Advance in Energy Harvesters/Nanogenerators and Self-Powered Sensors

Advance in Energy Harvesters/Nanogenerators and Self-Powered Sensors

Editors

Qiongfeng Shi
Jianxiong Zhu

MDPI • Basel • Beijing • Wuhan • Barcelona • Belgrade • Manchester • Tokyo • Cluj • Tianjin

Editors
Qiongfeng Shi
School of Electronic Science and Engineering
Southeast University
Nanjing
China

Jianxiong Zhu
School of Mechanical Engineering
Southeast University
Nanjing
China

Editorial Office
MDPI
St. Alban-Anlage 66
4052 Basel, Switzerland

This is a reprint of articles from the Special Issue published online in the open access journal *Nanomaterials* (ISSN 2079-4991) (available at: www.mdpi.com/journal/nanomaterials/special_issues/Energy_Harvesters).

For citation purposes, cite each article independently as indicated on the article page online and as indicated below:

LastName, A.A.; LastName, B.B.; LastName, C.C. Article Title. *Journal Name* **Year**, *Volume Number*, Page Range.

ISBN 978-3-0365-5884-4 (Hbk)
ISBN 978-3-0365-5883-7 (PDF)

Contents

About the Editors vii

Preface to "Advance in Energy Harvesters/Nanogenerators and Self-Powered Sensors" ix

Qiongfeng Shi and Jianxiong Zhu
Special Issue: Advance in Energy Harvesters/Nanogenerators and Self-Powered Sensors
Reprinted from: *Nanomaterials* **2022**, *12*, 3167, doi:10.3390/nano12183167 1

Taejun Park, Sangbin Park, Joon Hui Park, Ji Young Min, Yusup Jung and Sinsu Kyoung et al.
Temperature-Dependent Self-Powered Solar-Blind Photodetector Based on Ag_2O/β-Ga_2O_3 Heterojunction
Reprinted from: *Nanomaterials* **2022**, *12*, 2983, doi:10.3390/nano12172983 5

Ahmed Haroun, Mohamed Tarek, Mohamed Mosleh and Farouk Ismail
Recent Progress on Triboelectric Nanogenerators for Vibration Energy Harvesting and Vibration Sensing
Reprinted from: *Nanomaterials* **2022**, *12*, 2960, doi:10.3390/nano12172960 21

Enrique Delgado-Alvarado, Ernesto A. Elvira-Hernández, José Hernández-Hernández, Jesús Huerta-Chua, Héctor Vázquez-Leal and Jaime Martínez-Castillo et al.
Recent Progress of Nanogenerators for Green Energy Harvesting: Performance, Applications, and Challenges
Reprinted from: *Nanomaterials* **2022**, *12*, 2549, doi:10.3390/nano12152549 45

Yanzi Huang, Lingyu Wan, Jiang Jiang, Liuyan Li and Junyi Zhai
Self-Powered Resistance-Switching Properties of $Pr_{0.7}Ca_{0.3}MnO_3$ Film Driven by Triboelectric Nanogenerator
Reprinted from: *Nanomaterials* **2022**, *12*, 2199, doi:10.3390/nano12132199 77

Chan Wang, Qiongfeng Shi and Chengkuo Lee
Advanced Implantable Biomedical Devices Enabled by Triboelectric Nanogenerators
Reprinted from: *Nanomaterials* **2022**, *12*, 1366, doi:10.3390/nano12081366 87

Gang Tang, Zhen Wang, Xin Hu, Shaojie Wu, Bin Xu and Zhibiao Li et al.
A Non-Resonant Piezoelectric–Electromagnetic–Triboelectric Hybrid Energy Harvester for Low-Frequency Human Motions
Reprinted from: *Nanomaterials* **2022**, *12*, 1168, doi:10.3390/nano12071168 119

Madani Labed, Hojoong Kim, Joon Hui Park, Mohamed Labed, Afak Meftah and Nouredine Sengouga et al.
Physical Operations of a Self-Powered IZTO/-Ga_2O_3 Schottky Barrier Diode Photodetector
Reprinted from: *Nanomaterials* **2022**, *12*, 1061, doi:10.3390/nano12071061 133

Lei Zhai, Lingxiao Gao, Ziying Wang, Kejie Dai, Shuai Wu and Xiaojing Mu
An Energy Harvester Coupled with a Triboelectric Mechanism and Electrostatic Mechanism for Biomechanical Energy Harvesting
Reprinted from: *Nanomaterials* **2022**, *12*, 933, doi:10.3390/nano12060933 147

Haibing Hu, Baozhu Xie, Yangtian Lu and Jianxiong Zhu
Advances in Electrochemical Detection Electrodes for As(III)
Reprinted from: *Nanomaterials* **2022**, *12*, 781, doi:10.3390/nano12050781 159

Jie Lin, Yu-Jia Lv, Lei Han, Kuan Sun, Yan Xiang and Xiao-Xing Xing et al.
A Light-Driven Integrated Bio-Capacitor with Single Nano-Channel Modulation
Reprinted from: *Nanomaterials* **2022**, *12*, 592, doi:10.3390/nano12040592 . . . . **199**

About the Editors

Qiongfeng Shi

Qiongfeng Shi is currently a Professor in the School of Electronic Science and Engineering, Southeast University, Nanjing, China. He received his B.Eng. degree from the Department of Electronic Engineering and Information Science at the University of Science and Technology of China (USTC) in 2012, and received his Ph.D. degree from the Department of Electrical and Computer Engineering at the National University of Singapore (NUS) in 2018. He then worked as a Research Fellow in the same department at NUS from 2018 to 2022. His research interests include energy harvesting, self-powered sensing, flexible/wearable electronics, MEMS, human-machine interfaces, and intelligent systems. Prof. Shi has authored and co-authored more than 90 international journal papers and conference papers in his research fields, which have received more than 4800 citations with an H-index of 40 according to Google Scholar. Until now, Prof. Shi has served as a Guest Editor for several international journals, and as a reviewer for more than 20 peer-reviewed journals.

Jianxiong Zhu

Jianxiong Zhu is currently an Associate Professor of Southeast University, China. He received his B.Eng. degree in Electrical and Automation Engineering from the Hubei University of Technology in 2006. After that, he obtained his Master's degree from the University of Science and Technology of China and Ph.D. from the University of Missouri Columbia in 2015. His research interests are focused on triboelectric sensing, MEMS zero-power sensor, wearable flexible sensor, and gas sensor.

Preface to "Advance in Energy Harvesters/Nanogenerators and Self-Powered Sensors"

Preface written by: Xin Zheng, Jiangsu Youaiwei Intelligent Technology Co., Ltd.

As the CEO of Jiangsu Youaiwei Intelligent Technology Co., Ltd., which is a high-tech enterprise integrating independent research and the development of intelligent electronic engineering and smart sensing for airplanes and smart equipment, I am glad to have the chance to write this preface for this reprint. As we know, internet of things (IoT) technologies are greatly promoted by the rapidly developing 5G-and-beyond networks, which have spawned diversified applications, including smart homes, digital health, sports training, robotics, human-machine interactions, smart manufacturing, and Industry 4.0, etc.

In recent decades, I have witnessed the rapid advancement of energy harvesters and nanogenerators in terms of innovations in materials, mechanisms, structural designs, etc., leading to ever-increasing energy conversion efficiency and output power performance. Except for directly serving as energy harvesters and power sources, energy-harvesting devices can also be used as self-powered sensors to effectively lower the power consumption of a system, due to their self-generated signals corresponding to the external stimuli. Therefore, the advancement of energy harvesters and self-powered sensors could push forward the realization of self-sustained systems with continuous functional operations.

This reprint brings together 10 research and review articles on the above areas, showing the latest advances in energy harvesters, nanogenerators, and self-powered sensors for various applications. Engineers around the world can use this reprint to find new ideas to solve the problem of traditional energy, technology in practical application, and promote social development to achieve better economic efficiency.

I hope that the articles showcased here are interesting and helpful for readers and can inspire innovations in the field. I would like to thank all the authors for their great efforts in preparing their articles to contribute to this book.

Qiongfeng Shi and Jianxiong Zhu
Editors

nanomaterials

Editorial

Special Issue: Advance in Energy Harvesters/Nanogenerators and Self-Powered Sensors

Qiongfeng Shi [1,*] and Jianxiong Zhu [2,*]

1 Joint International Research Laboratory of Information Display and Visualization, School of Electronic Science and Engineering, Southeast University, Nanjing 210096, China
2 School of Mechanical Engineering, Southeast University, Nanjing 211189, China
* Correspondence: qiongfeng@seu.edu.cn (Q.S.); mezhujx@seu.edu.cn (J.Z.)

Internet of things (IoT) technologies are greatly promoted by the rapidly developed 5G-and-beyond networks, which have spawned diversified applications in the new era including smart homes, digital health, sports training, robotics, human–machine interaction, metaverse, smart manufacturing and industry 4.0, etc. To accommodate the enormous and widely distributed power demands needed for IoT nodes, self-sustained systems enabled by energy harvesting technologies (e.g., piezoelectric, triboelectric, electromagnetic, pyroelectric, thermoelectric, photovoltaic, etc.) have emerged as a promising solution. In recent decades, we have witnessed the rapid advancement of energy harvesters and nanogenerators in terms of innovations in materials, mechanisms, structural designs, etc., leading to ever-increasing energy conversion efficiency and output power performance. Except for directly serving as energy harvesters and power sources, energy harvesting devices can also be used as self-powered sensors to effectively lower the power consumption of a system, due to their self-generated signals corresponding to the external stimuli. Therefore, the advancement of energy harvesters and self-powered sensors could push forward the realization of self-sustained systems with continuous functional operations.

This Special Issue includes 10 research and review articles in the field that showcase the recent advances in energy harvesters, nanogenerators, and self-powered sensors for various applications, which should give readers a glimpse of the challenges, opportunities, and development trends of energy harvesting technologies.

There are six research articles from researchers across the world that explore device innovations in energy harvesting and self-powered sensing applications. To boost the conversion efficiency, Zhai et al. reported on an energy harvester using a triboelectric–electrostatic coupling mechanism that showed a 46-fold enhancement compared to a single triboelectric mechanism [1]. With this high output performance, a self-powered system was successfully demonstrated using gait recognition for access control. To effectively scavenge energy from low-frequency human motions, Tang et al. presented a non-resonant energy harvester that hybridized piezoelectric–electromagnetic–triboelectric in a synergic design [2]. In practical applications, the energy harvester was deployed on various body parts to show its energy generating capability for powering a wireless IoT sensor node. Then, Huang et al. investigated the self-powered resistance-switching ability of a triboelectric nanogenerator on resistive random access memory (RRAM) [3]. The RRAM exhibited excellent resistance-switching performance when driven by the triboelectric nanogenerator, with a switching ratio up to 2×10^5.

In terms of self-powered sensing, Labed et al. developed a self-powered photodetector using an InZnSnO/β-Ga_2O_3 Schottky barrier diode [4]. Operated in the self-powered mode at 0 V, the device achieved a high photo-to-dark current ratio (3.70×10^5) and a good photoresponsivity (0.64 mA/W). Based on a Ag_2O/β-Ga_2O_3 heterojunction, Park et al. proposed a self-powered deep ultraviolet photodetector [5]. When operated at 0 bias voltage, a high photoresponsivity of 12.87 mA/W was obtained, showing good potential for

Citation: Shi, Q.; Zhu, J. Special Issue: Advance in Energy Harvesters/Nanogenerators and Self-Powered Sensors. *Nanomaterials* **2022**, *12*, 3167. https://doi.org/10.3390/nano12183167

Received: 6 September 2022
Accepted: 9 September 2022
Published: 13 September 2022

use in ultraviolet sensing systems. Next, Lin et al. constructed a new light-driven integrated bio-capacitor based on bacteriorhodopsin and artificial nanochannels [6]. Microfluidic chips and a single nanopore structure were integrated to make the bio-capacitor more stable, and the photocurrent duration time was effectively regulated by varying the single nanopore's size.

Meanwhile, for those readers that are seeking an overview of the progress of energy harvesting technologies, four review articles covering different aspects of the field are also included in this Special Issue. Delgado-Alvarado et al. reported on the recent progress of nanogenerator technologies, including piezoelectric, electromagnetic, thermoelectric, and triboelectric, in terms of materials, applications, challenges, and future prospects for green energy harvesting [7]. Then, Wang et al. systematically reviewed the progress of implantable biomedical devices (i.e., biosensors, energy harvesters, and stimulation therapy devices) based on triboelectric nanogenerators, and discussed the remaining challenges and opportunities for the aspects of multifunctional materials and self-sustained close-looped systems [8]. Haroun et al. summarized the recent progresses in using triboelectric nanogenerators and hybridized generators for vibration energy harvesting and monitoring, with detailed analyses of the working mechanism, design principle, output performance, and important applications [9]. To manage the increasingly serious arsenic pollution in the living environment and maintain a healthy and beautiful ecosystem for human beings, Hu et al. reviewed the advances in electrode systems based on nanomaterials and their performance in arsenic detection [10]. By using silicon and its compounds, as well as novel polymers, for the detection of arsenic detection, such as noble metals, bimetals, other metals, and their compounds, such as carbon-, nano-, and biomolecules, they showed new routes for investigating novel nanomaterial sensing.

We would like to thank all the authors and reviewers for their contributions to this Special Issue. We also hope that the articles showcased here are interesting and helpful for readers and can inspire new innovations in the field.

Funding: This research received no external funding.

Acknowledgments: We would like to thank all the authors for their great efforts in preparing their articles to contribute to this Special Issue. Meanwhile, we would also like to thank all the reviewers for donating their precious time in helping in the peer-review processes, which is important to maintain the high standard of this Special Issue.

Conflicts of Interest: The authors declare no conflict of interest.

References

1. Zhai, L.; Gao, L.; Wang, Z.; Dai, K.; Wu, S.; Mu, X. An Energy Harvester Coupled with a Triboelectric Mechanism and Electrostatic Mechanism for Biomechanical Energy Harvesting. *Nanomaterials* **2022**, *12*, 933. [CrossRef] [PubMed]
2. Tang, G.; Wang, Z.; Hu, X.; Wu, S.; Xu, B.; Li, Z.; Yan, X.; Xu, F.; Yuan, D.; Li, P.; et al. A Non-Resonant Piezoelectric–Electromagnetic–Triboelectric Hybrid Energy Harvester for Low-Frequency Human Motions. *Nanomaterials* **2022**, *12*, 1168. [CrossRef] [PubMed]
3. Huang, Y.; Wan, L.; Jiang, J.; Li, L.; Zhai, J. Self-Powered Resistance-Switching Properties of $Pr_{0.7}Ca_{0.3}MnO_3$ Film Driven by Triboelectric Nanogenerator. *Nanomaterials* **2022**, *12*, 2199. [CrossRef] [PubMed]
4. Labed, M.; Kim, H.; Park, J.H.; Labed, M.; Meftah, A.; Sengouga, N.; Rim, Y.S. Physical Operations of a Self-Powered IZTO/β-Ga_2O_3 Schottky Barrier Diode Photodetector. *Nanomaterials* **2022**, *12*, 1061. [CrossRef] [PubMed]
5. Park, T.; Park, S.; Park, J.H.; Min, J.Y.; Jung, Y.; Kyoung, S.; Kang, T.Y.; Kim, K.; Rim, Y.S.; Hong, J. Temperature-Dependent Self-Powered Solar-Blind Photodetector Based on Ag_2O/β-Ga_2O_3 Heterojunction. *Nanomaterials* **2022**, *12*, 2983. [CrossRef] [PubMed]
6. Lin, J.; Lv, Y.; Han, L.; Sun, K.; Xiang, Y.; Xing, X.; Li, Y. A Light-Driven Integrated Bio-Capacitor with Single Nano-Channel Modulation. *Nanomaterials* **2022**, *12*, 592. [CrossRef] [PubMed]
7. Delgado-Alvarado, E.; Elvira-Hernández, E.A.; Hernández-Hernández, J.; Huerta-Chua, J.; Vázquez-Leal, H.; Martínez-Castillo, J.; García-Ramírez, P.J.; Herrera-May, A.L. Recent Progress of Nanogenerators for Green Energy Harvesting: Performance, Applications, and Challenges. *Nanomaterials* **2022**, *12*, 2549. [CrossRef] [PubMed]
8. Wang, C.; Shi, Q.; Lee, C. Advanced Implantable Biomedical Devices Enabled by Triboelectric Nanogenerators. *Nanomaterials* **2022**, *12*, 1366. [CrossRef]

9. Haroun, A.; Tarek, M.; Mosleh, M.; Ismail, F. Recent Progress on Triboelectric Nanogenerators for Vibration Energy Harvesting and Vibration Sensing. *Nanomaterials* **2022**, *12*, 2960. [CrossRef]
10. Hu, H.; Xie, B.; Lu, Y.; Zhu, J. Advances in Electrochemical Detection Electrodes for As(III). *Nanomaterials* **2022**, *12*, 781. [CrossRef] [PubMed]

nanomaterials

MDPI

Article

Temperature-Dependent Self-Powered Solar-Blind Photodetector Based on $Ag_2O/\beta\text{-}Ga_2O_3$ Heterojunction

Taejun Park [1], Sangbin Park [1], Joon Hui Park [2], Ji Young Min [2], Yusup Jung [3], Sinsu Kyoung [3], Tai Young Kang [3], Kyunghwan Kim [1], You Seung Rim [2,*] and Jeongsoo Hong [1,*]

1 Department of Electrical Engineering, College of IT Convergence, Gachon University, 1342, Seongnam-daero, Sujeong-gu, Seongnam-si 13120, Korea
2 Department of Intelligent Mechatronics Engineering and Convergence Engineering for Intelligent Drone, Sejong University, 209, Neungdong-ro, Gwangjin-gu, Seoul 05006, Korea
3 PowerCubeSemi, Inc., 686, Cheonggyesan-ro, Sujeong-gu, Seongnam-si 13105, Korea
* Correspondence: youseung@sejong.ac.kr (Y.S.R.); hongjs@gachon.ac.kr (J.H.)

Abstract: In this study, a high-photoresponsivity self-powered deep ultraviolet (DUV) photodetector based on an $Ag_2O/\beta\text{-}Ga_2O_3$ heterojunction was fabricated by depositing a p-type Ag_2O thin film onto an n-type $\beta\text{-}Ga_2O_3$ layer. The device characteristics after post-annealing at temperatures ranging from 0 to 400 °C were investigated. Our DUV devices exhibited typical rectification characteristics. At a post-annealing temperature of 300 °C, the as-fabricated device had a low leakage current of 4.24×10^{-11} A, ideality factor of 2.08, and a barrier height of 1.12 eV. Moreover, a high photoresponsivity of 12.87 mA/W was obtained at a 100 μW/cm^2 light intensity at a 254 nm wavelength at zero bias voltage, the detectivity was 2.70×10^{11} Jones, and the rise and fall time were 29.76, 46.73 ms, respectively. Based on these results, the $Ag_2O/\beta\text{-}Ga_2O_3$ heterojunction photodetector operates without an externally applied voltage and has high responsivity, which will help in the performance improvement of ultraviolet sensing systems.

Keywords: $Ag_2O/\beta\text{-}Ga_2O_3$; heterojunction; deep ultraviolet; photodetector; post-annealing

Citation: Park, T.; Park, S.; Park, J.H.; Min, J.Y.; Jung, Y.; Kyoung, S.; Kang, T.Y.; Kim, K.; Rim, Y.S.; Hong, J. Temperature-Dependent Self-Powered Solar-Blind Photodetector Based on $Ag_2O/\beta\text{-}Ga_2O_3$ Heterojunction. *Nanomaterials* **2022**, *12*, 2983. https://doi.org/10.3390/nano12172983

Academic Editors: Jianxiong Zhu and Qiongfeng Shi

Received: 3 August 2022
Accepted: 26 August 2022
Published: 29 August 2022

1. Introduction

Ultraviolet (UV) radiation from the sun has been used in a variety of applications such as disinfection of air, surfaces, and instruments [1]. Depending on the wavelength, UV light can be divided into UV-A (320–400 nm), UV-B (280–320 nm), and UV-C (200–280 nm) Among them, UV-C light does not reach the ground level owing to strong ozone layer absorption at the earth's surface. It is also known as solar-blind deep UV (DUV) radiation. In addition, DUV can catalyze chemical reactions, sterilize, and radiate from initial flame combustion. Owing to these characteristics, the need for DUV utilization has increased, and photodetector sensing DUV research has been extensively conducted in various fields, including biological and chemical analyses, solar UV monitoring, and flame sensors [2–6].

Proper materials for DUV photodetectors are ultra-wide bandgap semiconductors, such as ZnMgO and AlGaN [7,8]. However, the incorporation of significant amounts of Al in AlGaN deteriorates the crystal quality, and ZnMgO appears to be phase-separated [9,10]. Conversely, gallium oxide ($\beta\text{-}Ga_2O_3$), a well-known wide-bandgap semiconductor, has attracted considerable attention because it can fabricate high-quality thin films without other methods such as doping [11]. $\beta\text{-}Ga_2O_3$ has been extensively studied for DUV optoelectronics and power electronic devices because of its large direct bandgap energy (4.8 eV), high breakdown electric field ($E_{br} = 8 \times 10^6$ V/cm), and excellent chemical and thermal stability [12–14]. In addition, $\beta\text{-}Ga_2O_3$ exhibits n-type semiconductor properties owing to oxygen vacancies [15].

Most photodetectors require additional external power to promote the generation of photocurrents. This increases the size of the device, making it inconvenient to package and carry, and becomes overly dependent on external power supplies, limiting practical application in various negative complex environments. To complement these disadvantages, a new type of self-powered photodetector has been designed through extensive research. Small in size, economical, and featuring low energy consumption, these self-powered photodetectors do not depend on externally applied voltages and can be implemented independently of self-powered photodetectors. The response mechanism of this self-powered photodetector utilizes electron-hole pairs generated by light irradiation. i.e., It is based on the photovoltaic effect of semiconductors. The photodetector can be classified into a photoconductor and photodiode type according to the operating principle. However, the photodiode type is important because it implements a self-powered sensor with high responsivity depending on the material selection. According to photodiode construction mechanism, the current self-powered photodiode can be divided into two types: Schottky junction, and p-n heterojunction [16,17]. Schottky junctions have the advantages of fabrication simplicity and high-speed response. However, Schottky barriers built into the contacts of a Schottky barrier photodetector can make it difficult for the charge to transfer and degrades the device responsivity. The p-n heterojunction is important because of its low reverse saturation current, high breakdown voltage, and adjustment of the photodetection wavelength according to the heterojunction material [18]. Herein, suitable p-type semiconductors for the fabrication of photodetector devices as counterparts for n-type β-Ga_2O_3 are explored.

In this study, a photodetector based on a p-n junction was fabricated, and heterojunctions of n-type β-Ga_2O_3 and p-type Ag_2O were investigated. Compounds of Ag with oxygen have been extensively investigated. Silver oxide has various phases, such as AgO, Ag_2O, Ag_2O_3, Ag_4O_3, and Ag_4O_4 [19–23]. Among them, Ag_2O is the most thermodynamically stable, and Ag_2O thin films exhibit p-type semiconducting properties [24]. Various studies have been conducted on p-type Ag_2O thin films owing to their excellent electrical properties and transparent properties in the infrared and visible ranges [25]. In addition, the transition from Ag to Ag_2O by adjusting the oxygen flow ratio has previously been reported. Ag_2O thin films can be fabricated by various techniques, such as thermal oxidation of Ag films [26], thermal evaporation [27], pulsed laser deposition [28], and sputtering [29,30]. In this study, an Ag_2O layer was deposited using a facing target sputtering (FTS) system, which is a DC sputtering system. The FTS system consists of two targets facing each other within the chamber, and the substrate is separated from the plasma area, resulting in less surface damage compared to typical sputtering systems [31–34]. Based on the Ag_2O/β-Ga_2O_3 heterojunction, we fabricated a self-powered solar-blind photodetector. It exhibited high performance in terms of photoresponsivity and DUV detectivity at zero bias, and it was confirmed that the Ag_2O/β-Ga_2O_3 heterojunction photodetector operated with high performance without an external power supply.

2. Experimental Procedure

2.1. Materials

The solar-blind heterojunction photodetector was fabricated on an n-type single-crystal Sn-doped β-Ga_2O_3 wafer (thickness of 415 μm, N_d-N_a = 4.50 × 10^{18} cm^{-3}, Novel Crystal Technology, Inc.). Halide vapor phase epitaxy (HVPE) was used to grow the epitaxial layer of Si-doped β-Ga_2O_3 (thickness of 10 μm, N_d-N_a = 2.20 × 10^{16} cm^{-3}, μ_n = 300 cm^2/Vs, ρ = 0.94 Ω cm), which was used as the active layer in this solar-blind heterojunction photodetector. This layer is used due to its high purity and provides low resistance and on-resistance, and a high breakdown voltage [35]. Ti/Au electrodes (10/40 nm) were deposited by sputtering on the backside of the Sn-doped β-Ga_2O_3 wafer for the ohmic device contact. For the deposition of the p-type Ag_2O thin film, a 4–inch Ag targets (RND Korea, Gwangmyeong-si, Korea) and soda–lime glass substrate (75 × 25 cm, 1 mm thick, Marienfeld, Lauda-Königshofen, Germany) were used.

2.2. Fabrication of the Ag_2O/β-Ga_2O_3 Heterojunction Photodetector

Figure 1 shows a schematic illustration of the fabricated Ag_2O/β-Ga_2O_3 heterojunction photodetector. Ag is used as the front photodetector electrode because it is capable of multilayer Ag_2O deposition through continuous deposition. Ag_2O/Ag (50/50 nm) thin films were patterned into circles with a 300 μm radius using a shadow mask on the top-epitaxial surface of the n-type β-Ga_2O_3 using an FTS system to fabricate an Ag_2O/β-Ga_2O_3 heterojunction photodetector. The Ag_2O thin film was deposited by controlling the oxygen flow, and the Ag thin film was continuously deposited without an oxygen atmosphere. The sputtering conditions are summarized in detail in Table 1. In addition, to improve the device performance of the fabricated photodetector, a post-annealing process was performed in the range from room temperature (RT) to 400 °C using rapid thermal annealing (RTA) in an Ar condition at 100 mTorr for 1 min.

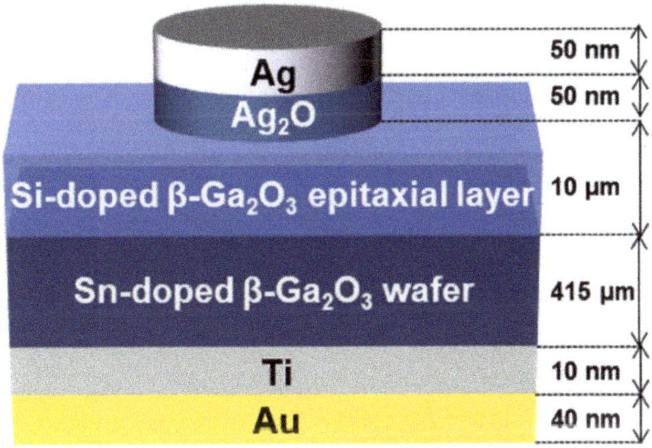

Figure 1. Schematic illustration of the fabricated Ag_2O/β-Ga_2O_3 heterojunction photodetector.

Table 1. Sputtering conditions of Ag_2O and Ag thin films.

Parameters	Sputtering Condition	
Layer	Ag_2O	Ag
Substrate	Soda-lime glass	Soda-lime glass
Targets	Ag (99.99%)	Ag (99.99%)
Base pressure	3×10^{-5} Torr	3×10^{-5} Torr
Working pressure	2 mTorr	2 mTorr
Gas flow	Ar: 10 sccm, O_2: 3 sccm	Ar: 10 sccm
Input power	50 W	50 W
Thickness	50 nm	50 nm

The crystallographic properties of the thin films were evaluated using X-ray diffraction (XRD; Bruker D8) at the Smart Materials Research Center for IoT, Gachon University. Scanning electron microscopy (SEM, S-4700, Hitachi, Tokyo, Japan) at the Smart Materials Research Center for IoT, Gachon University and a KLA-Tencor Alpha-step D-500 Stylus Profiler were used to observe the surface morphologies and thicknesses of the thin films, respectively. The optical properties of the as-deposited thin films were evaluated using a UV-vis spectrometer (Lambda 750 UV-vis-NIR, Perkin Elmer, Waltham, MA, USA). The electrical properties of the thin films were evaluated by Hall effect measurement system (HMS-5500, ECOPIA, Anyang, Korea). X-ray photoelectron spectroscopy (XPS, K-alpha +, Thermo Fisher Scientific, Waltham, MA, USA) was used to investigate the binding energy of Ag_2O before and after post-annealing.

2.3. Evaluation of the $Ag_2O/\beta\text{-}Ga_2O_3$ Heterojunction Photodetector

The electrical current density-voltage (J-V) characteristics were analyzed using a semiconductor analyzer (4200A-SCS parameter analyzer, Keithley, Cleveland, OSU, USA). A Keithley 2401 was used to measure the time-dependent photoresponse of the fabricated device. Time-dependent photoresponse measurements were performed using a UV-C lamp (TN-4LC, wavelength: 254 nm, Korea Ace Scientific) with varying light intensities ranging from 100 to 1000 $\mu W/cm^2$ at RT. The UV-C intensity was measured using a UV-C light meter (UVC-254SD, Lutron, PA, USA).

3. Results and Discussion

For the heterojunction photodetector fabrication, p-type Ag_2O thin-film deposition was conducted. In reactive sputtering, a transition region representing the change from various metals to oxides through discharge voltage changes has been observed in numerous studies. For example, by changing the discharge voltage of Si, the metal target surface switches to the oxide mode when oxygen reacts with the metal particles and target surface. A transition region was observed, and the result of the conversion to SiO_2 was reported in a previous study [36]. Therefore, to deposit p-type Ag_2O, the discharge voltage change under different oxygen flows in the Ag target was measured.

Figure 2 shows the variation in the discharge voltage with the oxygen flow rate in the Ag target. As the oxygen flow rate increased, the discharge voltage also increased; therefore, the transition region representing the change from metal to oxide was not observed. As a predictable cause of this trend, reactive gas such as oxygen can easily form negative ions in the plasma state. In the case of Si, because the emission coefficient of the secondary electrons emitted from the surface of the Si target is higher than that of oxygen gas, the discharge voltage decreases as the oxygen gas flow rate increases. However, in Ag, because the emission coefficient of secondary electrons emitted from the Ag metal target surface is lower than that of oxygen gas, the discharge voltage is expected to increase as the oxygen gas flow rate increases [37].

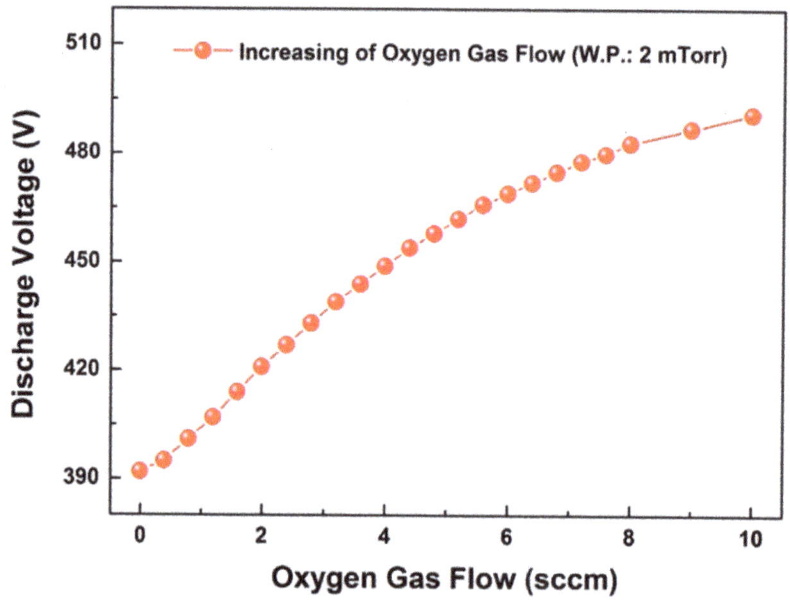

Figure 2. Variation of discharge voltage by changing the oxygen flow rate.

Figure 3A shows the XRD patterns of as-deposited thin film by changing oxygen gas flow in the range of 0 to 5 sccm. Without the oxygen gas, the Ag peaks appeared at 2θ = 38.2°, 44.4°, 64.6°, and 77.6° corresponding to the (111), (200), (220), and (311) planes, respectively, according to the reference data (ICDD card 01-087-0720). At 2 sccm, the intensity of Ag peaks decreased and the (100) planes of Ag_2O (ICDD card 01-072-2108) additionally appeared at 2θ = 33.6°, indicating a mixed phase of Ag_2O and Ag. Moreover, at 3 sccm, the broad peaks representing the (100) and (011) planes of Ag_2O appeared at 2θ = 33.6° and 38.4° without other impurities. The (011) peak of Ag_2O and the (111) peak of Ag may overlap. However, it could be sufficiently distinguished through the additional peaks of Ag and Ag_2O. As the oxygen gas flow increases from 4 to 5 sccm, the Ag_2O peaks were not seen, and peaks corresponding to AgO (JCPDS card 75-0969) were observed at 2θ = 32.3° and 35.6°. It is suggested that with an increased oxygen gas flow rate, an additional chemical bond between Ag and O is generated, resulting in AgO peaks [38]. The AgO demonstrated deteriorated electrical properties compared to Ag_2O, and its properties as a p-type are reduced [39]. Hence, our work was conducted at 3 sccm where only the Ag_2O peak exists. In the top-view SEM images of Figure 3B,C, the surface morphologies of the Ag and Ag_2O thin film indicate a dense surface.

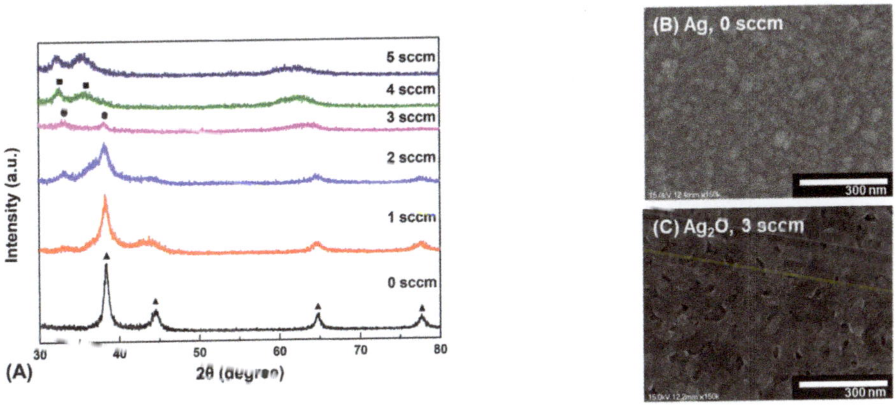

Figure 3. Structural properties of as-deposited thin film. (**A**) XRD patterns of as-deposited thin film under different oxygen gas flow rates. (•: Ag_2O, ▲: Ag, ■: AgO). (**B**) SEM image of Ag thin film deposited on a glass substrate. (**C**) SEM image of Ag_2O thin film.

Figure 4A shows the UV-vis spectra of the as-deposited thin film with the oxygen gas flow rate controlled from 0 to 5 sccm. The transmittance near 300 nm of the Ag thin film deposited at 0 and 1 sccm increase sharply than other conditions. The reason is the surface plasmon resonance (SPR) phenomenon that occurs on the surface of planar metals (typically gold or silver) or on the surface of metal nanoparticles. Therefore, it is a SPR phenomenon that appears on the surface of the Ag thin film through the light used for UV-vis, which increases the transmittance near 300 nm [40]. As the oxygen flow rate increased, the transition from Ag to Ag_2O occurred, increasing the transmittance in visible light of the thin film more than the Ag thin film [41]. Moreover, the average transmittance in visible light of the thin film deposited at 3 sccm was significantly increased to 42.5%. Additionally, when the oxygen gas flow was increased to 4–5 sccm, it was confirmed that the average visible light transmittance increased. However, as mentioned in the XRD study, the AgO phase appeared at the 4–5 sccm oxygen gas flow condition. Therefore, the optimal condition was chosen as 3 sccm.

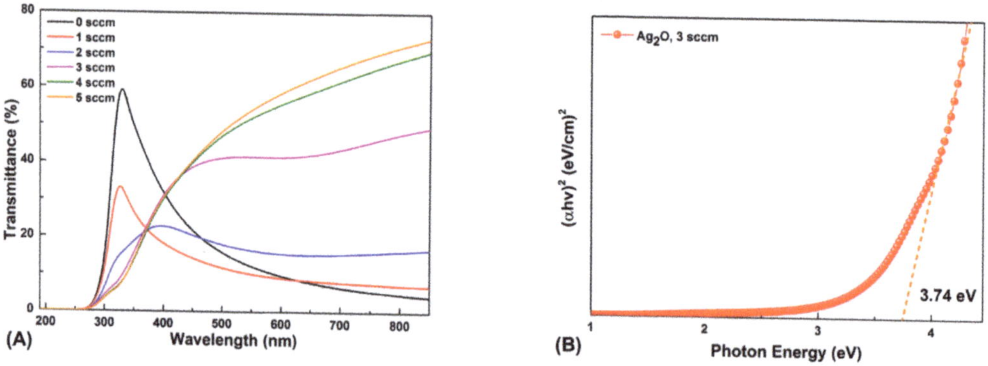

Figure 4. Optical properties of the as-deposited thin film prepared at different oxygen gas flow rates. (**A**) Transmittance (%) values of different samples at oxygen gas flow ranging from 0 to 5 sccm. (**B**) Optical bandgap energy of Ag_2O thin film.

Figure 4B shows the optical bandgap energy of the Ag_2O thin film calculated by the Tauc plot equation [42], given by

$$\alpha hv = \beta(hv - E_g)^{1/2} \tag{1}$$

where hv is the photon energy, α is the absorption coefficient, and β is a fixed constant. The Tauc plot shows that the optical bandgap energy can be estimated by extrapolating the linear section to the axis energy. The optical bandgap energy of the Ag_2O thin film deposited at 3 sccm was calculated to be 3.74 eV. Electrical properties of as-deposited Ag_2O thin film were evaluated using Hall measurement. The carrier concentration, hall mobility, and resistivity of as-deposited Ag_2O thin film are 6.2×10^{18} cm^{-3}, 41.5 cm^2/Vs, and 2.4×10^{-2} Ω cm, respectively. Moreover, the positive value of Hall coefficient of 1.0 cm^3/C was obtained, confirming that the as-deposited Ag_2O thin film was a p-type semiconductor.

The influence of post-annealing on the Ag_2O properties was also evaluated. It has been reported that an appropriate post-annealing temperature improves the photodetector characteristics owing to a reduction in interfacial defects [43]. The crystallinity quality, uniformity of the deposited thin film, and native defect correction at the heterojunction interface resulted in an increased photocurrent or apparent photoresponse for the fabricated device. Because the deposited Ag_2O thin films are affected by electrical and crystallographic properties depending on the post annealing temperature, the as-deposited Ag_2O thin films were studied in the temperature range RT−600 °C to evaluate the variation in Ag_2O crystallographic properties.

The XRD patterns of the post-annealed Ag_2O thin films are shown in Figure 5. The annealing temperature range was from RT to 600 °C. Broad peaks appeared in the Ag_2O thin film deposited at RT, and as the post-annealing temperature increased, the Ag_2O peak intensity increased. The intensity of (011) peak at 38.4°, the preferred growth plane, increased the most at 300 °C, without crystal transition from Ag_2O to Ag. The Scherrer equation was used to calculate the crystallite sizes of the Ag_2O (011) plane according to the annealing temperature [44]. It was confirmed that the crystallite size increase with annealing temperature (pristine: 13.3 nm, 100 °C: 14 nm, 200 °C: 15 nm, 300 °C: 16.5 nm). As the heat treatment temperature increases, it becomes a high-quality thin film with improved crystallinity, and the surface of the film becomes homogenous due to reduced roughness. The smooth surface exhibits good interfacial properties and can effectively suppress the formation of interfacial charge traps and reduce the carrier scattering centers to achieve greater electron mobility, which greatly improves the performance of the photodetector [45]. However, after heat treatment at 400 °C, the intensity of Ag_2O (100) peak was slightly decreased, and it was confirmed that the intensity of the Ag peaks (200), (220), and (311) increased. This means that the sample post-annealed at 400 °C presents both Ag_2O and Ag phases. These thin films have inhomogeneous surface which cause more traps and dislocations at the Ag_2O/Ag and Ag_2O/β-Ga_2O_3 interfaces. Since these defects acts as a resistance in the diode characteristics, it is expected that such as hump phenomenon will appear. The transition of Ag_2O into Ag is attributed to the increase in the Ag diffusion rate owing to the thermal effect, and the diffusion rate of oxygen atoms becomes relatively smaller than that of Ag. Therefore, owing to sufficient thermal energy, the Ag diffusion increases and the chemical bond with oxygen is broken [46]. Thus, at 500 °C or higher, the Ag_2O peaks were not observed and the Ag peaks appeared. Moreover, the core level binding energy was further investigated through XPS analysis of the post-annealed sample at 400 °C, in which both Ag_2O and Ag peaks exist.

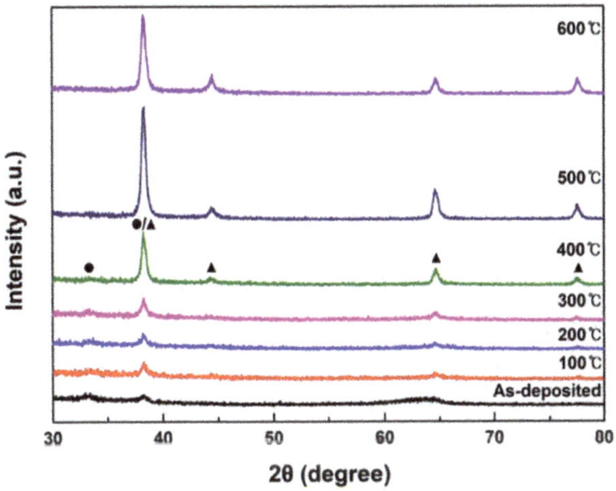

Figure 5. XRD patterns of the post-annealed Ag_2O thin films in the range of RT to 600 °C. (•: Ag_2O, ▲: Ag).

Figure 6 shows the XPS spectra of Ag_2O thin films with and without post-annealing. The XPS analysis was performed to further investigate the changes in Ag_2O film properties post-annealed at 400 °C via the core-level binding energy. The chemical composition of Ag_2O was determined by XPS. As presented in Figure 6A, the full XPS spectrum of Ag_2O reveals the presence of C, O, and Ag. Through the XPS analysis, the atom ratio between Ag and O was 2:1. However, after heat treatment at 400 °C, the atom ratio of Ag and O is changed to 3.5:1. Since the atom ratio of Ag becomes higher than that of O, it can be seen that Ag_2O is transitioning to Ag. The spectrum in Figure 6B shows the characteristic core-level binding energies at approximately 368 and 374 eV, corresponding to Ag $3d_{5/2}$ and Ag $3d_{3/2}$, respectively, related to the Ag_2O thin films. The core-level binding energies shown in the XPS spectra are consistent with those reported in other studies [47]. Moreover, after post-annealing at 400 °C, the core level binding energy of Ag $3d_{5/2}$ was shifted to 368.3 eV and Ag $3d_{3/2}$ shifted to 374.3 eV. In the O 1s XPS spectrum in Figure 6C, the two peaks at 529.5 and 531.4 eV are related to the core-level binding energy of the O ion in Ag_2O, whereas the higher-energy peak corresponds to hydroxyl groups (-OH). Compared with thin film post-annealed at 400 °C, the core-level binding energy of the O ion shifted to 529.4 eV and -OH shifted to 532.1 eV. This indicates that the transition from Ag_2O to Ag was caused by thermal decomposition [48]. As shown in Figure 6, the sample without post-annealing indicated Ag_2O. However, the post-annealing of the 400 °C sample showed an Ag_2O to Ag transition region. These XPS results are in good agreement with the XRD results. Hence, heat treatment was performed on the photodetector in the range of 0–400 °C, where the Ag_2O peaks were maintained.

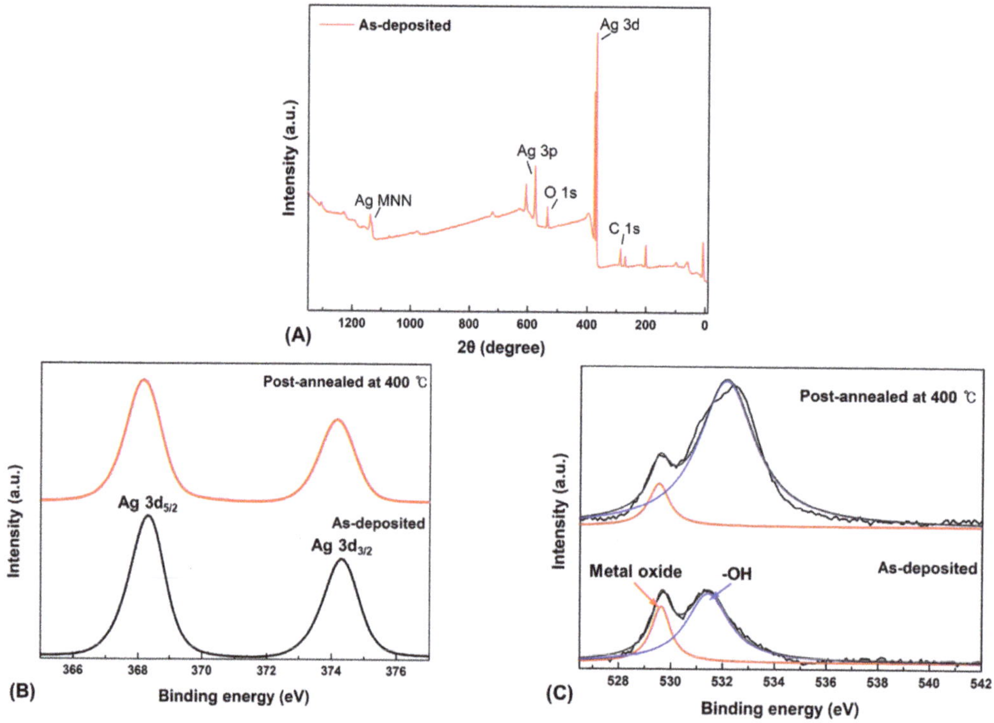

Figure 6. XPS spectra of the Ag_2O thin films with and without post-annealing. (**A**) Full XPS spectrum of Ag_2O thin film, (**B**) Ag 3d, (**C**) O 1s.

After optimizing the Ag_2O condition, an $Ag_2O/\beta\text{-}Ga_2O_3$ heterojunction photodetector was fabricated, and its structure is shown in Figure 7A. The J-V characteristics of the $Ag_2O/\beta\text{-}Ga_2O_3$ heterojunction photodetectors fabricated at various post-annealing temperatures were measured in the range of −2–4 V are shown in Figure 7B. All devices exhibited p-n junction rectification characteristics; in particular, a high on-off ratio of 1.40×10^8 was achieved through a low off-current of 4.24×10^{-11} A at 0 V and an on-current of 5.74×10^{-3} A at 1.86 V in the device post-annealed at 300 °C. Moreover, the J-V curve of the fabricated device without heat treatment indicates that the leakage current increases as the reverse bias voltage increases. However, the leakage current decreased as the post-annealing temperature increased. The change in leakage current with annealing temperature may influence the Ag_2O physical properties at different annealing temperatures. Post-annealing of the layers can improve the interface properties between the heterojunctions; thus, the homogeneity of the junction reduces carrier recombination and leakage current [49]. As shown in Figure 7C, it exhibited the variation of the breakdown voltage with post-annealing temperature. The breakdown voltage increased as the annealing temperature increased, and the device post-annealed at 300 °C exhibited the highest value at 238 V. The photodetector is primarily used at low current, but the as-fabricated device has high durability owing to its high breakdown voltage and low leakage current, such that it can be applied in harsh industries where a sudden short circuit occurs [50]. In addition, diode parameters such as the ideality factor and barrier height were investigated. The diode J-V characteristics can be expressed, correlating the current and voltage with the ideality factor and barrier height, as follows:

$$J = J_S \exp\{q(V - IR_S)/nkT\} \quad (2)$$

where J_S is the saturation current density, V is the voltage across the diode, T is the temperature in Kelvin, q is the electron charge, k is Boltzmann's constant, and n is an ideality factor. The value of the ideality factor determines the deviation from the ideal diode owing to the presence of the barrier inhomogeneity and tunneling component [51],

$$J_S = AA^*T^2\exp(-q\varphi_B/kT) \tag{3}$$

where A^* is Richardson's constant (41 A cm^{-2} K^{-2} for β-Ga_2O_3), A is the contact area, T is the temperature in Kelvin, q is the electron charge, k is Boltzmann's constant, and φ_B is the effective barrier height.

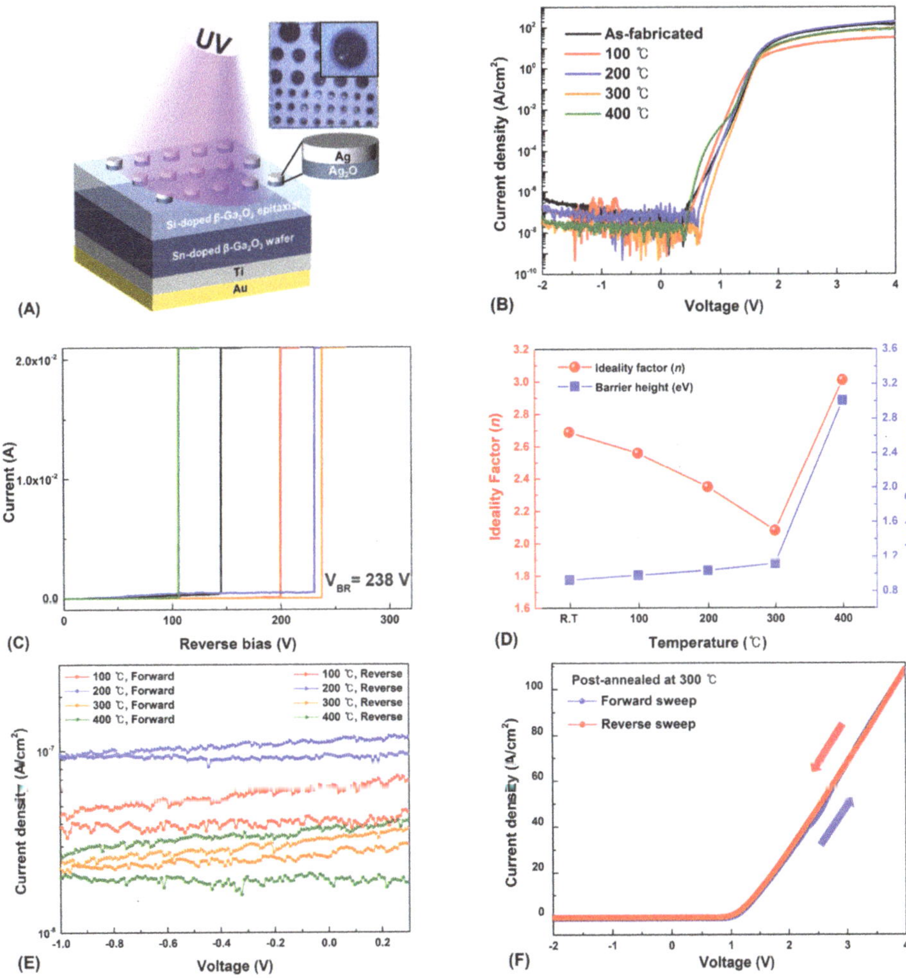

Figure 7. (**A**) Schematic illustration of the as-fabricated Ag_2O/β-Ga_2O_3 heterojunction photodetector. (**B**) J-V characteristics of Ag_2O/β-Ga_2O_3 heterojunction photodetectors. (**C**) Breakdown voltage of the as-fabricated devices. (**D**) Variation of the ideality factor (n) and barrier height (φ_B) with the post-annealing temperature. (**E**) Logarithmic plot of the J-V characteristics with difference in reverse saturation current according to post-annealing temperature. (**F**) J-V characteristics for the forward and reverse sweep of bias voltage.

Figure 7D shows the changes in n and φ_B with the post-annealing temperature. As the post-annealing temperature increased up to 300 °C, the value of n decreased and φ_B increased. The ideality factor of the device post-annealed at 300 °C had the lowest value of 2.08, and φ_B increased to 1.12 eV. This can be attributed to improvements in crystallization and reduced interfacial defects with fewer dislocations because of post-annealing [52]. However, the values of n and φ_B increased rapidly at 400 °C, this is due to the hump phenomenon observed in the J-V curve in Figure 7B. This was related to the transition of Ag_2O to Ag caused by the post-annealing process at 400 °C. As shown in the XRD study, the thin film post-annealed at 400 °C indicates a mixed state of Ag_2O and Ag. It is thought that the trap density increases owing to the trapping of charged species occurring from the difference in the resistivity and conductivity of the mixed crystallinity, causing the hump phenomenon [53]. Furthermore, the interface quality of the $Ag_2O/\beta\text{-}Ga_2O_3$ heterojunction was investigated. The occurrence of electrical hysteresis was analyzed by measuring the forward and reverse J-V characteristics. Figure 7E shows a logarithmic plot of the J–V characteristics with small electrical hysteresis at the reverse saturation current according to the post-annealing temperature. The presence of electrical hysteresis in the reverse saturation current exhibits the quality of interface states and surface defects for the $Ag_2O/\beta\text{-}Ga_2O_3$ heterojunction [54,55]. It was confirmed that the difference in reverse current and the reverse saturation current decreased as the post-annealing temperature increased, showing the narrowest gap at 300 °C. As the difference in reverse current and reverse saturation current is much smaller, charge trapping is reduced, indicating an enhanced interface state. Figure 7F shows the forward and reverse sweeps of the post-annealed device at 300 °C under dark condition. The measured J–V characteristics indicated almost no hysteresis effect compared to the forward and reverse, showing an excellent interface state.

Figure 8A shows the continuous time-dependent photoresponse according to the post-annealing temperature at a wavelength of 254 nm and a light intensity of 1000 $\mu W/cm^2$ at zero bias voltage. When the UV light was turned on, the photocurrent instantaneously flowed, and the photoresponse of all the samples remained approximately the same, indicating high reproducibility. Moreover, when the UV light was turned off, the photocurrent immediately returned to the dark current. This on/off operating characteristic of the UV irradiation clearly indicates that it behaves as a photodetector. Moreover, the effect on heat treatment was examined, the photocurrent increased with increasing post-annealing temperature, showing the highest photocurrent density of 9.70 $\mu A/cm^2$ at 300 °C. In the case of 400 °C, the photocurrent decreased owing to the hump phenomenon, in which Ag_2O transitions to Ag caused by post-annealing at 400 °C. The change in photocurrent density according to post-annealing temperature tends to be the same as the J-V analysis. The interfacial traps interfere the movement of electrons separated from a pair of electron holes generated by light irradiation. However, it can be seen that the interfacial traps reduced at 300 °C through the J-V analysis and photoresponse according to the heat treatment, thereby stabilizing the interface between Ag_2O and $\beta\text{-}Ga_2O_3$ and improving the quality. In addition, temperature-dependent external quantum efficiency (EQE) was calculated using the following formulas:

$$\eta_{\text{eff}} = (hvR/e) \times 100\% \quad (4)$$

where η_{eff} is the EQE, hv is the incident photon energy, R is the responsivity, and e is elementary charge.

Comparing the values of η_{eff} for each annealing temperature, it was confirmed that the device post-annealed at 300 °C achieved the highest η_{eff} of 6.3% (100 °C: 1.9%, 200 °C: 2.8%, 300 °C: 6.3%, 400 °C: 4.8%) at zero-bias under 254 nm irradiation at 100 $\mu W/cm^2$. Hence, it was confirmed that 300 °C is the optimal condition, the change in photocurrent of the 300 °C sample according to the intensity of light at 254 nm was investigated. Figure 8B shows the photoresponse of the post-annealed device at 300 °C under a UV light intensities ranging from 100 to 1000 $\mu W/cm^2$ under 254 nm irradiation. The photocurrent gradually increased with the intensity of light because the higher the light intensity, the more electron-hole pairs, which in turn produced a higher photocurrent. However, contrary to this tendency, the responsivity and detectivity of the photodetector exhibited different results. Figure 8C,D shows the change in the photodetector responsivity and detectivity according to the light intensity. The responsivity, which evaluates the photodetector sensitivity and the detectivity, which is the figure of merit for detecting the smallest signal, were calculated using the following formulas [56,57]:

$$R = (J_{Photo} - J_{Dark})/P \quad (5)$$

where R is the responsivity, J_{Photo} is the photocurrent density, J_{Dark} is the dark-current density, and P is the supplied light power. The detectivity value is determined by,

$$D = R/(2eJ)^{1/2} \quad (6)$$

where D is the detectivity, J is the dark-current density, and e is the elemental charge.

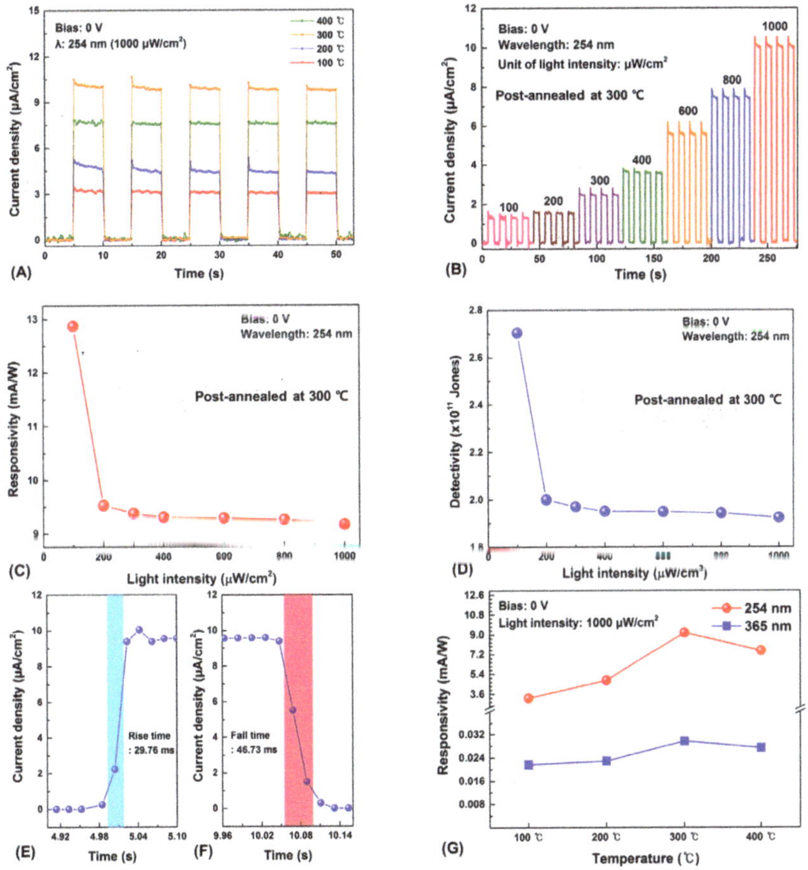

Figure 8. The photoresponse and parameters of the Ag_2O/β-Ga_2O_3 heterojunction photodetector under zero-bias at 254 nm irradiation. (**A**) Photoresponse of the devices with the Ag_2O thin films grown under different post-annealing temperatures. (**B**) Photoresponse of the device post-annealed at 300 °C with various light intensities. (**C**) Responsivity as a function of the light intensity. (**D**) Detectivity as a function of the light intensity. (**E**) Rise time and (**F**) Fall time. (**G**) Responsivity under zero-bias at 254 and 365 nm irradiation.

Consequently, the responsivity under 254 nm irradiation decreased with increasing light intensity. The maximum responsivity was obtained as 12.87 mA/W at 100 μW/cm^2. Moreover, the detectivity decreased with increasing light intensity, and the highest D was obtained at 2.70 × 10^{11} Jones under 254 nm irradiation at 100 μW/cm^2. The device exhibited high responsivity and detectivity at low light intensities. For this reason, it is suggested that as more electron-hole pairs are generated by high light intensity, scattering is increased because of the generated charge carriers, and the possibility of recombination increases, thereby reducing responsivity and detectivity [58]. Figure 8E,F show the rise and fall times of the Ag_2O/β-Ga_2O_3 heterojunction photodetector. It was measured to be 29.8 and 46.7 ms, respectively. As shown in Figure 8G, we confirmed that the rejection ratio of 254 nm (UVC) to 365 nm (UVA) at zero bias under the light intensity of 1000 μW/cm^2 is 3.1 × 10^2. The photoresponse parameters under zero bias of the Ag_2O/β-Ga_2O_3 heterojunction photodetector in this study showed higher responsivity and detectivity than those of other solar-blind photodetectors, as shown in Table 2 [59–65].

Table 2. Comparison of the self-powered solar-blind photodetector characteristics parameters with other reported studies.

Photodetector	Wavelength (nm)	Responsivity (mA/W)	Detectivity (Jones)	Rise Time/Fall Time	Ref.
Ag_2O/β-Ga_2O_3	254	12.87	2.70 × 10^{11}	29.76 ms/46.73 ms	This work
Diamond/β-Ga_2O_3	244	0.2			[59]
Au/β-Ga_2O_3	258	0.01		1 μs/100 μs	[60]
4H-SiC/β-Ga_2O_3	254	10.35	8.8 × 10^9	11 ms/19 ms	[61]
Polyanline/MgZnO	250	0.16	1.5 × 10^{11}	0.3 s/0.3 s	[62]
p-Gr/ZnS QDs/4H-SiC	250	0.29	1.41 × 10^{10}	28 μs/0.75 ms	[63]
MoS_2/β-Ga_2O_3	245	2.05	1.21 × 10^{11}		[64]
β-Ga_2O_3/Ga:ZnO	254	0.76		0.18 s/0.27 s	[65]

The photoresponse properties of the Ag_2O/β-Ga_2O_3 photodetector can be explained from the energy band diagram as shown in Figure 9. The mechanism of photocurrent generation is related to the depletion region. For p-type Ag_2O, in which positive holes are the major carriers, the Fermi level is located near the valance band, whereas n-type β-Ga_2O_3, in which free electrons are the major carriers, exists near the conduction band. When forming the p-n junction holes from p-type diffuse to n-type, and electrons from n-type diffuse to p-type. As a result, each Fermi level is matched, forming an electrically neutral depletion region. In addition, due to the large difference in optical bandgap, a strong built-in electric field will be created at the Ag_2O/β-Ga_2O_3 interface. When the device is irradiated to DUV light, electron–hole pairs are generated in the depletion region due to absorption of the incident DUV light, which is rapidly separated by the strong built-in electric field. Moreover, electrons are transferred to the electrode, leading to a fast photoresponse speed and an increase in photocurrent. Besides, the strong built-in electric field ensures that this heterojunction photodetector can operate at zero bias and proper heat treatment reduces defects between interfaces, resulting in low leakage currents and high photoresponse. In this study, we successfully fabricated a self-powered DUV photodetector through the heterojunction of Ag_2O and β-Ga_2O_3. The photodetector characteristics according to the post-annealing temperature were evaluated, and the sample that was post-annealed at 300 °C had the best performance. Our research provides a new route for the DUV sensing industry and shows potential for military applications, including flame sensing, spatial sensing, and image sensors.

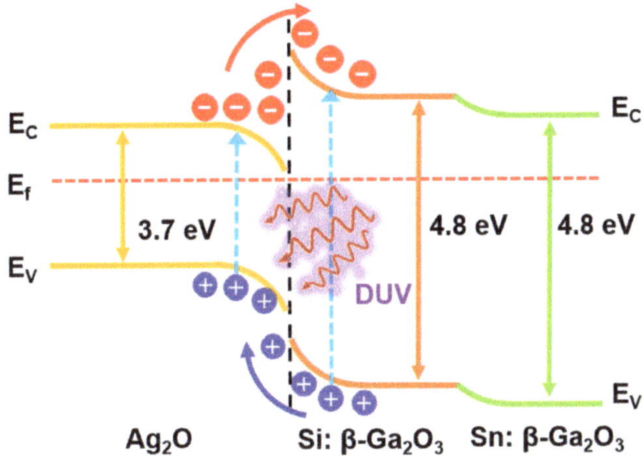

Figure 9. Energy band diagram of the $Ag_2O/\beta\text{-}Ga_2O_3$ heterojunction under DUV light irradiation.

4. Conclusions

A self-powered solar-blind photodetector was successfully fabricated with an $Ag_2O/\beta\text{-}Ga_2O_3$ heterojunction via deposition of a p-type Ag_2O thin film onto an n-type $\beta\text{-}Ga_2O_3$ layer using an FTS system. Ag_2O thin films were deposited by controlling the oxygen gas flow, with 3 sccm being the optimal condition for presenting p-type semiconductor characteristics. After the Ag/Ag_2O thin film deposition, the effects of post-annealing on the $Ag_2O/\beta\text{-}Ga_2O_3$ heterojunction photodetector were investigated. Typical rectification characteristics were observed in J-V characteristics, and as the post-annealing temperature increased to 300 °C, a low leakage current (4.24×10^{-11} A), high breakdown voltage of 238 V, ideality factor of 2.08, and barrier height of 1.12 eV were observed. A high responsivity of 12.87 mA/W, detectivity rate of 2.70×10^{11} Jones, and fast rise and fall times of 29.76 and 46.73 ms were observed under 254 nm irradiation at zero bias. From these results, it was confirmed that the device operates without an external applied voltage and has high responsivity in DUV. $Ag_2O/\beta\text{-}Ga_2O_3$ heterojunction photodetectors have confirmed their potential for application in DUV sensing systems, and further research is planned to improve the device characteristics

Author Contributions: Conceptualization, T.P. and J.H.; methodology, T.P., S.P., S.K. and J.H.; formal analysis, T.P., S.P., J.H.P. and J.Y.M.; investigation, T.P., S.P., Y.J., Y.S.R. and J.H.; resources, S.K. and T.Y.K.; writing—original draft preparation, T.P. and J.H.; writing—review & editing, K.K. and J.H.; supervision, J.H. All authors have read and agreed to the published version of the manuscript.

Funding: This research received no external funding.

Institutional Review Board Statement: Not applicable.

Informed Consent Statement: Not applicable.

Data Availability Statement: The data that support the findings of this study are available from the corresponding author upon reasonable request.

Acknowledgments: This research was supported by Korea Institute for Advancement of Technology(KIAT) grant funded by the Korea Government(MOTIE) (P0012451, The Competency Development Program for Industry Specialist) and was supported by the Gachon University research fund of 2021(GCU-202008450010).

Conflicts of Interest: The authors declare no conflict of interest.

References

1. Wolfe, R.L. Ultraviolet disinfection of potable water. *Environ. Sci. Technol.* **1990**, *24*, 768–773. [CrossRef]
2. Zhang, Z.-X.; Li, C.; Lu, Y.; Tong, X.-W.; Liang, F.-X.; Zhao, X.-Y.; Wu, D.; Xie, C.; Luo, L.-B. Sensitive deep ultraviolet photodetector and image sensor composed of inorganic lead-free Cs3Cu2I5 perovskite with wide bandgap. *J. Phys. Chem. Lett.* **2019**, *10*, 5343–5350. [CrossRef]
3. Yoshikawa, A.; Ushida, S.; Nagase, K.; Iwaya, M.; Takeuchi, T.; Kamiyama, S.; Akasaki, I. High-performance solar-blind Al0.6Ga0.4N/Al0.5Ga0.5N MSM type photodetector. *Appl. Phys. Lett.* **2017**, *111*, 191103. [CrossRef]
4. Arora, K.; Goel, N.; Kumar, M.; Kumar, M. Ultrahigh performance of self-powered β-Ga_2O_3 thin film solar-blind photodetector grown on cost-effective Si substrate using high-temperature seed layer. *ACS Photonics* **2018**, *5*, 2391–2401. [CrossRef]
5. Qian, L.; Wang, Y.; Wu, Z.; Sheng, T.; Liu, X. β-Ga_2O_3 solar-blind deep-ultraviolet photodetector based on annealed sapphire substrate. *Vacuum* **2017**, *140*, 106–110. [CrossRef]
6. Park, T.; Hur, J. Self-Powered Low-Cost UVC Sensor Based on Organic-Inorganic Heterojunction for Partial Discharge Detection. *Small* **2021**, *17*, 2100695. [CrossRef]
7. Wang, L.K.; Ju, Z.G.; Zhang, J.Y.; Zheng, J.; Shen, D.Z.; Yao, B.; Zhao, D.X.; Zhang, Z.Z.; Li, B.H.; Shan, C.X. Single-crystalline cubic MgZnO films and their application in deep-ultraviolet optoelectronic devices. *Appl. Phys. Lett.* **2009**, *95*, 131113. [CrossRef]
8. Zhao, Z.; Chu, C.; Zhang, G.; Tian, K.; Zhang, Y.; Zhang, Z.-H. Tuning the Plasmonic Resonance Peak for Al Nanorods on AlGaN Layer to Deep Ultraviolet Band. *IEEE Photonics J.* **2021**, *13*, 1–7.
9. Knigge, A.; Brendel, M.; Brunner, F.; Einfeldt, S.; Knauer, A.; Kueller, V.; Weyers, M. AlGaN photodetectors for the UV-C spectral region on planar and epitaxial laterally overgrown AlN/sapphire templates. *Phys. Status Solidi C* **2013**, *10*, 294–297. [CrossRef]
10. Chen, X.; Ren, F.; Gu, S.; Ye, J. Review of gallium-oxide-based solar-blind ultraviolet photodetectors. *Photonics Res.* **2019**, *7*, 381–415.
11. Shiojiri, D.; Fukuda, D.; Yamauchi, R.; Tsuchimine, N.; Koyama, K.; Kaneko, S.; Matsuda, A.; Yoshimoto, M. Room-temperature laser annealing for solid-phase epitaxial crystallization of beta-Ga_2O_3 thin films. *Appl. Phys. Express* **2016**, *9*, 105502.
12. Park, T.; Kim, K.; Hong, J. Effects of Drying Temperature and Molar Concentration on Structural, Optical, and Electrical Properties of β-Ga_2O_3 Thin Films Fabricated by Sol–Gel Method. *Coatings* **2021**, *11*, 1391.
13. Pearton, S.; Yang, J.; Cary, P.H., IV; Ren, F.; Kim, J.; Tadjer, M.J.; Mastro, M.A. A review of Ga_2O_3 materials, processing, and devices. *Appl. Phys. Rev.* **2018**, *5*, 011301.
14. Lin, T.-Y.; Han, S.-Y.; Huang, C.-Y.; Yang, C.-F.; Wei, S. Using Different Methods to Measure the Optical Energy Bandgap of Un-annealed and Annealed Ga_2O_3 Films. *AFM* **2021**, *1*, 25–30.
15. Jiao, S.; Lu, H.; Wang, X.; Nie, Y.; Wang, D.; Gao, S.; Wang, J. The structural and photoelectrical properties of gallium oxide thin film grown by radio frequency magnetron sputtering. *ECS J. Solid State Sci. Technol.* **2019**, *8*, Q3086.
16. Mukhopadhyay, P.; Schoenfeld, W.V. High responsivity tin gallium oxide Schottky ultraviolet photodetectors. *J. Vac. Sci.* **2020**, *38*, 013403.
17. Bae, H.; Charnas, A.; Sun, X.; Noh, J.; Si, M.; Chung, W.; Qiu, G.; Lyu, X.; Alghamdi, S.; Wang, H. Solar-Blind UV Photodetector Based on Atomic Layer-Deposited Cu_2O and Nanomembrane β-Ga_2O_3 pn Oxide Heterojunction. *ACS Omega* **2019**, *4*, 20756–20761.
18. Chiou, Y.-Z.; Su, Y.-K.; Chang, S.-J.; Gong, J.; Lin, Y.-C.; Liu, S.-H.; Chang, C.-S. High detectivity InGaN-GaN multiquantum well pn junction photodiodes. *IEEE J. Quantum Electron.* **2003**, *39*, 681–685.
19. Pierson, J.; Rousselot, C. Stability of reactively sputtered silver oxide films. *Surf. Coat. Technol.* **2005**, *200*, 276–279.
20. Fukaya, T.; Buchel, D.; Shinbori, S.; Tominaga, J.; Atoda, N.; Tsai, D.P.; Lin, W.C. Micro-optical nonlinearity of a silver oxide layer. *J. Appl. Phys.* **2001**, *89*, 6139–6144. [CrossRef]
21. Standke, B.; Jansen, M. Ag_2O_3, a novel binary silver oxide. *Angew. Chem. Int. Ed. Engl.* **1985**, *24*, 118–119. [CrossRef]
22. Mansour, A. Evidence for an Ag_4O_3 phase of silver oxide. *J. Phys. Chem.* **1990**, *94*, 1006–1010.
23. Dellasega, D.; Facibeni, A.; Di Fonzo, F.; Bogana, M.; Polissi, A.; Conti, C.; Ducati, C.; Casari, C.S.; Bassi, A.L.; Bottani, C.E. Nanostructured Ag_4O_4 films with enhanced antibacterial activity. *Nanotechnology* **2008**, *19*, 475602.
24. Barik, U.K.; Srinivasan, S.; Nagendra, C.; Subrahmanyam, A. Electrical and optical properties of reactive DC magnetron sputtered silver oxide thin films: Role of oxygen. *Thin Solid Films* **2003**, *429*, 129–134.
25. Hammad, A.; Abdel-Wahab, M.; Alshahrie, A. Structural and morphological properties of sputtered silver oxide thin films: The effect of thin film thickness. *Dig. J. Nanomater. Bio.* **2016**, *11*, 1245–1252.
26. Fakhri, M.A. Annealing effects on opto-electronic properties of Ag_2O films growth using thermal evaporation techniques. *Int. J. Nanoelectron. Mater.* **2016**, *9*, 93–102.
27. Yahia, K.Z. Study Optoelectronic Properties of Ag_2O Heterojunction Prepered by Thermal Oxidation Technique. *J. Eng. Technol.* **2008**, *26*, 570–578.
28. Dellasega, D.; Facibeni, A.; Di Fonzo, F.; Russo, V.; Conti, C.; Ducati, C.; Casari, C.S.; Bassi, A.L.; Bottani, C.E. Nanostructured high valence silver oxide produced by pulsed laser deposition. *Appl. Surf. Sci.* **2009**, *255*, 5248–5251.
29. Wu, Q.; Si, M.; Zhang, B.; Zhang, K.; Li, H.; Mi, L.; Jiang, Y.; Rong, Y.; Chen, J.; Fang, Y. Strong damping of the localized surface plasmon resonance of Ag nanoparticles by Ag_2O. *Nanotechnology* **2018**, *29*, 295702.
30. Narayana Reddy, P.; Sreedhar, A.; Hari Prasad Reddy, M.; Uthanna, S.; Pierson, J. The effect of oxygen partial pressure on physical properties of nano-crystalline silver oxide thin films deposited by RF magnetron sputtering. *Cryst. Res. Technol.* **2011**, *46*, 961–966.

31. Park, S.; Yoon, Y.; Lee, S.H.Y.; Park, T.; Kim, K.; Hong, J. Thermoinduced and Photoinduced Sustainable Hydrophilic Surface of Sputtered-TiO_2 Thin Film. *Coatings* **2021**, *11*, 1360.
32. Hong, J.S.; Matsushita, N.; Kim, K.H. Investigation of the effect of oxygen gas on properties of GAZO thin films fabricated by facing targets sputtering system. *Semicond. Sci. Technol.* **2014**, *29*, 075007.
33. Hong, J.; Matsushita, N.; Kim, K. Effect of dopants and thermal treatment on properties of Ga-Al-ZnO thin films fabricated by hetero targets sputtering system. *Thin Solid Films* **2013**, *531*, 238–242.
34. Hong, J.S.; Jang, K.W.; Park, Y.S.; Choi, H.W.; Kim, K.H. Preparation of ZnO Based Thin Films for OLED Anode by Facing Targets Sputtering System. *Mol. Cryst. Liq. Cryst.* **2011**, *538*, 103–111.
35. Labed, M.; Sengouga, N.; Labed, M.; Meftah, A.; Kyoung, S.; Kim, H.; Rim, Y.S. Modeling a Ni/beta-Ga_2O_3 Schottky barrier diode deposited by confined magnetic-field-based sputtering. *J. Phys. D Appl. Phys.* **2021**, *54*, 115102.
36. Hong, J.S.; Kim, S.M.; Kim, K.-H. Preparation of SiO_2 passivation thin film for improved the organic light-emitting device life time. *Jpn. J. Appl. Phys.* **2011**, *50*, 08KE02.
37. Anders, A. Tutorial: Reactive high power impulse magnetron sputtering (R-HiPIMS). *J. Appl. Phys.* **2017**, *121*, 171101.
38. Raju, N.R.C.; Kumar, K.J.; Subrahmanyam, A. Physical properties of silver oxide thin films by pulsed laser deposition: Effect of oxygen pressure during growth. *J. Phys. D Appl. Phys.* **2009**, *42*, 135411. [CrossRef]
39. Chen, C.-W.; Hsieh, P.-Y.; Chiang, H.-H.; Lin, C.-L.; Wu, H.-M.; Wu, C.-C. Top-emitting organic light-emitting devices using surface-modified Ag anode. *Appl. Phys. Lett.* **2003**, *83*, 5127–5129. [CrossRef]
40. Mukhtar, W.M.; Shaari, S.; Menon, P.S. Propagation of surface plasmon waves at metal thin film/air interface using modified optical waveguiding assembly. *Optoelectron. Adv. Mater.* **2013**, *7*, 9–13.
41. Abe, Y.; Hasegawa, T.; Kawamura, M.; Sasaki, K. Characterization of Ag oxide thin films prepared by reactive RF sputtering. *Vacuum* **2004**, *76*, 1–6. [CrossRef]
42. Hong, J.; Katsumata, K.-I.; Matsushita, N. Fabrication of Al-Doped ZnO Film with High Conductivity Induced by Photocatalytic Activity. *J. Electron. Mater.* **2016**, *45*, 4875–4880. [CrossRef]
43. Kim, H.; Seok, H.-J.; Park, J.H.; Chung, K.-B.; Kyoung, S.; Kim, H.-K.; Rim, Y.S. Fully transparent InZnSnO/β-Ga_2O_3/InSnO solar-blind photodetectors with high schottky barrier height and low-defect interfaces. *J. Alloys Compd.* **2022**, *890*, 161931. [CrossRef]
44. Yoon, Y.; Katsumata, K.-I.; Park, S.; Fujishima, A.; Hong, J. Enhanced Hydrogen Production at Optimum pH for the Recovery Cycle of β-FeOOH. *ACS Omega* **2022**, *7*, 16049–16054. [CrossRef]
45. Zhong, W.; Li, G.; Lan, L.; Li, B.; Chen, R. Effects of annealing temperature on properties of InSnZnO thin film transistors prepared by Co-sputtering. *RSC Adv.* **2018**, *8*, 34817–34822. [CrossRef]
46. Waterhouse, G.I.; Bowmaker, G.A.; Metson, J.B. The thermal decomposition of silver (I, III) oxide: A combined XRD, FT-IR and Raman spectroscopic study. *Phys. Chem. Chem. Phys.* **2001**, *3*, 3838–3845. [CrossRef]
47. Tjeng, L.-H.; Meinders, M.B.; van Elp, J.; Ghijsen, J.; Sawatzky, G.A.; Johnson, R.L. Electronic structure of Ag_2O. *Phys. Rev. B* **1990**, *41*, 3190–3199. [CrossRef]
48. Reddy, P.N.; Reddy, M.; Pierson, J.; Uthanna, S. Characterization of silver oxide films formed by reactive RF sputtering at different substrate temperatures. *Int. Sch. Res. Not.* **2014**, *2014*, 684317. [CrossRef]
49. Ghosh, B.K.; Rani, A.I.; Mohamad, K.A.; Saad, I. Low Leakage Current by Solution Processed PTAA-ZnO Transparent Hybrid Hetero-Junction Device. *Electron. Mater. Lett.* **2020**, *16*, 457–465. [CrossRef]
50. Alema, F.; Hertog, B.; Mukhopadhyay, P.; Zhang, Y.; Mauze, A.; Osinsky, A.; Schoenfeld, W.V.; Speck, J.S.; Vogt, T. Solar blind Schottky photodiode based on an MOCVD-grown homoepitaxial β-Ga_2O_3 thin film. *APL Mater.* **2019**, *7*, 022527. [CrossRef]
51. Kalita, G.; Dzulsyahmi Shaarin, M.; Paudel, B.; Mahyavanshi, R.; Tanemura, M. Temperature dependent diode and photovoltaic characteristics of graphene-GaN heterojunction. *Appl. Phys. Lett.* **2017**, *111*, 013504. [CrossRef]
52. Li, Y.; Li, X.; Gao, X. Effects of post-annealing on Schottky contacts of Pt/ZnO films toward UV photodetector. *J. Alloys Compd.* **2011**, *509*, 7193–7197. [CrossRef]
53. Kim, Y.; Ha, T.-K.; Cho, Y.-J.; Kang, Y.-S.; Yu, S.; Kim, G.; Jeong, H.; Park, J.K.; Kim, O. Severe hump phenomenon induced by increased charge trapping and suppression of electron capture effect in amorphous In-Ga-Zn-O thin-film transistors under unipolar pulsed drain bias with static positive gate bias stress. *Solid-State Electron.* **2020**, *167*, 107785. [CrossRef]
54. Michaelson, H.B. The work function of the elements and its periodicity. *J. Appl. Phys.* **1977**, *48*, 4729–4733. [CrossRef]
55. Venkata Krishna Rao, R.; Ranade, A.K.; Desai, P.; Kalita, G.; Suzuki, H.; Hayashi, Y. Temperature-dependent device properties of γ-CuI and β-Ga_2O_3 heterojunctions. *SN Appl. Sci.* **2021**, *3*, 1–9. [CrossRef]
56. Guo, D.; Li, P.; Wu, Z.; Cui, W.; Zhao, X.; Lei, M.; Li, L.; Tang, W. Inhibition of unintentional extra carriers by Mn valence change for high insulating devices. *Sci. Rep.* **2016**, *6*, 24190. [CrossRef]
57. Zhao, B.; Wang, F.; Chen, H.; Wang, Y.; Jiang, M.; Fang, X.; Zhao, D. Solar-blind avalanche photodetector based on single ZnO–Ga_2O_3 core–shell microwire. *Nano Lett.* **2015**, *15*, 3988–3993. [CrossRef]
58. Guo, D.; Su, Y.; Shi, H.; Li, P.; Zhao, N.; Ye, J.; Wang, S.; Liu, A.; Chen, Z.; Li, C. Self-powered ultraviolet photodetector with superhigh photoresponsivity (3.05 A/W) based on the GaN/Sn: Ga_2O_3 pn junction. *ACS Nano* **2018**, *12*, 12827–12835. [CrossRef]
59. Chen, Y.-C.; Lu, Y.-J.; Lin, C.-N.; Tian, Y.-Z.; Gao, C.-J.; Dong, L.; Shan, C.-X. Self-powered diamond/β-Ga_2O_3 photodetectors for solar-blind imaging. *J. Mater. Chem. C* **2018**, *6*, 5727–5732. [CrossRef]
60. Chen, X.; Liu, K.; Zhang, Z.; Wang, C.; Li, B.; Zhao, H.; Zhao, D.; Shen, D. Self-powered solar-blind photodetector with fast response based on Au/β-Ga_2O_3 nanowires array film Schottky junction. *ACS Appl. Mater. Interfaces* **2016**, *8*, 4185–4191. [CrossRef]

61. Yu, J.; Dong, L.; Peng, B.; Yuan, L.; Huang, Y.; Zhang, L.; Zhang, Y.; Jia, R. Self-powered photodetectors based on β-Ga_2O_3/4H–SiC heterojunction with ultrahigh current on/off ratio and fast response. *J. Alloys Compd.* **2020**, *821*, 153532. [CrossRef]
62. Chen, H.; Yu, P.; Zhang, Z.; Teng, F.; Zheng, L.; Hu, K.; Fang, X. Ultrasensitive self-powered solar-blind deep-ultraviolet photodetector based on all-solid-state polyaniline/MgZnO bilayer. *Small* **2016**, *12*, 5809–5816. [CrossRef]
63. Kan, H.; Zheng, W.; Lin, R.; Li, M.; Fu, C.; Sun, H.; Dong, M.; Xu, C.; Luo, J.; Fu, Y. Ultrafast photovoltaic-type deep ultraviolet photodetectors using hybrid zero-/two-dimensional heterojunctions. *ACS Appl. Mater. Inter.* **2019**, *11*, 8412–8418. [CrossRef]
64. Zhuo, R.; Wu, D.; Wang, Y.; Wu, E.; Jia, C.; Shi, Z.; Xu, T.; Tian, Y.; Li, X. A self-powered solar-blind photodetector based on a MoS_2/β-Ga_2O_3 heterojunction. *J. Mater. Chem. C* **2018**, *6*, 10982–10986. [CrossRef]
65. Wu, Z.; Jiao, L.; Wang, X.; Guo, D.; Li, W.; Li, L.; Huang, F.; Tang, W. A self-powered deep-ultraviolet photodetector based on an epitaxial Ga_2O_3/Ga: ZnO heterojunction. *J. Mater. Chem. C* **2017**, *5*, 8688–8693. [CrossRef]

nanomaterials

MDPI

Review

Recent Progress on Triboelectric Nanogenerators for Vibration Energy Harvesting and Vibration Sensing

Ahmed Haroun *, Mohamed Tarek, Mohamed Mosleh and Farouk Ismail

Department of Mechanical Design and Production Engineering, Cairo University, Giza 12613, Egypt
* Correspondence: eng.ahmedharoun@cu.edu.eg

Abstract: The triboelectric nanogenerator (TENG) is a recent technology that reforms kinetic energy generation and motion sensing. A TENG comes with variety of structures and mechanisms that make it suitable for wide range of applications and working conditions. Since mechanical vibrations are abundant source of energy in the surrounding environment, the development of a TENG for vibration energy harvesting and vibration measurements has attracted a huge attention and great research interest through the past two decades. Due to the high output voltage and high-power density of a TENG, it can be used as a sustainable power supply for small electronics, smart devices, and wireless sensors. In addition, it can work as a vibration sensor with high sensitivity. This article reviews the recent progress in the development of a TENG for vibration energy harvesting and vibration measurements. Systems of only a TENG or a hybrid TENG with other transduction technologies, such as piezoelectric and electromagnetic, can be utilized for vibrations scavenging. Vibration measurement can be done by measuring either vibration displacement or vibration acceleration. Each can provide full information about the vibration amplitude and frequency. Some TENG vibration-sensing architectures may also be used for energy harvesting due to their large output power. Numerous applications can rely on TENG vibration sensors such as machine condition monitoring, structure health monitoring, and the Internet of things (IoT).

Keywords: triboelectric nanogenerator; vibration energy harvesting; hybrid generators; triboelectric displacement sensors; triboelectric acceleration sensors

Citation: Haroun, A.; Tarek, M.; Mosleh, M.; Ismail, F. Recent Progress on Triboelectric Nanogenerators for Vibration Energy Harvesting and Vibration Sensing. *Nanomaterials* **2022**, *12*, 2960. https://doi.org/10.3390/nano12172960

Academic Editors: Jyh-Ming Wu and Dong-Joo Kim

Received: 27 July 2022
Accepted: 22 August 2022
Published: 26 August 2022

1. Introduction

Energy harvesting, which is the process of capturing and converting dissipated energy from the surrounding into electricity [1], has become an attractive field of research. It is considered as an effective self-powering technique for wireless sensors. Several forms of wasted energy exist in the surrounding environment such as wind energy, solar energy, and vibrational energy. Mechanical vibrations are an abundant source of energy [2,3] that appear in everyday life (Figure 1) in vehicles [4,5], structures such as buildings and bridges [6], machines [7], etc. Vibration energy harvesting can be an effective powering strategy for wireless sensors that are placed in deep and dark locations, where there is no exposure to air or light [8,9]. Three main technologies of energy transduction have been used for years to convert mechanical energy into electricity, which are piezoelectric [10–17], electromagnetic [18–24], and electrostatic [25–30]. A newly developed electrostatic technology that can convert mechanical energy into electricity has been introduced by Wang's group in 2012, which is the triboelectric nanogenerator (TENG) [31–33]. TENG technology is based on combination between contact electrification and electrostatic induction [34]. It has been introduced as a promising technology to scavenge mechanical energy from many available sources [35] in the ambient environment such as vibrations [36–39], human motion [39,40], waves [41,42], and wind [43,44]. The triboelectric nanogenerator shows some advantages when compared with other technologies, as it exhibits a large output voltage [45], high efficiency [46], low cost [47], a simple fabrication method [48], great broadband behavior, excellent robustness, reliability [49], and eco-friendliness, as well as efficient low-frequency vibration energy scavenging [32–35,47]. According to those features of a TENG as well as the low power consumption of modern wireless sensors, a TENG can be a sustainable power source (Figure 1) for many wireless-sensing applications [50,51].

Figure 1. TENG can convert mechanical vibrations into electricity, which can be used for electrical powering or vibration sensing applications.

Coupling a TENG with other generator types, such as piezoelectric (PEG) and electromagnetic (EMG) generators, to form a hybrid is also introduced [52]. A hybrid generator can take the advantages of multiple generator types and enhance the output performance at different input motion conditions. For example, an EMG—TENG can provide higher output voltage, higher output current, and wider bandwidth than each generator separately [53].

The same concept of converting mechanical vibrations into electricity is also applied for vibration sensing (Figure 1). A self-powered vibration sensor can detect vibration signals without the need for a battery as a power unit. Vibration measurement is essential for many applications. Machine-condition monitoring is one of the important applications that allows the detection/prediction of machine damages, abnormal stoppings, and disasters. The accurate monitoring of a machine's condition and process is very essential to enhance the feedback system and optimize the operation. Some transduction technologies have been developed for vibration sensing, such as magnetoresistive [54,55], piezoresistive [55], piezoelectric [56], etc. Each can be fitted to specific applications [57].

A TENG can transfer vibrations or other kinds of motion such as triggering, rotational motion, linear displacement, or physical motion into a high output voltage. Thus, it can be used as a self-powered vibration sensor with high sensitivity [57–59]. Mechanical vibrations are specified by frequency and amplitude. Vibration frequency can be easily detected by most vibration sensors, however, measuring vibration amplitude is more difficult. Measuring one of the vibration physical characteristics such as displacement, velocity, and acceleration with time provides information about the vibration amplitude and frequency. Some TENGs have been developed to measure one of the vibration physical characteristics [57–59]. In this article, we summarize the recent progress of utilizing a TENG for energy harvesting and sensing the mechanical vibration of physical objects in the last 10 years. First, the working mechanism and some examples of a TENG, as well as a hybrid TENG for vibration energy scavenging, are presented. Then, measuring mechanical vibrations using a TENG by measuring either vibration displacement or vibration acceleration are reviewed. Finally, some TENGs that can work effectively for both vibration energy harvesting and vibration sensing are described.

2. Working Mechanism of a TENG

There are four operational modes of a triboelectric nanogenerator (TENG): vertical contact-separation mode, in-plane sliding mode, single-electrode mode, and free-standing triboelectric-layer mode, as shown in Figure 2 [31].

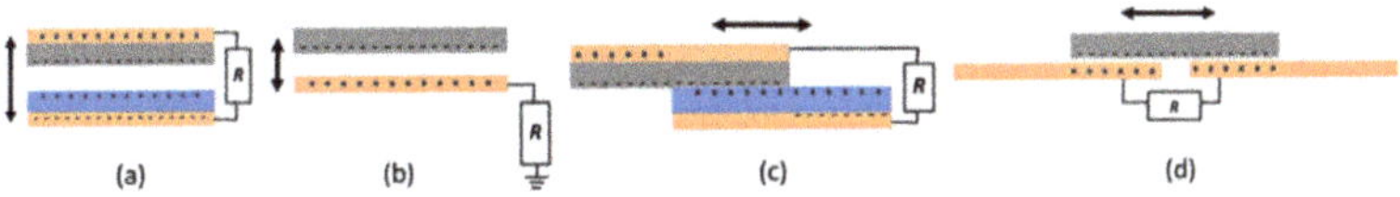

Figure 2. TENG modes: (**a**) vertical contact-separation mode, (**b**) single-electrode mode, (**c**) lateral-sliding mode, and (**d**) freestanding triboelectric-layer mode.

In the vertical contact-separation mode (Figure 2a), two surfaces from dissimilar materials come into contact and then are vertically separated. As a result, oppositely electrostatic charges are generated on the surfaces. Repeated contact-separation and varying the gap between surfaces by mechanical vibrations or another repeated motion generate voltage by electrostatic induction. Both surfaces are attached to electrodes for electrons transfer when a TENG is included in a closed circuit. Thus, mechanical energy is converted to electrical energy [31].

In the single-electrode mode, the triboelectrification is accomplished by contact and separation, as in the contact-separation mode. However, the ground is taken as a reference electrode. This mode is capable of harvesting energy from a freely moving object that is difficult to be attached to a lead or wire (Figure 2b).

The lateral-sliding mode (Figure 2c) has the same structure and starting position as the vertical contact-separation mode. However, triboelectric charges are created on both surfaces by the relative parallel sliding. The lateral polarization along the sliding direction generates triboelectric charges on the contact surfaces, which drive the electrons on the electrode to flow. With repeated sliding and closing, an AC output voltage is generated. This sliding can be in the form of planar motion, disc rotation, or cylindrical rotation [5].

The basic idea of freestanding triboelectric-layer mode (Figure 2d) is that a pair of symmetric electrodes are connected to the external circuit, and the movement of the freestanding triboelectric layer between them allows the current to flow from one electrode to another. The current flow direction depends on the motion direction of the freestanding layer. Usually, the freestanding layer is made from dielectric material, or both electrodes are placed underneath a dielectric layer [60]. The freestanding triboelectric layer mode can take one of two configurations: a sliding-freestanding TENG or a contact-freestanding TENG [61].

In most of vibration harvesting devices and sensors, the environmental vibrations are transmitted into internal oscillations inside the device via a spring-mass system. Such internal relative oscillations are converted into electricity by electromechanical transduction. For example, the relative oscillation between a permanent magnet and a coil generates EMF by electromagnetic induction, and the relative oscillation between two connected bodies by a piezoelectric element causes deformation/stress on the piezoelectric element, which in response generates electrostatic voltage. However, when triboelectricity is utilized for energy transduction, sort of periodic contact-separation or sliding movements between two surfaces should be implemented. A spring or a kind of elastic/elastomeric mechanism that can provide a restoring force is introduced in the TENG as well, to realize such movements. It is worth mentioning that, when the TENG is utilized for harvesting relative vibration motion between two external surfaces, no mechanical springs are required. The external relative motion can be directly converted into electricity by one of the TENG's working modes based on the nature of the relative motion. The TENG in that case becomes a non-resonant harvester, by which increasing the input frequency increases the frequency of the electrical output.

3. Triboelectric Nanogenerators for Vibration Energy Harvesting

The triboelectric nanogenerator (TENG) is considered as an effective and powerful technology to scavenge electricity from repetitive movements of objects, especially at low frequencies [62]. Mechanical vibration is a kind of repetitive movement that appears in everyday life. TENG technology shows great advantages that are represented in high output voltage, low cost, multiple optional materials, simplicity, and efficient response to low frequency—unsteady motion [63]. Those advantages promote the TENG for motion energy harvesting, especially from random vibrations. Harvesting mechanical vibrations can serve many important applications such as machine and engine monitoring, structure health monitoring, IoT, etc. Thus, researchers turn to take the advantages of the TENG for vibration energy harvesting. A TENG can be used on its own for energy transduction in the harvesting device, or in hybrid with other transduction technologies such as piezoelectric and electromagnetic. Here, we reviewed some of the TENG and hybrid TENG vibration energy harvesters. Table 1 also shows a summary of the reviewed harvesting systems showing their main features and output performance.

Table 1. Summary of the reviewed TENGs for vibration energy harvesting showing their main features, dimensions, and output performance.

TENG Energy Harvester	Only TENG/Hybrid	TENG Mode	TENG Materials	Dimensions	Input Excitation	Electrical Output
3D Stack Integrated (Yang et al.) [64]	TENG	Contact-separation	PTFE/Al	231.67 cm^3	40 Hz	I_{sc} = 1.14 mA, V_{oc} = 303 V, P_d = 104.6 W m^{-2} @ 2 MΩ
QZS-TENG (Wang et al.) [65]	TENG	Sliding	Bonded glass/metal	≅ 471.24 cm^3	3 Hz	P ≅ 4.06 mW
3D-TENG (Yang et al.) [66]	TENG	Contact-separation + in plane-sliding	PTFE/Al	≅23.56 cm^3	6 m/s^2 and 36 Hz	V = 123 V, I = 21 μA
CIT-TENG (Bhatia et al.) [67]	TENG	Contact-separation (impact)	PTFE/Al	25 × 7 × 5.5 cm^3	0.2g m/s^2@ 40 Hz range	5 V and 0.75 μA from air compressor
Gas-enhanced (Lv et al.) [68]	TENG	Contact-separation	FEP/conductive fabric	5 × 5 × 5 cm^3	3 Hz	V_o = 500 V, I_{sc} = 11 μA, Max. P = 0.8 W @ 8 MΩ
Marine pipe EH (Li et al.) [69]	TENG	Contact-separation	PTFE and nylon/nylon	5 × 5 × 1 cm^3	8 Hz and 4 mm	V = 140 V, Max. P = 14 μW @ 200 MΩ
PNG-EMG-TENG (Hemojit et al.) [70]	Hybrid	Contact-separation	PTFE/gold	6 × 6 × 6 cm^3	ND	V_{oc} = 192 V, I_{sc} = 2.78 mA
PET (He et al.) [71]	Hybrid	Contact-separation	Patterned PDMS/steel	Ø 51 × H 37 mm^2	0.5 g m/s^2 and 20 Hz	P= 41 μW, V = 12.6 V @ 800.1 KΩ
TPE (He et al.) [72]	Hybrid	Contact-separation	SR + CNT	Ø 48 × H 27 mm^2	20 Hz	V = 2V, I = 1 mA.
MFKI-HNG (Toyabur et al.) [40]	Hybrid	Contact-separation	PTFE/Al	Ø 26 × H 50.5 mm^2	1g m/s^2 and 5 Hz	P_d = 3.67 mW/cm^3
EMG-TENG (Zhu et al.) [73]	Hybrid	Contact-separation	PTFE/CB film	6 × 2 × 1 cm^3	1g m/s^2 and 18 Hz	TENG: 30 μW and EMG: 53 μW
Polymer-based EMG-TENG (Kumar et al.) [74]	Hybrid	Contact-separation	PTFE/ITO	2 × 2 × 1 cm^3	0.1–2 g m/s^2	TENG: 4.8 $\mu W/cm^3$, EMG: 6 $\mu W/cm^3$ (max)
DETEG (Haroun et al.) [75]	Hybrid	Contact-separation	PTFE/Al	8 × 7 × 6 cm^3	5 Hz	P_d= 500 W/m^2

3.1. TENG-Based Vibration Energy Harvesting

Figure 3 shows sample prototypes of recently developed TENG-based vibration energy harvesters. According to the type and style of input motion, a certain mode of the TENG is employed. Yang et al. [64] proposed a 3D stack integrated energy harvester (Figure 3A), which utilizes a multilayer triboelectric nanogenerator. It consists of a 3D acrylic structure with pinned and movable fingers that are installed on a movable acrylic base. The fingers are made from 3 mm acrylic sheets and are placed parallel to each other. The movable base is attached to the structure by eight identical helical springs. The function of those springs is to allow the movement of the movable fingers. This structure is based on the contact-separation TENG mode to harvest motion in a vertical direction.

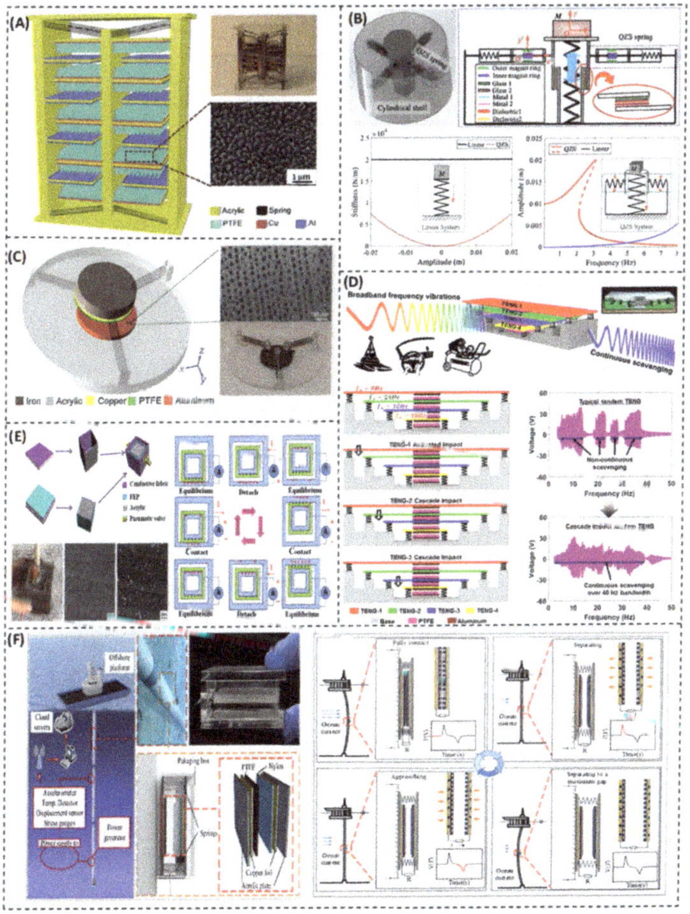

Figure 3. Vibration energy harvesting based on only TENG. (**A**) A 3D stack integrated triboelectric nanogenerator with a wide bandwidth of 36 Hz (Reprinted with permission [64], copyright 2014, John Wiley and Sons). (**B**) An energy harvester based on sliding mode triboelectric nanogenerator with a QZS system to gain an ultra-low stiffness in large displacement region (Reprinted with permission [65], copyright 2021, Elsevier). (**C**) A 3D TENG-based energy harvester that combines two operational TENG modes: vertical contact-separation mode and in-plane sliding mode (Reprinted with permission [66], copyright 2014, John Wiley and Sons). (**D**) A triboelectric nanogenerator with cascade structure based on impact to scavenge a wide range of input frequencies (Reprinted with permission [67], copyright 2019, Scientific Reports). (**E**) A gas-enhanced triboelectric nanogenerator based on fully enclosed structure that replaced the air with SF6 so that the output voltage and output current could be increased by 67% and 120%, respectively (Reprinted with permission [68], copyright 2018, Elsevier). (**F**) A contact-mode TENG to harvest vibrations of marine pipes (Reprinted with permission [69], copyright 2021, MDPI).

Thin films of aluminum are affixed to both sides of the pinned fingers to act as an electrode as well as a contact surface. Each side of the movable fingers has a layer of polytetrafluoroethylene (PTFE) with a layer of copper as another electrode. The purpose of the PTFE nanowires, as depicted in the SEM image of Figure 3A, is to increase the triboelectric charges and, thus, the overall electrical output.

The harvester was fabricated and tested through a wide range of vibration frequencies, varying from 2 Hz to 54 Hz. The resonant frequency appears at 40 Hz. Compared to other vibration energy harvesters, it shows a wide bandwidth of 36 Hz in the low frequency range. At the resonance frequency of 40 Hz, the open circuit voltage (Voc) and the short circuit current (Isc) are, respectively, 303 V and 1.14 mA. It has been noticed that with the same number of charges transported back and forth, a faster approaching produces larger current peaks than that during a slower separation. One of the important features of this 3D-TENG harvester is that it can serve different applications based on specific dimensional design. For example, it can be integrated into a ball with arbitrary size and be used to harvest human body kinetic energy while doing some sports such as basketball, football, baseball, etc. Moreover, it can efficiently harvest ocean waves by weaving together a large amount of those self-powered balls [64].

Wang et al. [65] introduced an energy harvester based on the unique idea of combining a quasi-zero-stiffness (QZS) mechanism and a sliding-mode triboelectric nanogenerator, as shown in Figure 3B. The main advantage of the QZS-TENG is to enhance the harvesting performance in the ultra-low frequency region (less than 4 Hz). The QZS-TENG shows a maximal output power of 6 mW at ultra-low frequencies [65]. The main purpose of the QZS mechanism is to create a system with a low resonant frequency (Figure 3B). The stiffness of the linear system is much larger than that of the QZS system. As a result of the low stiffness feature, the vibration response amplitude in the ultra-low frequency region of the linear system is lower than that of the QZS system.

As illustrated in Figure 3B, the QZS spring consists of two magnet rings (with green and blue colors) and two sliding bearings. The sliding bearings are fixed to the QZS spring to reduce the friction between the rod and an axial spring, while its stiffness is neutralized with the pair of magnetic rings. When the QZS-TENG is exposed to an external excitation, the QZS spring will rotate around the hinge, which is fixed onto the cylindrical shell, and the inner magnet ring moves from the equilibrium position in the horizontal direction by a distance. This will result in a thrust force along the axial direction. So, a negative stiffness will be provided by the QZS spring for the QZS-TENG along the vertical direction. Under the platform of mass M, there is a sleeve with four bonded glasses fixed on the frame of the QZS-TENG and four metal electrodes bonded with four dielectrics. Both the metal electrode and the dielectric are bonded on the glass. There are four similar configurations, which are fixed onto the platform, so that a sliding-mode triboelectric nanogenerator is formed, as shown in Figure 3B. This unique idea can be used in several applications of energy harvesting such as environmental and structure health monitoring [65].

Yang et al. [66] developed a multiple direction broadband TENG-based vibration energy harvester (Figure 3C). The developed system has an innovative design which could operate by combining two modes of the TENG: the vertical contact-separation mode and the in-plane sliding mode. This hybrid mode system allows harvesting random vibrations with different directions and with a wide bandwidth of relatively high frequencies. As schematically shown in Figure 3C, this 3D-TENG system consists of an acrylic base, a cylindrical core that lies in the center of the base, and three identical springs separated by an angle of 120° between each other. Those springs are connected to the cylindrical mass and the acrylic supporting base. The symmetrical design of the system allows it to have a constant resonant frequency in in-plane directions. There is a layer of polytetrafluoroethylene (PTFE) thin film pasted to the bottom side of the cylindrical mass as well as a copper electrode that acts as a back electrode. Moreover, there is a thin film of aluminum adhered to the acrylic base that has a dual function of a contact surface and a conductive electrode. The vertical contact-separation mode shows a wide working bandwidth of up to 75 Hz, while the in-plane sliding mode has a relatively lower bandwidth of 14.4 Hz [66]. This 3D-TENG has been tested, and the results were very promising for use in several practical applications such as self-powered sensors, infrastructure monitoring, and charging portable electronics. It has a maximum power density of 1.35 W/m^2 for the out-of-plane excitations and 1.45 W/m^2 for the in-plane excitations. These results exhibit the influence of hybridization between the two modes of the TENG.

For unsteady random vibrations, Bhatia et al. [67] proposed a tandem triboelectric nanogenerator with cascade impact structure (CIT-TENG) to continuously scavenge broadband vibrations (Figure 3D). The design concept of this harvester is based on resonance. It consists of four TENGs; each one has a different resonant frequency that enables the harvester to operate at a wide frequency range from 0 to 40 Hz. It can also work and produce remarkable output power away from resonance, which is due to the cascade impact structure of the harvester. Simulation and experimental results show that the impact dynamics of the CIT-TENG enables scavenging energy even under low input accelerations of 0.2–0.5 G m/s^2. It can also produce a continuous output voltage of 15 V over 40 Hz bandwidth. This CIT-TENG can be utilized in several applications without the need for redesign validation, such as a car dashboard, water waves, and air compressor machines [67].

A gas-enhanced triboelectric nanogenerator based on a fully enclosed structure (Figure 3E) was developed by Lv et al. [68]. They came up with the idea of using SF6 atmosphere instead of air. This increases the output voltage and current by 67% and 120%, respectively, compared to normal atmospheric air. The proposed cubic TENG is composed of two boxes inside each other. The inner box dimensions are $3 \times 3 \times 3$ cm^3, while the outer box dimensions are $5 \times 5 \times 5$ cm^3. The outer surfaces are covered with the etched fluorinated ethylene propylene (FEP) film with a thickness of 0.1 mm, except the top and bottom surface of the inner box. Two small springs are attached to the bottom center of the outer box, while the inner box was fixed with a glue on the top of the two springs. The box is completely sealed with a waterproof sealant.

This cubic TENG can harvest vibrational energy from four sides, as the triboelectric material covers the cubic faces. When the box is subjected to mechanical vibrations, the FEP film on the inner box makes contact-separation with the conductive fabric on the outer box. Thus, the two surfaces are positively and negatively charged, and a current is generated through the external circuit by electrostatic induction. The cubic TENG was optimized when filled with SF6, which produced a maximum output voltage of 500 V and a current of 11 μA. Those values were not affected by the ambient humidity.

Some TENG-based vibration energy harvesters are designed for specific applications. For example, Li et al. [69] introduced a contact-separation mode TENG to harvest marine pipe vibrations (Figure 3F). There are many sensors for structural monitoring of marine pipes, which need to be powered. This TENG exploits the vibration of the pipes, which comes from the ocean, to power the sensors. The device is composed of dielectric material films adhered to a simple mass-spring system, to ensure the contact-separation motions of the triboelectric materials [69]. The working cycle of this device has four stages, which occur in sequence: fully contacting, separating, separating to a maximal distance, and approaching. Each stage is shown in Figure 3F. The experimental tests have proven that the electrical output of the TENG device could meet the requirements of the monitoring systems in pipes, since the output power of the TENG has a maximum value of 14 μW with a power density of 5.56 mW/m^2 at a resistance of 200 MΩ.

3.2. *Hybrid TENG-Based Vibration Energy Harvesting*

The hybridization of a TENG with other transduction mechanisms, such as piezoelectric and electromagnetic, is proposed to improve the output performance. A hybrid generator is usually developed for higher output power, higher power density, higher energy conversion efficiency, wider bandwidth, better environmental adaptability, and/or better applicability. Singh et al. [70] have designed a harvester that utilizes the three transduction techniques to get the most benefit from the size and achieve the highest power density in a cuboid shape (Figure 4A). The harvester can produce an open-circuit output voltage of 192 V because of the hybridization. It uses a ZnO–PVDF film for the combination of the piezoelectric and the triboelectric effects. The harvester design allows the acquisition of the simultaneous output power of each effect from a single vibration direction. It consists of two acrylic bases connected by four identical springs, which allow the excitation of the system in the vertical direction. The vertical excitation creates a contact-separation motion on the TENG as well as pressure on the piezoelectric material, as shown in Figure 4A. Due to the dual nature of the ZnO–PVDF material, the triboelectric effect and the piezoelectric effect have been combined [70]. The electromagnetic harvesting effect is implemented by a copper coil attached to the upper acrylic base, and a permanent magnet fixed to the other base. When the two bases approach each other, the magnetic flux changes across the coil and generates electricity by electromagnetic induction (EMG). The performance of this hybrid generator is checked with some applications just like lighting several LEDs connected in series, powering a digital calculator, and a screw gauge meter [70].

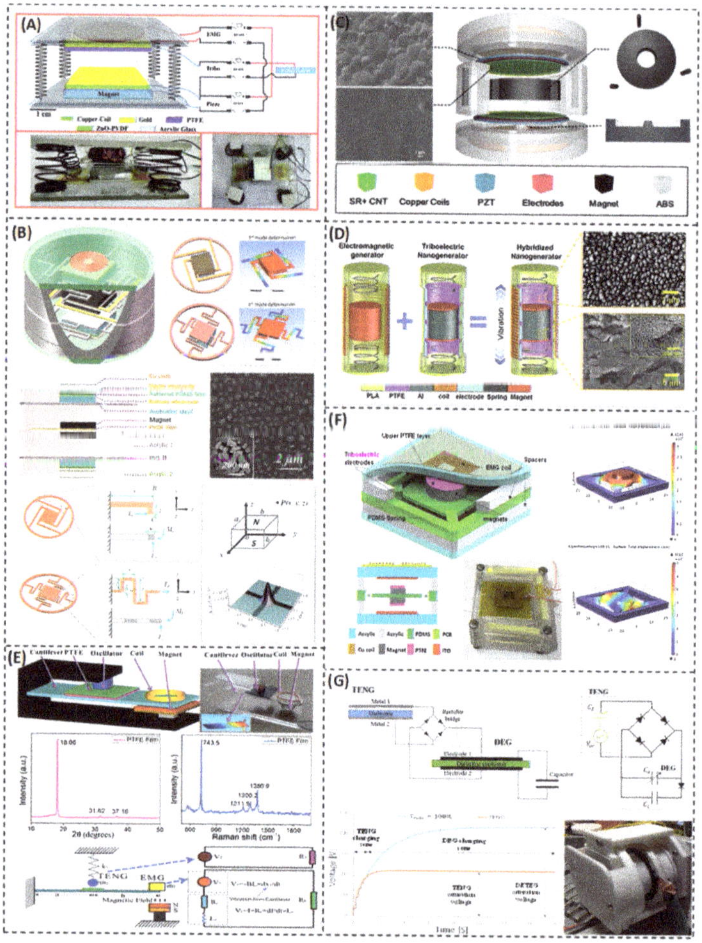

Figure 4. Hybrid TENG-based vibration energy harvesting. (**A**) Vibration energy harvester based on combining different devices in order to increase the efficiency and applicability (Reprinted with permission [70], copyright 2021, Royal Society of Chemistry). (**B**) A piezoelectric–electromagnetic–triboelectric hybrid energy harvester that can scavenge vibrations with frequency ranged from 3.5 Hz

to 10 Hz (Reprinted with permission [71], copyright 2017, Elsevier). (**C**) Cylindrical triboelectric–piezoelectric–electromagnetic hybrid nanogenerator that is mainly based on a magnetic levitating structure (Reprinted with permission [72], copyright 2017, Elsevier). (**D**) A hybridized electromagnetic–triboelectric nanogenerator for harvesting human vibrations that are less than 5 Hz (Reprinted with permission [40], copyright 2020, Elsevier). (**E**) An electromagnetic–triboelectric nanogenerator based on multi-impact technique for harvesting wideband frequency vibrations (Reprinted with permission [73], copyright 2017, MDPI). (**F**) Broadband energy harvester utilizing non-linear polymer spring and electromagnetic–triboelectric hybrid harvesting mechanism (Reprinted with permission [74], copyright 2021, Scientific Reports). (**G**) Hybrid dielectric elastomer–triboelectric nanogenerator with enhanced output performance (Reprinted with permission [75], copyright 2021, Elsevier).

The hybrid generators can have the advantage of broadband performance, which can work effectively at a wide range of input frequencies. He et al. [71] introduced a piezoelectric–electromagnetic–triboelectric (PET) hybrid generator (Figure 4B), which can efficiently scavenge vibrational energy with frequency ranging from 5 to 40 Hz. It has an output voltage of 12.6 V at 20 Hz and 13.3 V at 24 Hz [71]. The structure of the generator is in a form of cylindrical acrylic case with three layers of hybrid energy harvesters, as shown schematically in Figure 4B. The middle layer is a piezoelectric–electromagnetic hybrid harvester. It has a unique structure, which consists of four L-shaped PET beams with flexible PVDF piezo film pasted on them, two permanent magnets, and other two magnetic coils. This structure is called the picking up vibration structure (PVS) I. The top and bottom layers of the harvester are the same and are placed symmetrically. Each one is a triboelectric energy harvester, which is based on the picking up vibration structure (PVS) II. (PVS) II includes four PET folded beams. Each harvester of the three previously is considered as a lumped spring-mass-damper system. Each system has a resonant frequency according to its structure. Both structures have two close resonant frequencies. In addition, the excitation of any of these harvester causes the other harvester to be excited, due to them hitting against each other. This allows the cylindrical PET energy harvester to operate at a very low and wide range of frequencies [71].

He et al. [72] presented a compact hybrid generator (Figure 4C) with high power density that is able to work efficiently at low frequencies. The proposed harvester has a cylindrical compact design with dimensions of 48 mm × 27 mm. It has a light weight due to the low-density materials and is capable of scavenging vibrational energy under 20 Hz. This low frequency operating range is a result of combining the three mechanisms of transduction: piezoelectric, triboelectric, and electromagnetic. The design of this PET harvester is so simple and has an innovative idea for the mechanism of movement. It is all about the movable core, which is a permanent magnetic core levitating in the middle of three cylindrical magnets in a triangle-shape distribution. According to the layout design of the harvester (Figure 4C), the levitated annular magnet has three roles in generating electricity from the PEG, EMG, and TENG. The first role is generating electricity from the piezoelectric effect, due to the impact between the PZT sheets and PEG units in the harvester. The second one is changing the magnetic flux in the two copper coils existing in the top and bottom of the system. This results in generating electricity by electromagnetic induction. The third role is from there being a triboelectric electrode in the TENG unit, which delivers electrons between the two rubber layers at the top and bottom of the harvester during the up and down vertical oscillation [72]. The presented harvester can generate 78.4 μW output power due to the triboelectric nanogenerator (TENG), 36 mW and 38.4 mW due to the electromagnetic generators in the top and bottom of the system, respectively, and 122 mW and 105 mW from the piezoelectric generator in the top and bottom of the harvester, respectively [72].

Energy harvesting of ultra-low frequency vibrations such as human walking can serve important applications. It can be a sustainable power source for portable electronics and wearable sensors [45,76,77]. Rahman et al. [40] invented a miniaturized freestanding kinetic-impact-based hybridized nanogenerator (MFKI-HNG) that is capable of harvesting energy from human body motion in the ultra-low frequency range (Figure 4D). This MFKI-HNG is designed to generate electricity simultaneously from an electromagnetic generator (EMG) and a triboelectric nanogenerator (TENG) in a simple way. The system configuration is like a battery, as they both have almost the same shape and size. The harvester consists of a cylindrical aluminum mass oscillating in a cylindrical tube, and two cylindrical magnets are fixed to the top and the bottom of the mass. Moreover, there are two identical springs; each has a fixed end to the end of the tube, while the other end is free. The copper coil is wound over the outer surface of the tube, where the magnetic flux changes due to the magnets' movement up and down and, consequently, generates electricity by electromagnetic induction. A layer of triboelectric material partially covers the internal surface of the tube, which creates, with the cylindrical mass, a sliding mode TENG and generates electricity by the triboelectric effect [40]. Figure 4D shows the results of the harvester output. By optimizing the parameters, it can generate an output voltage of 18 V under 6 Hz. It also generates a maximum power of 102.29 mW and a power density of 3.67 mW/cm^3. This power value can be sufficient for powering smartphones and smartwatches.

Due to the convenience and effectiveness of a device like that invented by Rahman et al. [40], a few hybrid generators were presented with the same concept but with different designs. One was introduced by Seol et al. [78] in 2016. A magnetic-based spring was utilized instead of a mechanical one. Two permanent magnets were fixed to the ends of the tube, which provides repulsive forces, and two copper coils were wound over the tube. The operation range of frequencies was under 10 Hz and the output power from the EMG and the TENG was 128 W/Kg m^3 and 130 W/Kg m^3, respectively. Another energy harvester was that presented by Salauddin et al. [79] in 2017. It was composed of two rectangular-shaped cylinders inside each other with a magnetic stiffness between them and copper coils. The fabricated hybrid energy harvester gave an output power of 10.07 mW and an output current of 3.74 mA at a resonant frequency of 4.5 Hz, which corresponded to a 344 W/m^3 volume power density.

Zhu et al. [73] presented a hybrid broadband TENG—EMG device. As shown in Figure 4E, the device consists of a PTFE cantilever, where the TENG and EMG are placed at the middle and the end of the cantilever, respectively. It has a wide operating low frequency bandwidth (10–45 Hz) due to the multi-impact approach, which introduces a nonlinearity in the system. It also has a small-size device, which could be suitable for powering small electronics just like smartphones and smartwatches. From the results shown in Figure 4E, the proposed harvester can generate an average power from the TENG and the EMG up to 30 μW/m^2 and 53 μW/m^2, respectively [73].

A way of combining non-linear stiffening effect with the dual-way energy harvesting (TENG and EMG) is introduced by Gupta et al. [74]. They presented a TENG—EMG hybrid generator with a wide bandwidth of 23–68 Hz over a wide range of input acceleration (0.1–2.0g) (Figure 4F). There are two working modes of this device, as shock and resonant. Harvesting energy in resonant mode is achieved at 82 Hz with output voltage of 20 mV and 55 mV from the TENG and EMG, respectively. Whereas energy harvesting in the non-linear mode or hand tapping mode reaches a maximum power density of 4.8 μW/cm^3 from the TENG and 6 μW/cm^3 from the EMG.

Hybridization between TENG technology and the other technologies such as EMG and PEG is proven to be an innovative strategy to boost the output power, increase the operating bandwidth, and efficiently harvest low and ultra-low frequency vibrations. Though, there are other transduction technologies that can benefit from coupling with a TENG, such as dielectric elastomers (DE). Haroun and Lee [75] developed a DE—TENG hybrid generator, namely DETEG (Figure 4G). A dielectric elastomer has the advantages of stretch ability, high energy density, and the ability to match low frequency motion. It deforms with nonlinear stiffening behavior, which can work as a nonlinear self-spring and be utilized for broadband vibration energy harvesting. However, a dielectric elastomer cannot convert mechanical energy to electrical energy on its own. It needs a prime voltage or an external source of prime electrical charges. It converts mechanical energy to electrical energy by amplifying the prime voltage to higher voltage levels. Therefore, tethering a TENG with a DEG can provide the necessary prime voltage to the DEG, which in turn is amplified by the external input mechanical energy. The hybrid DETEG system was introduced to gain use of the DEG advantages mentioned before, for portable applications. In addition, the DE membrane has an electric field breakdown strength larger than air. A higher voltage can be sustained across the DE membrane. Thus, DETEG can be more convenient with high voltage applications such as the powering of dielectric elastomer actuators (DEA) and the self-powering of electrospinning systems.

4. Triboelectric Nanogenerators for Vibration Measurements

Vibration measurement is a key component of machine-condition monitoring, which enables early detection of machine damages, faults, and abnormal stoppings. A TENG can convert mechanical motion in the form of contact-separation or sliding between two surfaces into electrical output with high voltage. Therefore, a TENG can be used as a vibration sensor with high sensitivity, by converting vibration motion into one of a TENG's modes of contact. A TENG for vibration sensing uses the same concept as a TENG for vibration energy harvesting. However, other important features are considered in case of vibration sensing such as sensitivity, resolution, measuring range, and linearity. Vibration motion is mainly defined by vibration amplitude and frequency. Vibration frequency can be easily detected by most vibration sensors, but the amplitude is more difficult. Measuring one of the vibration physical characteristics such as displacement velocity or acceleration with time provides information about vibration amplitude and frequency. Some TENGs have been developed to measure either vibration displacement or vibration acceleration, which are reviewed in this work.

4.1. Triboelectric Sensors for Displacement Measurements

Vibration displacement is an important characteristic that represents the strength of mechanical vibrations. Some TENGs are developed to measure only the displacement amplitude, while others measure the vibration displacement over time, which can provide information about both vibration amplitude and frequency. Wang et al. [80] proposed a contact separation-enabled freestanding TENG (CF-TENG) for self-powered displacement amplitude measurements. The configuration of the system is schematically shown in Figure 5A. Two Al-deposited acrylic plates are supported in a parallel setup with a gap of 2 cm, which act as the stationary electrodes. An acrylic sheet is placed between them and supported with eight springs through its corners, which acts as a vibration resonator. Two 50-μm fluorinated ethylene propylene (FEP) films are laminated onto the two sides of the acrylic sheet. They are purposely chosen as the freestanding triboelectric layers due to their distinctively opposite triboelectric polarity with Al. When the system is excited with input vibrations, the resonating acrylic sheet allows the two triboelectric layers to alternatively reach the electrodes. When the vibration amplitude is high, the generation of triboelectric charge will occur due to contact between the FEP layers.

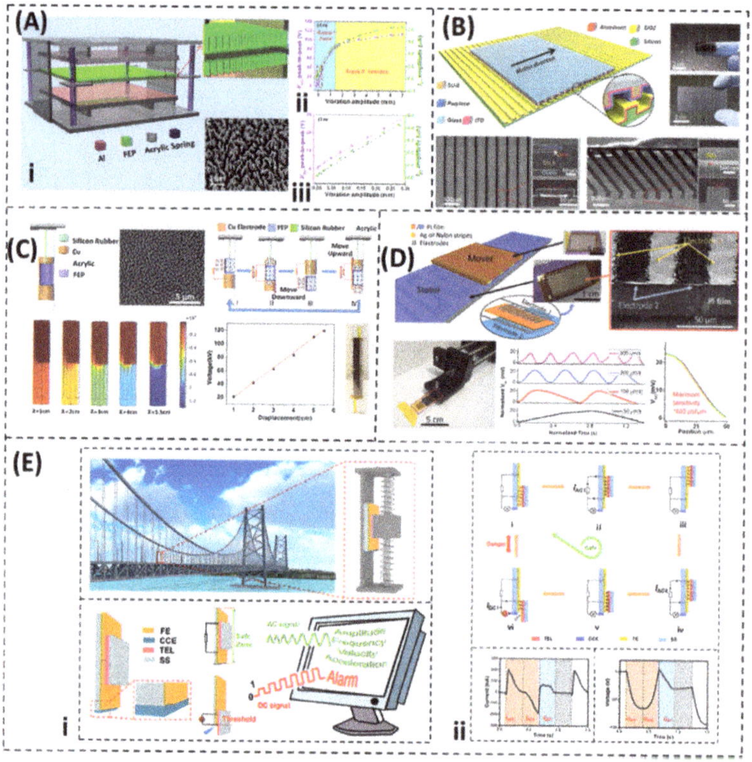

Figure 5. Triboelectric sensors for vibration-displacement measurements. (**A**) A contact-separation-enabled freestanding TENG sensor for measuring vibration displacement amplitude, (Reprinted with permission [80], copyright 2014, American Chemical Society). (**B**) A single dimensional displacement and speed sensor technology based on two layers of micro-grated triboelectric layers (Reprinted with permission [81], copyright 2013, John Wiley and Sons). (**C**) An integrated self-powered dynamic displacement monitoring system for structural health monitoring (Reprinted with permission [82], copyright 2017, John Wiley and Sons). (**D**) An aerosol jet printing (AJP) that is used to create triboelectric sensors from different materials, including polymers, which may be directly printed into finely detailed grating structures for increased sensitivity to displacement and speed of motion (Reprinted with permission [83], copyright 2018, John Wiley and Sons). (**E**) A self-powered vibration sensing system that can detect vibration properties in real time, continuously using a dual-mode triboelectric nanogenerator that can create either alternating current (AC) or direct current (DC) within various operation zones (Reprinted with permission [84], copyright 2020, American Chemical Society).

The resonator plate reaches the largest vibration distance at resonance. The resonant frequency in this system is around 15 Hz, at which the best sensitivity of the system can be found. When the system works at 15 Hz, as the amplitude increases, the open circuit voltage and the short circuit current increase almost linearly, until the amplitude reaches 1.5 mm. This is called region I, as depicted by Figure 5A(ii). After 1.5 mm, the acrylic plate starts to impact with the electrodes, which is called region II. Figure 5A(iii) shows a linear relationship between the open-circuit voltage and the vibration amplitude over a wide range of displacements. This linearity indicates the performance enhancement of the CF-TENG over simple contact-mode TENGs that generate remarkable signals only within narrow limits of separation distance. It also gives an advantage to the CF-TENG when utilized as an energy harvester, since it would have the capability to harvest large amplitude vibrations. This sensor can detect small displacement amplitudes down to 3.5 μm at the resonant frequency and up to 15 mm at a frequency of 21 Hz. This refers to a good sensitivity of the sensor.

Wang et al. [81] proposed a nanometer-resolution self-powered static and dynamic sensor, based on micro-grated triboelectrification, which consists of a pair of micro-grating structures with relative motion between them that leads to the periodic separation of two grates that have opposite charges, as shown in Figure 5B. The sensor can detect the vibration displacement by the open circuit voltage and the short circuit current generated by the sensor. The motion range of this sensor is 200 μm, and the resolution is 173 nm, with a linearity error of 0.02%. The displacement can be calculated by identifying the number of peaks in the output signal. One peak is an indication of 200 μm displacement. When the sensor is placed on a vibration generator and the number of zero-crossings of the output voltage are counted, each of which represents a 100 μm step of displacement, then the real-time displacement can be derived.

Yu et al. [82] presented a self-powered dynamic displacement monitoring system, as illustrated in Figure 5C. It consists of an outer tube made of acrylic with an inertial mass inside, which is suspended by a stretchable silicone rubber wire. Two nylon films coated with copper are glued to the inner surface of the tube, which provide an optimized gap of 2 mm inside the tube. The fluorinated ethylene propylene (FEP) film, as the freestanding triboelectric layer, is coated on the cylindrical mass that has a diameter of 15 mm and weight of 32 g. The external input motion causes a relative motion between the mass and the tube. The resulted open circuit voltage shows an acceptable linearity with the displacement of the inertial mass. The natural frequency of the proposed sensor is important, which can be determined by selecting the appropriate size and material parameters, such as wire elastic modulus, wire length, and cylinder diameter. The natural frequency of the proposed prototype is designed to be 434.8 Hz. In this sensor, the dynamic displacement monitoring system automatically sends an alarm signal when the displacement amplitude exceeds a pre-set limit. The testing results of the sensor show that it behaves accurately below 5 Hz, compared to a traditional piezoelectric accelerometer. Moreover, it has a high sensitivity of 0.391 V s^2/m when it operates at 3 Hz.

Jing et al. [83] use the aerosol jet printing (AJP) technique to develop novel triboelectric sensors for high-resolution motion sensing, including displacement, as shown in Figure 5D. This is done by direct and rapid printing of both metallic and polymeric inks into the desired finely grated structures. The fabrication of an ultra-fine grated and patterned interdigitated electrode structure is done by printing silver and polyimide inks onto flexible polyimide substrates. These make up the main functional parts for the mover and the stator of the sensor. The stator contains a pair of interdigitated electrodes at a pitch of 50 μm. The centers of the comb electrodes on either side are displaced by half a pitch to form a pair of interdigitated electrodes. For the mover, a grated pattern with a period of 100 μm was printed on polyimide film with Ag ink. Material selection of triboelectric layers is critical when designing this system. So, in this work, optimization was done by printing three ink movers with different combinations of metals as follow: Ag on Kapton, PVDF-TrFE on Kapton, Kapton on nylon, Kapton on Teflon, and Kapton on aluminum foil. When the layers are tested under the same condition, the highest output voltage is acquired from the Ag–Kapton combination, which is an interesting result. This sensor has a high resolution of up to 50 μm, with a maximum sensitivity of around 630 mV/μm. Aerosol jet printing shows a high resolution, a low prototyping time, wide material options, and a large scale.

To achieve a real time monitoring of structures and predict vibrations behavior with a warning system, a self-powered vibration monitoring system is introduced by Li et al. [84], as shown in Figure 5E. The proposed system can detect vibration characteristics (e.g., displacement amplitude, frequency, and acceleration) as well as detect the working state of the construction with an alarm system, when reaching the safety threshold. The configuration of the AC/DC-TENG consists of a stator and a mover, as shown in Figure 5E. The stator is made of two friction electrodes, a charge-collecting electrode, and an acrylic layer that acts as a frame. The two friction electrodes are equal in size (20 mm × 20 mm), and are pasted side by side on the acrylic layer with a gap of 0.5 mm. The collecting electrode is put on the bottom edge of the acrylic layer. The slider (20 mm × 20 mm) is attached to a fluorinated ethylene film as a triboelectric layer. When the mover oscillates between the two friction electrodes without exceeding the edge of the acrylic layer (the safe zone), an AC signal is generated by triboelectrification and electrostatic induction, and, hence, the vibration characteristic is obtained.

The working mechanism of the device is shown in Figure 5E(ii). Initially, the mover is fully overlapped with the friction electrode. In the safe region, the case (vi) will not occur. If part of the slider moves off the edge of the friction electrode, the case (vi) occurred. Due to the high electrostatic field between the collecting charge electrode and the negatively charged friction electrodes surface, air breakdown occurs in the gap, causing a DC in the external circuit. This was confirmed by the same author in previous research [85]. In the sage zone, the open circuit voltage is proportional to vibration displacement amplitude. The device was tested for more than 10,000 cycles, to prove a high stability. The system also shows a linear behavior in both the safe and danger zones.

4.2. Triboelectric Sensors for Acceleration Measurements

Acceleration sensor is an important tool in vibration condition monitoring and others. It plays an essential role in many applications such as GPS devices, biomedical devices, artificial intelligence products, automotive, environmental monitoring, and equipment monitoring. In recent years, the triboelectric nanogenerator has been developed as a promising technology for acceleration sensing [86]. The most important features of the reviewed TENG displacement and acceleration sensors are summarized in Table 2.

Table 2. Summary of the reviewed TENG vibration sensors, which shows their main features and performance.

Sensor	Sensor Type	TENG Mode	Triboelectric Materials	Dimensions	Input Excitation	Performance
CF-TENG (Wang et al.) [80]	Displacement	Contact-separation	FEP/Al	200 cm^3	15 Hz and 3.5 μm	I_{sc} = 10 nA, V_{oc} = 0.54 V
Micro-grating motion sensor (Zhou et al.) [81]	Displacement	Sliding	Parylene/SiO_2	≅ 0.18 cm^3	1 Hz	Resolution = 173 nm
FTENG (Yu et al.) [82]	Displacement	Sliding	FEP/Cu	Ø25 mm × ≅ 65 mm	3 Hz	Sensitivity = 0.391 $Vs^2\ m^{-1}$
AJP motion sensor (Jing et al.) [83]	Displacement	Sliding	PVDF-TrFE/Ag	(25 × 7 × 5.5) cm^3	ND	Max. sensitivity = 630 μV/μm
AC/DC-TENG (Li et al.) [84]	Displacement	Sliding	FEP/NA	Electrodes: (20 × 20) mm^2	2 Hz and 5 mm	V_o = 3 V
Liquid-metal-based TENG (Zhang et al.) [87]	Acceleration	Contact-separation	PVDF/Hg	(30 × 30 × 6) mm^3	60 m/s^2	I_{sc} = 300 nA, V_{oc} = 15.5 V, sensitivity = 0.26 $V{\cdot}s/m^2$
MFMS (Wu et al.) [88]	Acceleration	Sliding	PTFE/NdFeB	(Ø 40 × H 19) mm^2. V = 23.88 cm^3	ND	ND
High-g sensor (Dai et al.) [89]	Acceleration	Contact-separation	PDMS/Cu	(14 × 14 × 8) mm^3	ND	Sensitivity = 1.8 mV/g @ 200 μm beam thickness
MMS (Wang et al.) [90]	Acceleration	Contact-separation	PU/PTFE	ND	50 Hz and 3 m s^{-2}	V_o = 330 mV
Elastic-beam TENG (Chen et al.) [91]	Acceleration	Contact-separation	PTFE/stainless steel foil (SSF)	(7.5 × 60) mm^2		Sensitivity = 900 V N^{-1}
Inertial sensor (Adly et al.) [92]	Acceleration	Contact-separation	Teflon/Al	(200 × 50 × 20) mm^3	30 Hz	V_o = 6.2 V
V-Q-a TENG (Liu et al.) [93]	Acceleration	Contact-separation	ITO/PET	(3.5 × 6) cm^2	1–11 m/s^2	P_d = 371.8 mW/m^2 sensitivity = 20.4 $V/(m/s^2)$

Based on a liquid metal triboelectric nanogenerator, a self-powered acceleration sensor is introduced by Zhang et al. [87], as shown in Figure 6A. The sensor was tested for over 200,000 cycles. It shows a high sensitivity of 26 Vs/m^2 and a wide working range from 0 to 60 m/s^2, with excellent stability and durability. The high stability and durability are achieved due to the usage of liquid metal mercury. The sensor also shows a good linear relationship between the measured acceleration, and the output voltage and current. The graph in Figure 6A shows the effect of different volume ratios of the mercury droplet and acrylic cavity on the measured accelerations.

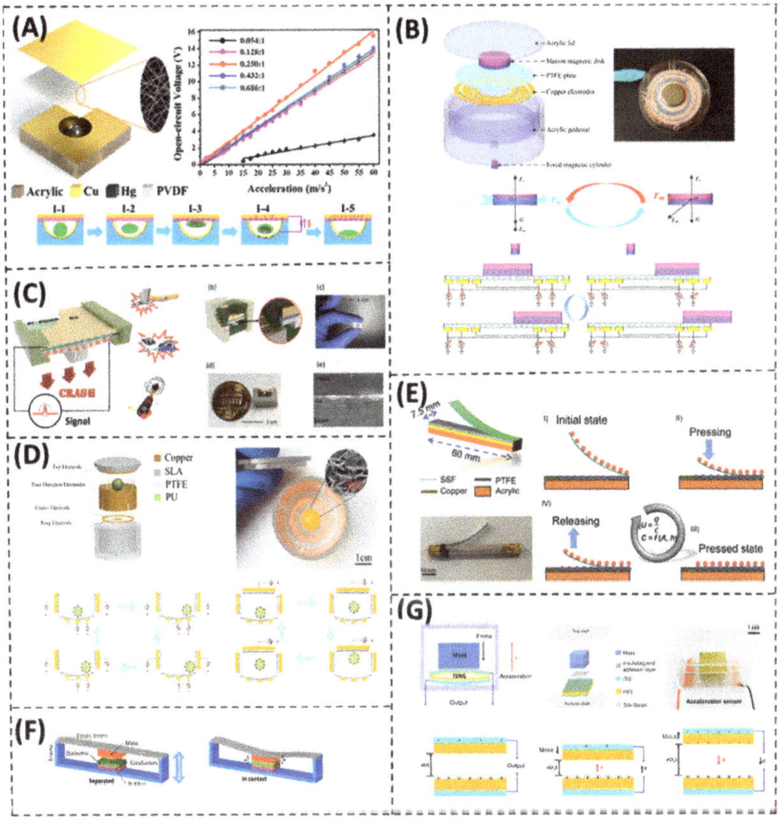

Figure 6. Triboelectric sensors for acceleration measurements. (**A**) Self-powered and extremely sensitive triboelectric acceleration sensor that is made of a liquid metal mercury droplet and a nanofiber-networked polyvinylidene fluoride film (Reprinted with permission [87], copyright 2017, American Chemical Society). (**B**) Self-powered multifunctional triboelectric motion sensor (MFMS), which is capable of concurrently sensing motion characteristics such as direction, speed, and acceleration of linear and rotating motions (Reprinted with permission [88], copyright 2018, American Chemical Society). (**C**) Self-powered triboelectric high-g acceleration sensor (Reprinted with permission [89], copyright 2017, Elsevier). (**D**) Self-powered multifunctional triboelectric sensor based on PTFE and PU for vibration and motion sensing (Reprinted with permission [90], copyright 2020, John Wiley and Sons). (**E**) Multifunction sensor based on elastic-beam triboelectric nanogenerator. It consists of copper and polytetrafluoroethylene layers that are glued on an acrylic plate (Reprinted with permission [91], copyright 2018, John Wiley and Sons). (**F**) Inertial triboelectric motion sensor with optimized parameters (Reprinted with permission [92], copyright 2021, Elsevier). (**G**) Self-powered and high-sensitivity triboelectric acceleration sensor. It is composed of transparent outer shell and inner mass and silk fibroin patches (Reprinted with permission [93], copyright 2019, Elsevier).

This self-powered acceleration sensor consists of an internal mercury droplet and external acrylic shell. In the acrylic face, there is a hemisphere cavity with a deposited copper film that acts as an electrode. The nonstick mercury droplet was put in the cavity as a free moving electrode, and the face of the cavity is made from a copper layer coated with nn-PVDF which acts as a TENG layer. The working mechanism of the self-powered acceleration sensor is shown in Figure 6A. At the initial state, the mercury sticks on the bottom of the cavity with no electric charge on the surface of the nn-PVDF layer. When the sensor moves up, the mercury will contact the nn-PVDF layer. Since, they are different in their triboelectric polarity, positive and negative charges are generated on the mercury and on the PVDF layer surface, respectively. When there is no contact between the mercury and the PVDF layer while the sensor is moving down, the equilibration of the electric field is changed, and then an electric difference is created between the electrodes. The electrons will flow from the top electrode to the bottom electrode until the equilibrium of the collected charges.

Wu et al. [88] presented a self-powered multifunctional motion sensor that can detect acceleration of linear and rotary motions, as shown in Figure 6B. The dimension of the device are 40 mm diameter and 19 mm height. The body is made from acrylic. The TENG is composed of a magnetic disk and PTFE as a friction layer, with copper electrodes stuck onto it. There are an inner circle electrode, an outer circle electrode, and arc electrodes between them. By increasing the number of electrodes, the sensitivity increases. For the sake of design, the number of electrodes in this device are stated as four. The four arc electrodes are referring to as north, east, south, and west.

Measuring the linear acceleration is experimentally checked. When the magnetic disk moves linearly in any direction, crossing the inner and outer circle electrodes, the electrostatic potential of the two circle electrodes reach each peak. The voltage waveforms of the inner and outer circle electrodes show the relationship between the voltage and the motion position. When the magnetic disk is at the center, the open-circuit voltages of the two electrodes equal 0 V. When the magnetic disk is moving far from the inner circle electrode, its open-circuit voltage will reach a certain value. After that, with the magnetic disk moving away from the inner circle electrode and toward the outer circle electrode, the open-circuit voltage of the inner circle electrode will reach a peak value. To determine the acceleration, the time is recorded, and the derivation of its value can be done.

In Figure 6C, a self-powered high-g acceleration sensor based on a triboelectric nanogenerator is proposed for the first time [89]. It has a volume of $(14 \times 14 \times 8)$ mm^3, a measuring range of up to 1.8×10^4 g, and sensitivity of 1.8 mV/g with a good linearity. The structure of this sensor is shown in Figure 6C. The positive electrode is made of aluminum foil in a beam shape having hinged weight. The negative electrode is made of a PDMS film that is bonded to the face of the copper plate. When there is no high-g acceleration, a contact between the positive and negative electrodes will occur and there is no voltage output. As high-g acceleration occurs, the aluminum beam will deform away from the positive electrode due to the hinged mass. When the deformation reaches the maximum, a potential difference between positive and negative electrodes occurs and reaches its peak due to the distance between charges. After the impact of acceleration, the deformation of the aluminum electrode is restored, which, in turn, causes the voltage to drop back to zero.

A multifunction sensor based on an elastic-beam triboelectric nanogenerator is introduced [90]. As shown in Figure 6D, the sensor consists of a copper (Cu) layer and a polytetrafluoroethylene layer. Both are glued to an acrylic plate. One end of an arc-SSF is fixed on the acrylic plate that holds the copper and the polytetrafluoroethylene layers. The working principle of the sensor is as follows: by contact-separation between the SSF and the PTFE that is caused by external force, the SSF gains a positive charge and the PTFE gains a negative charge. These charges stay on both layers due to the open-circuit state and the dielectric properties of the PTFE layer. Increasing the pressing would not change the amount of charge, but it changes the capacitance, which is related to the contacting area and the separation distance. At the full pressing, the maximum contact area and the minimum separated distance occur. When releasing, the SSF will return to the initial position to complete the cycle. This sensor can be used as a force sensor, acceleration sensor, sensitive scale, and intelligent keyboard with good performance with sensitivity of 30–900 V/N [90].

Another self-powered multifunctional triboelectric sensor based on the PTFE and PU for vibration and motion sensing is proposed [91]. It is made up of a hollow cylindrical structure with a movable cover and a resin frame (Figure 6E). Seven copper electrodes with 0.1 mm thickness are stuck to the inner walls of the frame. One electrode is placed on the top cover. A center electrode and a ring electrode are placed on the bottom, and four directional electrodes are attached to the side wall of the cylinder. All electrodes are coated with the PTFE layer. A PU ball is inserted in the inner hallow cylinder. The working flow of the sensor (Figure 6E) appears in two working modes. The sensor can achieve multiple sensing functions, due to the different combinations of the ball and the electrodes. The sensor can sense speed, acceleration, and direction.

Adly et al. [92] presented a study on the modeling and optimization of an inertial triboelectric motion sensor. The proposed sensor consists of a U-shaped frame with an elastic beam mounted on its ends (Figure 6F). A counterweight is attached to the beam, with aluminum layer pasted on it as a triboelectric layer. The other triboelectric layer is made of Teflon and is fixed on the bottom of the U-shaped frame, with a rubber layer in between. This device is based on the contact-separation mode between the aluminum layer and the dielectric Teflon layer. As shown in Figure 6F, the system is selected to have a resonant frequency of 30 Hz based on an optimization, and it has been tested under different accelerations ranging from 0.4 g to 1.2 g. The focus of this work is to optimize all the device parameters.

Liu et al. [93] came up with an idea for a self-powered and high-sensitivity TENG acceleration sensor. It is composed of a proof mass attached to an arch-shaped triboelectric layer (Figure 6G) The sensor has a transparent outer shell with the mass inside and silk fibroin patches. The arch layer supports the mass. The working principle of the sensor is based on a contact-separation mode TENG. It has two working phases: when the acceleration is upwards, and when the acceleration is downwards. For upwards acceleration, the proof mass tends to move downwards, forcing the triboelectric arch layers to be in full contact. The electron charges will then transfer from the ITO/PET/SF layer to the ITO/PET layer and meet electrostatic equilibrium. For downward acceleration, the mass will then separate the triboelectric layers, and a potential drop will occur, forcing electrical signal to be generated through the two electrodes. This TENG is highly recommended to serve as wearable devices with an alarm and vibration monitoring systems, which require high sensitivity. It exhibits a high sensitivity of 20.4 $V/(m/s^2)$ and a power density of 371.8 mW/m^2.

5. TENGs for Both Vibration Energy Harvesting and Vibration Sensing

Some TENGs can work effectively as self-powered vibration sensors as well as energy harvesters. Their designs are suitable for certain sensing applications with sufficient sensitivity. In addition, they have an enhanced output power, which allows them to work effectively as energy harvesters. One example is that presented by Xu et al. [38]. They proposed a bidirectional spring-based triboelectric nanogenerator (S-TENG), as shown in Figure 7A. The spring-based TENG was previously presented, but it can benefit from motion in a single direction. Benefiting from motion in more than one direction at the same time remains a big challenge. Xu et al. succeeded in integrating a spring into a TENG and gaining use of two-directional vibration modes, which come from its bendable property. It has great potential to serve as a self-powered active sensor for acceleration and in harvesting arbitrary vibrational energy generated from buildings, ocean waves, wind, etc. As shown in Figure 7A, the helical structure of the S-TENG is composed of two elastomeric conductive layers, which serve as electrodes. With the aid of wire spring and silicone rubber and through external vibration triggering, both vertical and horizontal vibrations will decrease the adjacent distance between the helical coil, forming a contact-separation mode in the TENG. The S-TENG shows great performance under vertical and horizontal resonant frequencies of 16 Hz and 8.5 Hz, respectively. The average maximum power density was found to be 240 and 45 mW/m^2, respectively.

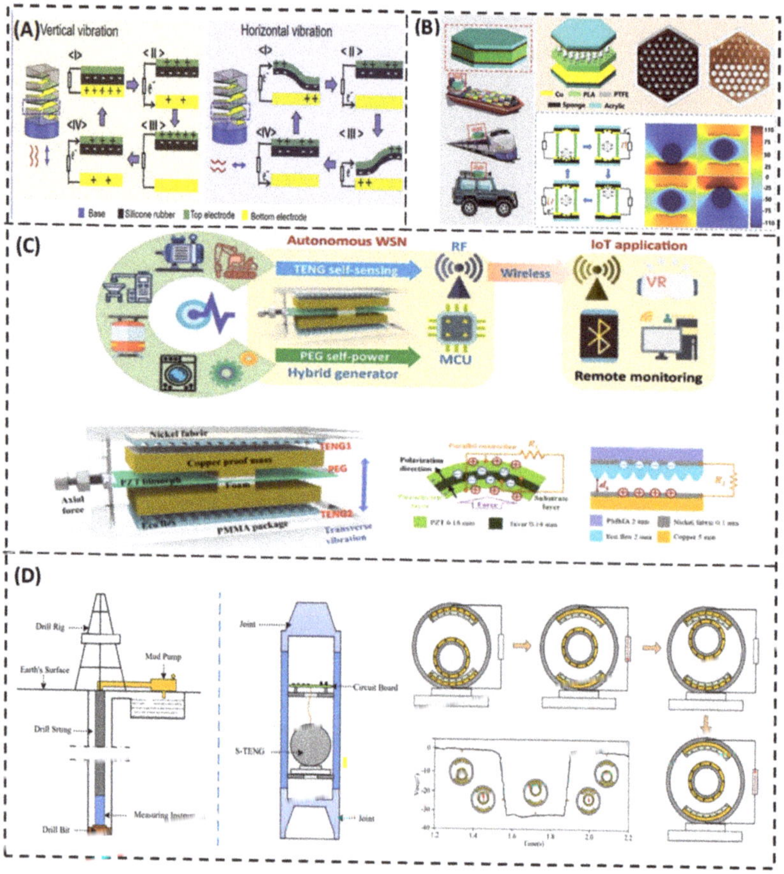

Figure 7. TENGs that can work as energy harvesters as well as self-powered sensors. (**A**) Soft and resilient spring-based triboelectric nanogenerator that scavenges vibrational energy from horizontal and vertical excitations (Reprinted with permission [38], copyright 2019, Advanced Energy Materials). (**B**) Honeycomb-structure vibration sensor for monitoring motors conditions (Reprinted with permission [94], copyright 2019, Advanced Energy Materials). (**C**) Hybridized triboelectric and piezoelectric nanogenerator with a tunable resonant frequency (Reprinted with permission [95], copyright 2020, Nano Energy). (**D**) The structure and mechanism of spherical triboelectric nanogenerator, which is used in drilling process (Reprinted with permission [36], copyright 2020, MDPI).

Another example of enhancing sensing and harvesting performance is presented by Xiao et al. [94]. They proposed a honeycomb-structure-inspired triboelectric nanogenerator (HIS-TENG). It is composed of two flexible copper electrodes and an intermediate honeycomb frame, which is fabricated using 3D printing (Figure 7B). The overall structure is then framed by two acrylic plates. Each groove in the honeycomb structure, has a poly tetrafluoroethylene ball, which acts as the triboelectric layer. Compared to the square grid structure, the honeycomb structure increases the maximum power density by 43.2% and has an advantage of strong frequency adaptability compared to a spring-assisted TENG. By applying this HIS-TENG on the block of a diesel engine of a real sailing ship, it was found that the HIS-TENG can work as an effective self-powered sensor used for monitoring engine condition; in addition, it has a remarkable performance and great potential in vibration energy harvesting.

The working mechanism of the HIS-TENG simply depends on the motion of the small ball inside each groove. When the PTFE ball is in contact with the lower copper film, electrons are transferred from the bottom electrode to the PTFE ball, which in return will achieve positive and negative triboelectric charges on the PTFE ball and the surface of the electrode layer. As the structure is subjected to any external source of vibrations, the PTFE ball will move upward and separate from the bottom electrode, while hitting the upper one. During this separation, electrons are transfered from the bottom electrode to the top electrode, resulting in a current and charged signals transfer to be generated in the external circuit. This work proved that the HIS-TENG exhibited a robust performance, and it can serve as a self-powered engine condition monitor and may have a great impact on machine monitoring in the future [94].

We all know that autonomous wireless sensor nodes play an important role in the field of industry and Internet of things. That's why a self-sustained power supply for sensors is highly recommended. Wang et al. [95] proposed a self-sustained autonomous wireless sensing based on a hybridized TENG and PEG vibration mechanism. Their proposed mechanism is shown in Figure 7C. It has a broadband behavior and tunable resonant frequency, which can be applied for various ranges of vibration frequencies. What makes the hybridization of a TENG and a PEG so perfect in this device is that the PEG generates a high current and is more suitable to act as energy harvester, while the TENG has a high voltage output and is preferred to be a self-powered accelerometer, as it has good sensing capability. This hybrid WSN is composed of one PEG and two TENGs. There is a Lead Zicronate Titanate (PZT) bimorph, which is hinged-hinged mounted. There are two T-shaped copper proof masses, in which they are bonded with the PZT bimorph through a piece of foam. This piece of foam will reduce the concentrated stress. The bolt and nut fixed at the side of the outer package give an axial force to tune the resonant frequency of the PZT bimorph. The two TENGs work as energy harvesters as well as act as stoppers for PEG overload, protecting it. The triboelectric positive electrode is a nickel fiber, while ecoflex with a pyramidal synaptic array is selected to act as a negative electrode. This material is selected for its soft stiffness, which ensures sufficient contact in the contact-separation mode and generates more triboelectric charges. Its working mechanism depends on the strength of the vibration excitation. As with stronger vibration and acceleration, more output voltage and current would be generated. When the outer package receives an ambient vibration, the proof mass will force the bimorph PZT to strain, and this will generate electricity to power the acceleration sensor. When the proof mass hits the nickel fabric and full contact occurs, triboelectric charges are transferred to the ecoflex material through the contact-separation mode. This separation is translated to an electric signal through the triboelectrification process. This mechanism can be applied on a train cabin to monitor the conditions and overload status through the vibrations exerted by the moving cabins. It produces an output power of 6.5 mW at 25 Hz and 1.0 g [95].

Another TENG for vibration measurements and energy harvesting from mechanical drilling was invented by Wu et al. [36]. They proposed a spherical triboelectric nanogenerator (S-TENG) for scavenging and measuring mechanical vibrations during the traditional downhole drilling process (Figure 7D), which can be used for exploitation applications such as the exploitation of petroleum, gas, or other mineral resources. Drill string vibrations inevitably occur during any drilling process. With excessive vibrations, the drilling tool would be damaged, and accidents may occur. To avoid that hazard, there must be a method for real-time vibration measurements of the downhole drill string. The traditional methods used for drilling vibrations measurement (using cables and batteries) increase the drilling cost and decrease the drilling efficiency. However, the S-TENG brings hope for solving these problems. The S-TENG consists of a fixed ball, a freely moveable ball, and a supporting seat. The fixed ball is attached to the supporting seat and the moveable ball is embedded inside the fixed ball, as shown in Figure 7D. On the upper and lower sections of the fixed ball there are copper, and polytetrafluoroethylene (PTFE) pasted layers. The Cu serves as the electrode, while the PTFE is used as the friction layer for generating electric charges. When the drill string vibrates, the moveable ball moves up and down relative to the axial vibration of the drill string, generating charges due to the triboelectric effect and electrostatic induction. When the moveable ball separates from the lower seat/lower friction layer, electrons flow through the external load and form electric current. When the moveable ball contacts the upper friction layer (negatively charged) and separates again, electrons move through the external load in reverse, and a reverse current is generated. This S-TENG generates an electrical output of 70 V, 33 μA, and 10.9 nW at a maximum frequency value of 8 Hz [36].

6. Conclusions

Mechanical vibrations are an abundant source of motion in the surrounding environment. They appear almost everywhere: in manufacturing machines, cars, trains, household equipment, turbines, etc. That is why researchers have thought about mechanical vibrations as a sustainable energy source for wireless sensors associated with IoT and sensor network applications. Other renewable energy sources are employed such as solar and wind energies. Nevertheless, mechanical vibrations are preferred when the sensors are placed in deep and dark environments. In the past few decades, EMGs and PEGs have shown success in vibration energy harvesting for a wide range of applications. However, the prosperous development of the TENG shows its great potential for vibration scavenging. The TENG has a high energy conversion efficiency, low cost, simple fabrication method, and the ability to match low frequency vibrations. For vibration to electrical energy conversion, the TENG utilizes a spring-mass system or other elastomeric mechanisms that can provide a restoring force. Such a restoring force mechanism guarantees a sustained internal oscillation with the input vibration, which is converted into electricity through one or more of the TENG modes of contact. TENG-based energy harvesters can rely on only triboelectric as a mechanism of energy transduction or a hybrid with other transduction technologies such as electromagnetic and piezoelectric. Hybridization of the TENG is proposed to increase the system power density, the output power, the operating bandwidth, and/or provide a better environmental adaptability and applicability. Some recent developed TENGs and hybrid TENGs are reviewed in this paper. The development of the TENG for vibration scavenge depends on increasing the system output power, widening the frequency bandwidth, having better applicability with certain applications, and/or matching lower frequencies. Those are achieved by utilizing innovative system designs, and, in some cases, by boosting the triboelectrification performance [68]. There is still a potential for the future development of the TENG for vibration scavenging. Some vibration mechanisms, such as a bistable compliant mechanism [96], and variable stiffness [97] mechanisms can be employed with the TENG to enhance its harvesting performance.

The effectiveness of the TENG for vibration energy harvesting encourages researchers to consider it as a self-powered vibration sensor. The TENG converts a mechanical vibration trigger into an electrical signal. So, it can sense vibrations directly without applying a power unit to the device. The TENG shows many advantages when utilized as a vibration sensor, such as high output voltage, high efficiency, low cost, stability, and robustness. Due to the four different TENG modes of contact, it has also great design adaptability and different design possibilities. One of the most important applications of vibration measurement is the machine condition monitoring. Machine condition monitoring helps to detect early signs of deterioration, and it is the core element of machine predictive maintenance. In this work, the recent progress in the development of a TENG for vibration measurements is reviewed as well. Vibration measurements can be achieved by measuring either vibration displacement or vibration acceleration. Unlike a TENG for vibration energy harvesting, other criteria are considered when a TENG is utilized for vibration measurement. Sensitivity, resolution, measuring range, and linearity are among the main criteria. Researchers are trying to improve the performance of a TENG as a self-powered vibration sensor, according to those criteria. Among challenges that are faced when utilizing a TENG as a vibration sensor are the nonlinearity and low resolution. To overcome the nonlinearity problem, the system should be designed with a lateral dimension much larger than their vertical-separation distance. For example, the electricity generation has a linear relationship with the moving distance of the freestanding layer when working in contact-separation mode. Moreover, using grated structures can introduce high-resolution vibration sensing.

Although, high output power is the main criterion for a TENG as a vibration scavenger, some TENG vibration sensors can produce high output power. They can work as effective self-powered sensors as well as energy harvesters. Some of them are reviewed in this work.

Author Contributions: Conceptualization, A.H.; resources, A.H., M.T., M.M. and F.I.; writing—original draft preparation, A.H., M.T., M.M. and F.I.; writing—review and editing, A.H., M.T. and M.M.; visualization, A.H.; supervision, A.H. All authors have read and agreed to the published version of the manuscript.

Funding: This research received no external funding.

Institutional Review Board Statement: Not applicable.

Informed Consent Statement: Not applicable.

Data Availability Statement: No new data were created or analyzed in this study. Data sharing is not applicable to this article.

Conflicts of Interest: The authors declare no conflict of interest.

References

1. Sojan, S.; Kulkarni, R.K. A Comprehensive Review of Energy Harvesting Techniques and Its Potential Applications. *Int. J. Comput. Appl.* **2016**, *139*, 14–19. [CrossRef]
2. Beeby, S.P.; Tudor, M.J.; White, N.M. Energy harvesting vibration sources for microsystems applications. *Meas. Sci. Technol.* **2006**, *17*, r175. [CrossRef]
3. Fang, L.H.; Hassan, S.I.S.; Rahim, R.B.A.; Malek, M.F.A. A study of vibration energy harvester. *ARPN J. Eng. Appl. Sci.* **2016**, *11*, 5028–5041.
4. Kim, W.; Bhatia, D.; Hwang, H.J.; Choi, K.; Choi, D. Double impact triboelectric nanogenerators for harvesting broadband vibrations from vehicles. *Funct. Compos. Struct.* **2019**, *1*, 035003. [CrossRef]
5. Zhu, Q.; Guan, M.; He, Y. Vibration energy harvesting in automobiles to power wireless sensors. In Proceedings of the IEEE International Conference on Information and Automation, ICIA, Shenyang, China, 6 June 2012; pp. 349–354. [CrossRef]
6. Carder, D.S. Observed vibrations of bridges. *Bull. Seismol. Soc. Am.* **1937**, *27*, 267–303. [CrossRef]
7. Xabibulla, G.; Guzulov, J.; Nishonova, G. Causes and extinguishing equipment of vibrations occurred by machinery and mechanisms. *Sci. Prog.* **2021**, *2*, 950.
8. Homayounfar, S.Z.; Andrew, T.L. Wearable Sensors for Monitoring Human Motion: A Review on Mechanisms, Materials, and Challenges. *SLAS Technol.* **2019**, *25*, 9–24. [CrossRef]
9. Choi, Y.M.; Lee, M.G.; Jeon, Y. Wearable biomechanical energy harvesting technologies. *Energies* **2017**, *10*, 1483. [CrossRef]
10. Safaei, M.; Sodano, H.A.; Anton, S.R. A review of energy harvesting using piezoelectric materials: State-of-the-art a decade later (2008–2018). *Smart Mater. Struct.* **2019**, *28*, 113001. [CrossRef]
11. Xu, J.; Tang, J. Multi-directional energy harvesting by piezoelectric cantilever-pendulum with internal resonance. *Appl. Phys. Lett.* **2015**, *107*, 213902. [CrossRef]
12. Wang, L.; Tong, X.; Yang, H.; Wei, Y.; Miao, Y. Design and analysis of a hollow triangular piezoelectric cantilever beam harvester for vibration energy collection. *Int. J. Pavement Res. Technol.* **2019**, *12*, 259–268. [CrossRef]
13. Chilabi, H.J.; Salleh, H.; Supeni, E.E.; As'arry, A.B.; Rezali, K.A.M.; Atrah, A.B. Harvesting energy from planetary gear using piezoelectric material. *Energies* **2020**, *13*, 223. [CrossRef]
14. Wang, C.; Zhang, Q.; Wang, W.; Feng, J. A low-frequency, wideband quad-stable energy harvester using combined nonlinearity and frequency up-conversion by cantilever-surface contact. *Mech. Syst. Signal Process.* **2018**, *112*, 305–318. [CrossRef]
15. Song, H.C.; Kumar, P.; Maurya, D.; Kang, M.G.; Reynolds, W.T.; Jeong, D.Y. Ultra-low resonant piezoelectric MEMS energy harvester with high power density. *J. Microelectromech. Syst.* **2017**, *26*, 1226–1234. [CrossRef]
16. Wu, J.; Hu, Z.; Gao, X.; Cheng, M.; Zhao, X.; Su, W.; Li, X.; Wang, Z.; Zhou, Z.; Dong, S.; et al. Electrode shape dependence of the barbell-shaped magneto-mechano-electric energy harvester for low-frequency applications. *Sens. Actuators A Phys.* **2019**, *297*, 111535. [CrossRef]
17. Wang, C.; Zhang, Q.; Wang, W. Low-frequency wideband vibration energy harvesting by using frequency up-conversion and quin-stable nonlinearity. *J. Sound Vib.* **2017**, *399*, 169–181. [CrossRef]
18. Fan, K.; Cai, M.; Liu, H.; Zhang, Y. Capturing energy from ultra-low frequency vibrations and human motion through a monostable electromagnetic energy harvester. *Energy* **2019**, *169*, 356–368. [CrossRef]

19. Fan, K.; Zhang, Y.; Liu, H.; Cai, M.; Tan, Q. A nonlinear two-degree-of-freedom electromagnetic energy harvester for ultra-low frequency vibrations and human body motions. *Renew. Energy* **2019**, *138*, 292–302. [CrossRef]
20. Haroun, A.; Yamada, I.; Warisawa, S. Micro electromagnetic vibration energy harvester based on free/impact motion for low frequency-large amplitude operation. *Sens. Actuators A Phys.* **2015**, *224*, 87–98. [CrossRef]
21. Haroun, A.; Yamada, I.; Warisawa, S. Study of electromagnetic vibration energy harvesting with free/impact motion for low frequency operation. *J. Sound Vib.* **2015**, *349*, 389–402. [CrossRef]
22. Chae, S.H.; Ju, S.; Choi, Y.; Chi, Y.E.; Ji, C.H. Electromagnetic linear vibration energy harvester using sliding permanent magnet array and ferrofluid as a lubricant. *Micromachines* **2017**, *8*, 288. [CrossRef]
23. Ambrozkiewicz, B.; Litak, G.; Wolszczak, P. Modelling of electromagnetic energy harvester with rotational pendulum using mechanical vibrations to scavenge electrical energy. *Appl. Sci.* **2020**, *10*, 671. [CrossRef]
24. Kecik, K.; Mitura, A. Energy recovery from a pendulum tuned mass damper with two independent harvesting sources. *Int. J. Mech. Sci.* **2020**, *174*, 105568. [CrossRef]
25. Basset, P.; Galayko, D.; Paracha, A.M.; Marty, F.; Dudka, A.; Bourouina, T. A batch-fabricated and electret-free silicon electrostatic vibration energy harvester. *J. Micromech. Microeng.* **2009**, *19*, 115025. [CrossRef]
26. Sidek, O.; Khalid, M.A.; Ishak, M.Z.; Miskam, M.A. Design and simulation of SOI-MEMS electrostatic vibration energy harvester for micro power generation. In Proceedings of the ECCE—International Conference on Electrical, Control and Computer Engineering, Pahang, Malaysia, 21 June 2011. [CrossRef]
27. Dudka, A.; Basset, P.; Cottone, F.; Blokhina, E.; Galayko, D. Wideband electrostatic vibration energy harvester (e-VEH) having a low start-up voltage employing a high-voltage integrated interface. *J. Phys.* **2013**, *476*, 012127. [CrossRef]
28. Dragunov, V.; Dorzhiev, V. Electrostatic vibration energy harvester with increased charging current. *J. Phys.* **2013**, *476*, 3–8. [CrossRef]
29. Dorzhiev, V.; Karami, A.; Basset, P.; Dragunov, V.; Galayko, D. MEMS electrostatic vibration energy harvester without switches and inductive elements. *J. Phys.* **2014**, *557*, 012126. [CrossRef]
30. Basset, P.; Galayko, D.; Cottone, F.; Guillemet, R.; Blokhina, E.; Marty, F.; Bourouina, T. Electrostatic vibration energy harvester with combined effect of electrical nonlinearities and mechanical impact. *J. Micromech. Microeng.* **2014**, *24*, 035001. [CrossRef]
31. Wu, C.; Wang, A.C.; Ding, W.; Guo, H.; Wang, Z.L. Triboelectric Nanogenerator: A Foundation of the Energy for the New Era. *Adv. Energy Mater.* **2019**, *9*, 1802906. [CrossRef]
32. Bera, B. Literature Review on Triboelectric Nanogenerator. *Imp. J. Interdiscip. Res.* **2016**, *2*, 1263–1271.
33. Zhang, H.; Yao, L.; Quan, L.; Zheng, X. Theories for triboelectric nanogenerators: A comprehensive review. *Nanotechnol. Rev.* **2020**, *9*, 610–625. [CrossRef]
34. Niu, S.; Wang, S.; Lin, L.; Liu, Y.; Zhou, Y.S.; Hu, Y.; Wang, Z.L. Theoretical study of contact-mode triboelectric nanogenerators as an effective power source. *Energy Environ. Sci.* **2013**, *6*, 3576–3583. [CrossRef]
35. Lin, Z.; Chen, J.; Yang, J. Recent Progress in Triboelectric Nanogenerators as a Renewable and Sustainable Power Source. *J. Nanomater.* **2016**, *2016*, 5651613. [CrossRef]
36. Wu, C.; Huang, H.; Li, R.; Fan, C. Research on the potential of spherical triboelectric nanogenerator for collecting vibration energy and measuring vibration. *Sensors* **2020**, *20*, 1063. [CrossRef]
37. Zhang, H.; Yang, Y.; Su, Y.; Chen, J.; Adams, K.; Lee, S.; Hu, C.; Wang, Z.L. Triboelectric nanogenerator for harvesting vibration energy in full space and as self-powered acceleration sensor. *Adv. Funct. Mater.* **2014**, *24*, 1401–1407. [CrossRef]
38. Xu, M.; Wang, P.; Wang, Y.C.; Zhang, S.L.; Wang, A.C.; Zhang, C.; Wang, Z.; Pan, X.; Wang, Z.L. A Soft and Robust Spring Based Triboelectric Nanogenerator for Harvesting Arbitrary Directional Vibration Energy and Self-Powered Vibration Sensing. *Adv. Energy Mater.* **2018**, *8*, 1702432. [CrossRef]
39. Zhang, Z.; Cai, J. High output triboelectric nanogenerator based on PTFE and cotton for energy harvester and human motion sensor. *Curr. Appl. Phys.* **2020**, *22*, 1–5. [CrossRef]
40. Rahman, M.T.; Sohel Rana, S.; Salauddin, M.; Maharjan, P.; Bhatta, T.; Kim, H.; Park, J.Y. A highly miniaturized freestanding kinetic-impact-based non-resonant hybridized electromagnetic-triboelectric nanogenerator for human induced vibrations harvesting. *Appl. Energy* **2020**, *279*, 115799. [CrossRef]
41. Wang, X.; Wen, Z.; Guo, H.; Wu, C.; He, X.; Lin, L.; Wang, Z.L. Fully Packaged Blue Energy Harvester by Hybridizing a Rolling Triboelectric Nanogenerator and an Electromagnetic Generator. *ACS Nano* **2016**, *10*, 11369–11376. [CrossRef]
42. Hu, Y.; Yang, J.; Jing, Q.; Niu, S.; Wu, W.; Wang, Z.L. Triboelectric nanogenerator built on suspended 3D spiral structure as vibration and positioning sensor and wave energy harvester. *ACS Nano* **2013**, *7*, 10424–10432. [CrossRef]
43. Zeng, Q.; Wu, Y.; Tang, Q.; Liu, W.; Wu, J.; Zhang, Y.; Yin, G.; Yang, H.; Yuan, S.; Tan, D.; et al. A high-efficient breeze energy harvester utilizing a full-packaged triboelectric nanogenerator based on flow-induced vibration. *Nano Energy* **2020**, *70*, 104524. [CrossRef]
44. Lu, S.; Gao, L.; Chen, X.; Tong, D.; Lei, W.; Yuan, P.; Mu, X.; Yu, H. Simultaneous energy harvesting and signal sensing from a single triboelectric nanogenerator for intelligent self-powered wireless sensing systems. *Nano Energy* **2020**, *75*, 104813. [CrossRef]
45. He, W.; Fu, X.; Zhang, D.; Zhang, Q.; Zhuo, K.; Yuan, Z.; Ma, R. Recent progress of flexible/wearable self-charging power units based on triboelectric nanogenerators. *Nano Energy* **2021**, *84*, 105880. [CrossRef]

46. Li, G.Z.; Wang, G.G.; Cai, Y.W.; Sun, N.; Li, F.; Zhou, H.L.; Zhao, H.X.; Zhang, X.N.; Han, J.C.; Yang, Y. A high-performance transparent and flexible triboelectric nanogenerator based on hydrophobic composite films. *Nano Energy* **2020**, *75*, 104918. [CrossRef]
47. Bertacchini, A.; Larcher, L.; Lasagni, M.; Pavan, P. Ultra-low cost triboelectric energy harvesting solutions for embedded sensor systems. In Proceedings of the IEEE-NANO—15th International Conference on Nanotechnology, Rome, Italy, 27 July 2015. [CrossRef]
48. Jang, S.; Kim, H.; Oh, J.H. Simple and rapid fabrication of pencil-on-paper triboelectric nanogenerators with enhanced electrical performance. *Nanoscale* **2017**, *9*, 13034–13041. [CrossRef]
49. Jing, T.; Xu, B.; Yang, Y.; Li, M.; Ga, Y. Organogel electrode enables highly transparent and stretchable triboelectric nanogenerators of high power density for robust and reliable energy harvesting. *Nano Energy* **2020**, *78*, 105373. [CrossRef]
50. Zhu, J.; Zhu, M.; Shi, Q.; Wen, F.; Liu, L.; Dong, B.; Haroun, A.; Yang, Y.; Vachon, P.; Guo, X.; et al. Progress in TENG technology—A journey from energy harvesting to nanoenergy and nanosystem. *EcoMat* **2020**, *2*, e12058. [CrossRef]
51. Haroun, A.; Le, X.; Gao, S.; Dong, B.; He, T.; Zhang, Z.; Wen, F.; Xu, S.; Lee, C. Progress in micro/nano sensors and nanoenergy for future AIoT-based smart home applications. *Nano Express* **2021**, *2*, 022005. [CrossRef]
52. Xie, L.; Zhai, N.; Liu, Y.; Wen, Z.; Sun, X. Hybrid Triboelectric Nanogenerators: From Energy Complementation to Integration. *Research* **2021**, *2021*, 9143762. [CrossRef]
53. Vidal, J.V.; Slabov, V.; Kholkin, A.L.; dos Santos, S.M.P. Hybrid Triboelectric-Electromagnetic Nanogenerators for Mechanical Energy Harvesting: A Review. *Nano-Micro Lett.* **2021**, *13*, 1–58. [CrossRef]
54. Ha, M.; Canon Bermudez, G.S.; Kosub, T.; Mönch, I.; Zabila, Y.; Oliveros Mata, E.S.; Illing, R.; Wang, Y.; Fassbender, J.; Makarov, D. Printable and Stretchable Giant Magnetoresistive Sensors for Highly Compliant and Skin-Conformal Electronics. *Adv. Mater.* **2021**, *33*, 2005521. [CrossRef]
55. Dionisio, R.; Torres, P.; Ramalho, A.; Ferreira, R. Magnetoresistive sensors and piezoresistive accelerometers for vibration measurements: A comparative study. *J. Sens. Actuator Netw.* **2021**, *10*, 22. [CrossRef]
56. Qian, S.; Qin, L.; He, J.; Zhang, N.; Qian, J.; Mu, J.; Geng, W.; Hou, X.; Chou, X. A lead-free stretchable piezoelectric composite for human motion monitoring. *Mater. Lett.* **2020**, *261*, 127119. [CrossRef]
57. Gao, Q.; Cheng, T.; Wang, Z.L. Triboelectric mechanical sensors—Progress and prospects. *Extreme Mech. Lett.* **2021**, *42*, 101100. [CrossRef]
58. Du, T.; Zuo, X.; Dong, F.; Li, S.; Mtui, A.E.; Zou, Y.; Zhang, P.; Zhao, J.; Zhang, Y.; Sun, P.; et al. A self-powered and highly accurate vibration sensor based on bouncing-ball triboelectric nanogenerator for intelligent ship machinery monitoring. *Micromachines* **2021**, *12*, 218. [CrossRef]
59. Chen, J.; Wang, Z.L. Reviving Vibration Energy Harvesting and Self-Powered Sensing by a Triboelectric Nanogenerator. *Joule* **2017**, *1*, 480–521. [CrossRef]
60. Lin, Z.; Long, W.; Chen, L.J.; Niu, S.; Zi, Y. *Green Energy and Technology Triboelectric Nanogenerators*, 1st ed.; Springer: Berlin/Heidelberg, Gemany, 2016; p. 925.
61. Niu, S.; Liu, Y.; Chen, X.; Wang, S.; Zhou, Y.S.; Lin, L.; Xie, Y.; Wang, Z.L. Theory of freestanding triboelectric-layer-based nanogenerators. *Nano Energy* **2015**, *12*, 760–774. [CrossRef]
62. Liu, C.; Zhang, N.; Li, J.; Dong, L.; Wang, T.; Wang, Z.; Wang, G.; Zhou, X.; Zhang, J. Harvesting ultralow frequency (<1 Hz) mechanical energy using triboelectric nanogenerator. *Nano Energy* **2019**, *65*, 104011. [CrossRef]
63. He, J.; Fan, X.; Mu, J.; Wang, C.; Qian, J.; Li, X.; Hou, X.; Geng, W.; Wang, X.; Chou, X. 3D full-space triboelectric-electromagnetic hybrid nanogenerator for high-efficient mechanical energy harvesting in vibration system. *Energy* **2020**, *194*, 116871. [CrossRef]
64. Yang, W.; Chen, J.; Jing, Q.; Yang, J.; Wen, X.; Su, Y.; Zhu, G.; Bai, P.; Wang, Z.L. 3D stack integrated triboelectric nanogenerator for harvesting vibration energy. *Adv. Funct. Mater.* **2014**, *24*, 4090–4096. [CrossRef]
65. Wang, K.; Zhou, J.; Ouyang, H.; Chang, Y.; Xu, D. A dual quasi-zero-stiffness sliding-mode triboelectric nanogenerator for harvesting ultralow-low frequency vibration energy. *Mech. Syst. Signal Process.* **2021**, *151*, 107368. [CrossRef]
66. Yang, J.; Chen, J.; Yang, Y.; Zhang, H.; Yang, W.; Bai, P.; Su, Y.; Wang, Z.L. Broadband vibrational energy harvesting based on a triboelectric nanogenerator. *Adv. Energy Mater.* **2013**, *4*, 1301322. [CrossRef]
67. Bhatia, D.; Hwang, H.J.; Huynh, N.D.; Lee, S.; Lee, C.; Nam, Y.; Kim, J.G.; Choi, D. Continuous scavenging of broadband vibrations via omnipotent tandem triboelectric nanogenerators with cascade impact structure. *Sci. Rep.* **2019**, *9*, 8223. [CrossRef]
68. Lv, S.; Yu, B.; Huang, T.; Yu, H.; Wang, H.; Zhang, Q.; Zhu, M. Gas-enhanced triboelectric nanogenerator based on fully-enclosed structure for energy harvesting and sensing. *Nano Energy* **2018**, *55*, 463–469. [CrossRef]
69. Li, R.; Zhang, H.; Wang, L.; Liu, G. A Contact-Mode Triboelectric Nanogenerator for Energy. *Sensors* **2021**, *21*, 1514. [CrossRef] [PubMed]
70. Singh, H.H.; Kumar, D.; Khare, N. A synchronous piezoelectric-triboelectric-electromagnetic hybrid generator for harvesting vibration energy Sustain. *Energy Fuels* **2020**, *5*, 212–218. [CrossRef]
71. He, X.; Wen, Q.; Sun, Y.; Wen, Z. A low-frequency piezoelectric-electromagnetic-triboelectric hybrid broadband vibration energy harvester. *Nano Energy* **2017**, *40*, 300–307. [CrossRef]
72. He, J.; Wen, T.; Qian, S.; Zhang, Z.; Tian, Z.; Zhu, J.; Mu, J.; Hou, X.; Geng, W.; Cho, J.; et al. Triboelectric-piezoelectric-electromagnetic hybrid nanogenerator for high-efficient vibration energy harvesting and self-powered wireless monitoring system. *Nano Energy* **2018**, *43*, 326–339. [CrossRef]

73. Zhu, J.; Wang, A.; Hu, H.; Zhu, H. Hybrid electromagnetic and triboelectric nanogenerators with multi-impact for wideband frequency energy harvesting. *Energies* **2017**, *10*, 2024. [CrossRef]
74. Gupta, R.K.; Shi, Q.; Dhakar, L.; Wang, T.; Heng, C.H.; Lee, C. Broadband energy harvester using non-linear polymer spring and electromagnetic/triboelectric hybrid mechanism. *Sci. Rep.* **2017**, *7*, 41396. [CrossRef]
75. Haroun, A.; Lee, C. Dielectric-elastomer-enhanced triboelectric nanogenerator with amplified outputs. *Sens. Actuators A Phys.* **2022**, *333*, 113270. [CrossRef]
76. Hassani, F.A.; Shi, Q.; Wen, F.; He, T.; Haroun, A.; Yang, Y.; Feng, Y.; Lee, C. Smart materials for smart healthcare—Moving from sensors and actuators to self-sustained nanoenergy nanosystems. *Smart Mater. Med.* **2020**, *1*, 92–124. [CrossRef]
77. Haroun, A.; Yamada, I.; Warisawa, S. Investigation of Kinetic Energy Harvesting from Human Body Motion Activities Using Free/Impact Based Micro Electromagnetic Generator. *Diabetes Cholest. Metab.* **2016**, *1*, 12–16.
78. Seol, M.L.; Han, J.W.; Park, S.J.; Jeon, S.B.; Choi, Y.K. Hybrid energy harvester with simultaneous triboelectric and electromagnetic generation from an embedded floating oscillator in a single package. *Nano Energy* **2016**, *23*, 50–59. [CrossRef]
79. Salauddin, M.; Toyabur, R.M.; Maharjan, P.; Park, J.Y. High performance human-induced vibration driven hybrid energy harvester for powering portable electronics. *Nano Energy* **2018**, *45*, 236–246. [CrossRef]
80. Wang, S.; Niu, S.; Yang, J.; Lin, L.; Wang, Z.L. Quantitative measurements of vibration amplitude using a contact-mode freestanding triboelectric nanogenerator. *ACS Nano* **2014**, *8*, 12004–12013. [CrossRef]
81. Zhou, Y.S.; Zhu, G.; Niu, S.; Liu, Y.; Bai, P.; Jing, Q.; Wang, Z.L. Nanometer resolution self-powered static and dynamic motion sensor based on micro-grated triboelectrification. *Adv. Mater.* **2013**, *26*, 1719–1724. [CrossRef]
82. Yu, H.; He, X.; Ding, W.; Hu, Y.; Yang, D.; Lu, S.; Wu, C.; Zou, H.; Liu, R.; Lu, C.; et al. A Self-Powered Dynamic Displacement Monitoring System Based on Triboelectric Accelerometer. *Adv. Energy Mater.* **2017**, *7*, 1700565. [CrossRef]
83. Jing, Q.; Choi, Y.S.; Smith, M.; Ćatić, N.; Ou, C.; Kar-Narayan, S. Aerosol-Jet Printed Fine-Featured Triboelectric Sensors for Motion Sensing. *Adv. Mater. Technol.* **2019**, *4*, 1800328. [CrossRef]
84. Li, S.; Liu, D.; Zhao, Z.; Zhou, L.; Yin, X.; Li, X.; Gao, Y.; Zhang, C.; Zhang, Q.; Wang, J.; et al. A Fully Self-Powered Vibration Monitoring System Driven by Dual-Mode Triboelectric Nanogenerators. *ACS Nano* **2020**, *14*, 2475–2482. [CrossRef]
85. Liu, D.; Yin, X.; Guo, H.; Zhou, L.; Li, X.; Zhang, C.; Wang, J.; Wang, Z.L. A constant current triboelectric nanogenerator arising from electrostatic breakdown. *Sci. Adv.* **2019**, *5*, eaav6437. [CrossRef] [PubMed]
86. Liu, C.; Fang, L.; Zou, H.; Wang, Y.; Chi, J.; Che, L.; Zhou, X.; Wang, Z.; Wang, T.; Dong, L.; et al. Theoretical investigation and experimental verification of the self-powered acceleration sensor based on triboelectric nanogenerators (TENGs). *Extreme Mech. Lett.* **2021**, *42*, 101021. [CrossRef]
87. Zhang, B.; Zhang, L.; Deng, W.; Jin, L.; Chun, F.; Pan, H.; Wang, Z.L. Self-Powered Acceleration Sensor Based on Liquid Metal Triboelectric Nanogenerator for Vibration Monitoring. *ACS Nano* **2017**, *11*, 7440–7446. [CrossRef] [PubMed]
88. Wu, Z.; Ding, W.; Dai, Y.; Dong, K.; Wu, C.; Zhang, L.; Lin, Z.; Cheng, J.; Wang, Z.L. Self-Powered Multifunctional Motion Sensor Enabled by Magnetic Regulated Triboelectric Nanogenerator. *ACS Nano* **2018**, *12*, 5726–5733. [CrossRef]
89. Dai, K.; Wang, X.; Yi, F.; Jiang, C.; Li, R.; You, Z. Triboelectric nanogenerators as self-powered acceleration sensor under high-g impact. *Nano Energy* **2018**, *45*, 84–93. [CrossRef]
90. Chen, Y.; Wang, Y.-C.; Zhang, Y.; Zou, H.; Lin, Z.; Zhang, G.; Zou, C.; Wang, Z.L. Elastic-Beam Triboelectric Nanogenerator for High-Performance Multifunctional Applications: Sensitive Scale, Acceleration/Force/Vibration Sensor, and Intelligent Keyboard. *Adv. Energy Mater.* **2018**, *8*, 1802159. [CrossRef]
91. Wang, Z.; Zhang, F.; Li, N.; Yao, T.; Lv, D.; Cao, G. Self-Powered Multifunctional Triboelectric Sensor Based on PTFE/PU for Linear, Rotary, and Vibration Motion Sensing. *Adv. Mater. Technol.* **2020**, *5*, 2000159. [CrossRef]
92. Adly, M.A.; Arafa, M.H.; Hegazi, H.A. Modeling and optimization of an inertial triboelectric motion sensor. *Nano Energy* **2021**, *85*, 105952. [CrossRef]
93. Liu, C.; Wang, Y.; Zhang, N.; Yang, X.; Wang, Z.; Zhao, L.; Yang, W.; Dong, L.; Che, L.; Wang, G.; et al. A self-powered and high sensitivity acceleration sensor with V-Q-a model based on triboelectric nanogenerators (TENGs). *Nano Energy* **2020**, *67*, 104228. [CrossRef]
94. Xiao, X.; Zhang, X.; Wang, S.; Ouyang, H.; Chen, P.; Song, L.; Yuan, H.; Ji, Y.; Wang, P.; Li, Z.; et al. Honeycomb Structure Inspired Triboelectric Nanogenerator for Highly Effective Vibration Energy Harvesting and Self-Powered Engine Condition Monitoring. *Adv. Energy Mater.* **2019**, *9*, 1902460. [CrossRef]
95. Wang, L.; He, T.; Zhang, Z.; Zhao, L.; Lee, C.; Luo, G.; Mao, Q.; Yang, P.; Lin, Q.; Li, X.; et al. Self-sustained autonomous wireless sensing based on a hybridized TENG and PEG vibration mechanism. *Nano Energy* **2021**, *80*, 105555. [CrossRef]
96. Kim, G.W.; Kim, J. Compliant bistable mechanism for low frequency vibration energy harvester inspired by auditory hair bundle structures. *Smart Mater. Struct.* **2013**, *22*, 014005. [CrossRef]
97. Li, Z.; Bai, S.; Madsen, O.; Chen, W.; Zhang, J. Design, modeling and testing of a compact variable stiffness mechanism for exoskeletons. *Mech. Mach. Theory* **2020**, *151*, 103905. [CrossRef]

nanomaterials

Review

Recent Progress of Nanogenerators for Green Energy Harvesting: Performance, Applications, and Challenges

Enrique Delgado-Alvarado [1,2], Ernesto A. Elvira-Hernández [1], José Hernández-Hernández [3,4,*], Jesús Huerta-Chua [2], Héctor Vázquez-Leal [5], Jaime Martínez-Castillo [1], Pedro J. García-Ramírez [6] and Agustín L. Herrera-May [1,4,*]

1 Micro and Nanotechnology Research Center, Universidad Veracruzana, Boca del Río 94294, VER, Mexico; endelgado@uv.mx (E.D.-A.); aelvirah@hotmail.com (E.A.E.-H.); jaimartinez@uv.mx (J.M.-C.)
2 Departamento de Investigación, Instituto Tecnológico Superior de Poza Rica, Tecnológico Nacional de México, Poza Riza 93230, VER, Mexico; chua@itspozarica.edu.mx
3 Facultad de Ingeniería Mecánica y Ciencias Navales, Universidad Veracruzana, Boca del Río 94294, VER, Mexico
4 Maestría en Ingeniería Aplicada, Facultad de Ingeniería de la Construcción y el Hábitat, Universidad Veracruzana, Boca del Río 94294, VER, Mexico
5 Facultad de Instrumentación Electrónica, Universidad Veracruzana, Boca del Río 94294, VER, Mexico; hvazquez@uv.mx
6 Instituto de Ingeniería, Universidad Veracruzana, Boca del Río 94294, VER, Mexico; jagarcia@uv.mx
* Correspondence: josehernandez02@uv.mx (J.H.-H.); leherrera@uv.mx (A.L.H.-M.); Tel.: +52-2297752000 (J.H.-H.)

Citation: Delgado-Alvarado, E.; Elvira-Hernández, E.A.; Hernández-Hernández, J.; Huerta-Chua, J.; Vázquez-Leal, H.; Martínez-Castillo, J.; García-Ramírez, P.J.; Herrera-May, A.L. Recent Progress of Nanogenerators for Green Energy Harvesting: Performance, Applications, and Challenges. *Nanomaterials* **2022**, *12*, 2549. https://doi.org/10.3390/nano12152549

Academic Editors: Qiongfeng Shi and Jianxiong Zhu

Received: 11 June 2022
Accepted: 21 July 2022
Published: 25 July 2022

Publisher's Note: MDPI stays neutral with regard to jurisdictional claims in published maps and institutional affiliations.

Abstract: Natural sources of green energy include sunshine, water, biomass, geothermal heat, and wind. These energies are alternate forms of electrical energy that do not rely on fossil fuels. Green energy is environmentally benign, as it avoids the generation of greenhouse gases and pollutants. Various systems and equipment have been utilized to gather natural energy. However, most technologies need a huge amount of infrastructure and expensive equipment in order to power electronic gadgets, smart sensors, and wearable devices. Nanogenerators have recently emerged as an alternative technique for collecting energy from both natural and artificial sources, with significant benefits such as light weight, low-cost production, simple operation, easy signal processing, and low-cost materials. These nanogenerators might power electronic components and wearable devices used in a variety of applications such as telecommunications, the medical sector, the military and automotive industries, and internet of things (IoT) devices. We describe new research on the performance of nanogenerators employing several green energy acquisition processes such as piezoelectric, electromagnetic, thermoelectric, and triboelectric. Furthermore, the materials, applications, challenges, and future prospects of several nanogenerators are discussed.

Keywords: energy harvesting; green energy; hybrid nanogenerators; piezoelectric nanogenerator; thermoelectric nanogenerators; triboelectric nanogenerator

1. Introduction

The internet of things (IoT) gadgets, smart sensors, internet of medical things (IoMT) for healthcare systems, and consumer electronics devices have seen significant expansion in recent years. These devices often employ traditional batteries, which have drawbacks owing to their huge size, finite lifetime, and harmful components that contaminate the environment [1–3]. This issue with traditional batteries may restrict the efficiency of future IoT gadgets, smart sensors, and wearable devices. Thus, new eco-friendly alternative technologies to power these gadgets are current and future research challenges. Recent studies [4–8] have described nanogenerators capable of harvesting green energy by several transduction methods such as the piezoelectric, triboelectric, electromagnetic, and thermoelectric effects. The nanogenerators can harvest green energy from natural and artificial sources from wind, water, thermal, solar, mechanical vibrations, and motions of the human body [9–13]. These nanogenerators have unique features such as light weight, low-cost fabrication, tiny size, simple performance and signal processing, high power density, and a longer lifetime when compared to conventional batteries. Thus, nanogenerators provide a cost-effective alternative for powering future IoT devices, smart sensors, and consumer electronics products based on green energy harvesting from the environment. Furthermore, nanogenerators may be used to drive self-powered sensors for applications ranging from telecommunications to health monitoring, the automotive and military industries, agriculture, aerospace, and smart cities [14–19].

Most commercial low-power electronic devices require rectifier circuits to convert the variable output current of nanogenerators into direct current (DC). In addition, several researchers have used rectifiers coupled with antennas to design rectennas, which can harvest radio frequency (RF) energy and convert it to direct current [20–25]. Supercapacitors can also be integrated into nanogenerators to store their output power [26,27]. Thus, rectifier circuits and supercapacitors can enable the nanogenerators to have a consistent output power. In addition, hybrid nanogenerators may gather multiple green energy sources using two or more acquisition processes [28–31]. Due to this performance characteristic, hybrid nanogenerators can increase their output power densities in comparison to a single nanogenerator. The hybrid nanogenerators can power electronic devices for longer periods of time by utilizing various green energy sources (e.g., wind, heat, rain, solar radiation, and mechanical vibrations). These hybrid nanogenerators may be capable of harvesting a mix of green energies to continuously power electronics and sensors. This might enable the conversion of accessible green energy sources into electricity both during the day and at night, as well as in both indoor and outdoor environments.

More research is needed to increase the performance, stability, and reliability of nanogenerators. For instance, optimization methods may be utilized in the design of nano-generators for each individual application to forecast the best electrical and structural configurations and material selection. This optimized nanogenerator design can increase output power density and service time. Another idea is to employ wearable and flexible materials to create nanogenerators that are adaptive to the human body and gather biomechanical energy [32,33]. Additionally, effective packaging solutions for nanogenerators are necessary to improve their wear resistance and resistance to high temperature and humidity fluctuations. Better packing materials and the usage of long-lasting materials for nanogenerators can improve their reliability. The sensitivity of rectification circuits used in nanogenerators can be improved in the electronic section to produce a higher output DC power. Furthermore, these circuits may be manufactured utilizing microelectronic technology to reduce their size [34,35].

We present new research on nanogenerators that transform various green energy sources into electricity. This review looks at the principles of operation, materials, performance, and applications of several nanogenerators, including multiple green energy acquisition processes. The performance advantages of hybrid nanogenerators are also explored. We also consider the problems and views of nanogenerators, including their design phase, materials, energy storage, fabrication method, and dependability. Nanogenerators technology is an alternative solution for replacing traditional batteries and powering future electronic devices and sensors in the IoT and military industries, IoMT for healthcare systems, consumer electronics, telecommunications, automotive sector, robotics, wearable optoelectronics, and other fields.

2. Operation Principle

2.1. Vibration Energy

The vibration energy from the environment can be harvested using nanogenerators with transduction mechanisms such as piezoelectric, electromagnetic, triboelectric, and piezotronic effects. For instance, these nanogenerator types can convert mechanical vibrations caused by the wind effect, sound, water waves, human body motion, machines, and vehicles into electrical energy.

2.1.1. Piezoelectric Nanogenerators

The piezoelectric nanogenerators (PENGs) use the piezoelectric effect to capture green energy from ocean water waves, wind, biomechanical movements, and environmental mechanical vibrations. The output voltage of this type of nanogenerator is affected by mechanical deformations and the parameters of its piezoelectric layer. Mechanical vibrations in the environment can induce varied deformations in the piezoelectric nano-generators that generate the AC output voltage. A piezoelectric layer, a substrate, and two electrodes make up these nanogenerators. PENGs feature a basic structural design, easy performance, a simple construction method, high stability, and a low cost [36–44].

2.1.2. Electromagnetic Nanogenerators

Electromagnetic generators (EMGs) employ magnetic materials and coils to function according to the Faraday law. These generators may convert the kinetic energy of flowing water into electricity [45]. This wave flow is utilized to vary the location of the magnet material relative to the coil, resulting in a changing magnetic field that induces a voltage in the coil. However, as compared to triboelectric nanogenerators, these generators can have a larger volume and weight. Furthermore, EMGs require support structures that let them float on the water's surface [46]. The performance of electromagnetic nanogenerators is determined by the rate of change of the magnetic flux. EMGs can be made to function at frequencies comparable to those of ocean waves to scavenge energy from them. Ocean waves move randomly at low frequencies of roughly 1 Hz [47]. The EMGs' performance is limited by their low frequency. Due to wind sources and environmental mechanical vibrations, which may function at higher frequencies, EMGs are ideal for scavenging green energy.

2.1.3. Triboelectricity Nanogenerators

Triboelectric nanogenerators (TENGs) may gather green energy from irregular surroundings at low frequencies by connecting contact electrification with electrostatic induction. Blue energy, for example, may be extracted from ocean wave motion, which is fundamentally random and travels at low frequencies (near to 1 Hz) [48–60]. The benefits of triboelectric nanogenerators are their small weight, low cost, simple operation principle, and lack of sophisticated production [61–64]. To attain the highest performance, the triboelectric materials and electro-mechanical designs of the nanogenerators must be optimized [65–71]. As a result, optimizing the design of triboelectric nanogenerators is critical for improving the conversion of green energy into electric energy.

TENGs may be configured to function in four basic modes (Figure 1): vertical contact-separation (CS), lateral sliding (LS), single-electrode (SE), and freestanding triboelectric-layer (FSTL). TENGs usually require two triboelectric surfaces and two electrodes. Electron attraction between two triboelectric surfaces creates an electrostatic charge transfer from one surface to another in these operational modes. The displacement of the triboelectric layers changes their initial electrostatic state, resulting in an electric potential difference between the layers. The potential difference drives the current through the external load to balance the electrostatic state. The movement of the triboelectric layer in the opposite direction will generate a difference in the current flow. TENGs can therefore have alternating current (AC) voltages between their two output electrodes, depending on the triboelectric material type, operating mechanism, and green energy source.

2.1.4. Piezotronic Nanogenerators

The piezotronic nanogenerators harvest low-frequency vibration/friction energy into electricity by using the linked piezoelectric and semiconducting capabilities of nanowires/nanobelts, as well as the influence of a Schottky barrier at the metal-semiconductor [73,74].

These nanogenerators might be incorporated into textile strands to recycle energy generated by human movement. Thus, the piezoelectronic nanogenerator is a potentially useful technology for harvesting/recycling energy from the environment to power self-powered nanodevices that may be operated wirelessly and remotely. This technique will enable self-powered wireless nanosystems and nanodevices to have a sustained energy supply [75].

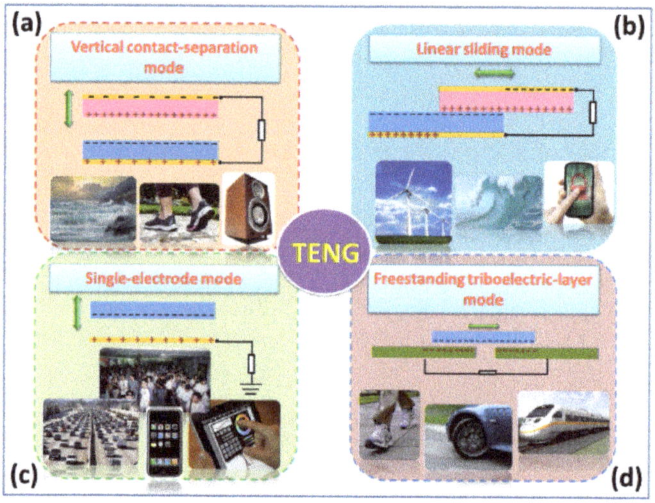

Figure 1. Several potential applications of TENGs using different operational modes: (**a**) vertical contact-separation, (**b**) linear sliding, (**c**) single-electrode, and (**d**) freestanding triboelectric-layer. Reprinted with permission from [72]. Copyright ©2014, Royal Society of Chemistry.

2.2. *Thermal Energy*

Thermoelectric and pyroelectric nanogenerators can transform thermal energy from the environment into electrical energy to power electronic devices.

Thermoelectric and Pyroelectric Nanogenerators

Another sort of green energy that may be obtained from the environment is thermal energy. This energy may be transformed into electric energy and used to power low-power electronic devices employing thermoelectric nanogenerators (TEGs) [76]. TEGs produce electricity by using the Seebeck effect to scavenge thermal energy caused by temperature differences between two thermoelectric (TE) materials (Figure 2). This temperature differential causes charge carriers to migrate from a high-temperature TE material to a low-temperature TE material [77,78]. A TEG's voltage output is proportional to the temperature gradient. TEGs, on the other hand, need significant temperature gradients across TE materials. TEGs are classified into two types: rigid thermoelectric nanogenerators and flexible thermoelectric nanogenerators, with the latter depending on their deformation properties. Stretchable, compressible, collapsible, lightweight, tiny in volume, affordable, and simple are advantages of TEGs [79–81]. Flexible TEGs have the potential to be employed in waste heat recovery [82–84], portable electronics [85–87], and human health monitoring due to their properties [88–90].

Pyroelectric nanogenerators (PyENGs) use the variation in spontaneous polarization inside pyroelectric materials to transform heat energy into electric energy. This is generated by oscillations of electric dipoles caused by a change in time-dependent temperature [91,92]. The creation of electric current through materials having a non-center symmetrical crystalline structure when subjected to a time-dependent temperature gradient is referred to as the pyroelectric effect [93,94].

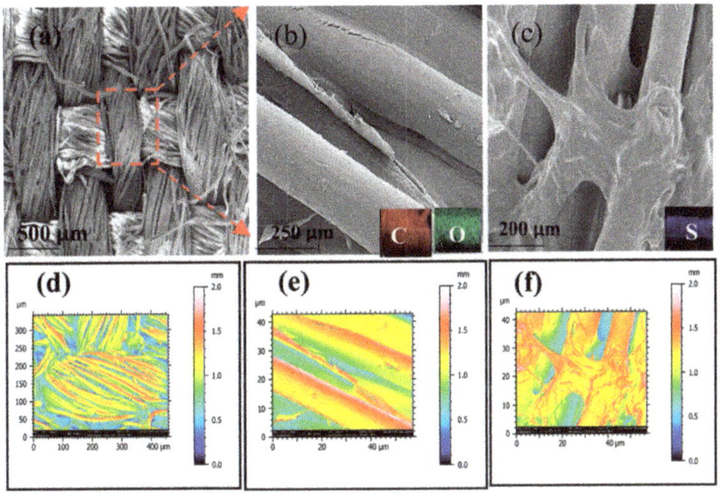

Figure 2. Reduced graphene oxide poly(3,4-ethylenedioxythiophene): poly (styrenesulfonate) (rGO-PEDOT:PSS) film-coated fabric of the flexible and washable thermoelectric nanogenerator fabricated by Khoso et al. [78]. This nanogenerator has potential application for harvesting green energy from human body heat. FESEM images with magnifications of (**a**) 500 μm and (**b**) 250 μm rGO-coated fabric and (**c**) 200 μm of rGO-PEDPT:PSS coated fabric. (**d–f**) Color mapping of SEM images' infrared rendering. Reprinted with permission from [78]. Copyright ©2021, Royal Society of Chemistry.

Pyroelectric nanogenerators have been identified as the energy collectors of the future, with the potential to be a viable energy technology for scavenging thermal energy in everyday life [94]. Thus, PyENGs and TEGs may have significant uses in powering future intelligent electronic sensors and IoT-connected wearable devices. More investigations on inorganic and organic materials, structure, performance, and reliability are required for the development of these nanogenerators.

2.3. Hybrid Nanogenerators

In the meantime, hybrid nanogenerators may harvest/recycle green energy from the environment by using several energy acquisition mechanisms or numerous connected nanogenerators with the same energy acquisition method (Figure 3). In hybrid nanogenerators, for example, piezoelectric, pyroelectric, triboelectric, and electromagnetic phenomena can be used. In comparison to individual nanogenerators, this nanogenerator type can provide high and efficient power density [95]. Recent research has led to the development of hybrid nanogenerators based on piezoelectric–pyroelectric [96–98], triboelectric–piezoelectric [31,99–117], electromagnetic–triboelectric [118–132], triboelectric–piezoelectric–pyroelectric [133–136], triboelectric–piezoelectric–electromagnetic [137–148], and photovoltaic–triboelectric effect [149–154].

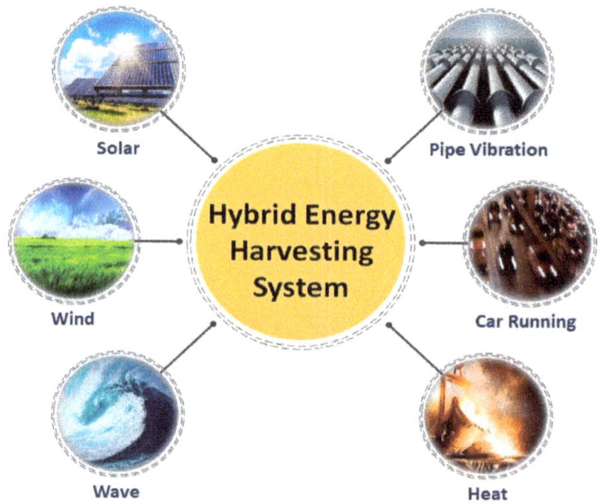

Figure 3. Potential applications of hybrid nanogenerators.

3. Performance and Applications

Hu et al. [155] designed an eco-friendly fabric-based TENG for converting biomechanical energy into electric energy, which can then be utilized to drive self-powered gadgets and wearable electronic sensors. This energy may be acquired by everyday human movements including leaping, jogging, walking, arms lifting, arms bending, and leg lifting. This TENG is made up of cellulose-based conductive macrofibers with key properties such as being super-strong, biodegradable, and washable. As illustrated in Figure 4, these microfibers were created by wet-stretching and wet-twisting bacterial cellulose (BC) hydro-gel with polypyrrole (PPy) and carbon nanotubes (CNTs). The microfibers were woven into a nylon fabric to generate the cellulose-based/nylon macrofiber. In this scenario, nylon serves as a positive triboelectric material, and a silver thin membrane is attached to a PDMS thin membrane to form a PDMS/silver film. As a result, the TENG features a cellulose-based/nylon macrofiber fabric that acts as a friction layer/electrode and a PDMS/silver layer that acts as a second friction film/electrode. The proposed microfibers demonstrated great tensile strength (449 MPa), strong electrical conductivity (5.32 Scm^{-1}), and good stability. The highest open-circuit voltage of the TENG is 170 V, the short-circuit current is 0.8 μA, and the output power is 352 μW. (Figure 5). Furthermore, these TENG may function as self-powered devices for tracking human body motions (Figure 6).

Zhao et al. [156] produced a triboelectric–electromagnetic hybrid nanogenerator (TEHG) that can gather wind energy while also powering electronic gadgets. This nanogenerator is made up of a TENG that operates in the sliding independent triboelectric-layer mode and an EMG that operates in the rotating mode. Figure 7 depicts the structure and materials of the TEHG, which is made up of a rotor and a stator. The stator has a cylindrical shell that is sealed, while the rotor has a disk and a projecting cylinder. This cylinder features an inside cylinder that can accommodate wind cups to convert environmental wind energy into mechanical energy. The cylindrical magnets of the EMG are positioned in ten cylindrical grooves on the upper surface of the rotor disk. The completed TEHG structure has an outside diameter of 80 mm and a height of 20 mm. The TENG employs PTFE and nylon as triboelectric layers that are in touch with one another. The PTFE functions as a 0.3 mm thick negative friction layer, while the nylon acts as a positive friction substance. Six aluminum electrodes are joined as interdigital electrodes on the nylon layer. Furthermore, the bottom of the shell has nine grooves for installing the copper coils of the EMG. These coils are wired in series to boost the output signal. The rotation of the TEHG structure caused by the wind source causes surface charge transfer between the two triboelectric layers. During the TEHG rotation process, an alternating current with a changing direction is produced. Figure 8 displays the TEHG's output open-circuit voltage and short-circuit current readings at various rotation speeds. Peak-to-peak voltage and peak-to-peak current of the TENG grow from 106 V to 190 V and 2.27 μA to 14.6 μA, respectively, with rotation speeds ranging from 100 rpm to 900 rpm (14 m/s of wind speed). The output response of the EMG is determined by the relative rotation of the magnet and coil based on electromagnetic induction (Faraday's law). The peak-to-peak voltage and peak-to-peak current rise from 5 V to 38 V and 3.3 mA to 20 mA, respectively, as the rotation speed increases from 100 rpm (5 m/s of wind speed) to 900 rpm. The TENG has a maximum average output power of 0.33 mW at an ideal load resistance of 12 MΩ. The EMG, on the other hand, has a maximum average output power of 32.87 mW and a maximum load resistance of 1.25 KΩ. The TEHG was evaluated for its ability to provide energy to wireless sensor network nodes. For this, Zhao et al. designed a circuit that incorporates a test device for measuring voltage changes and a cell phone for receiving data from the node (Figure 9). The TEHG was used to light up 200 LEDs in tandem and power an ambient humidity and temperature sensor at a rotation speed of 400 rpm (9 m/s of wind speed).

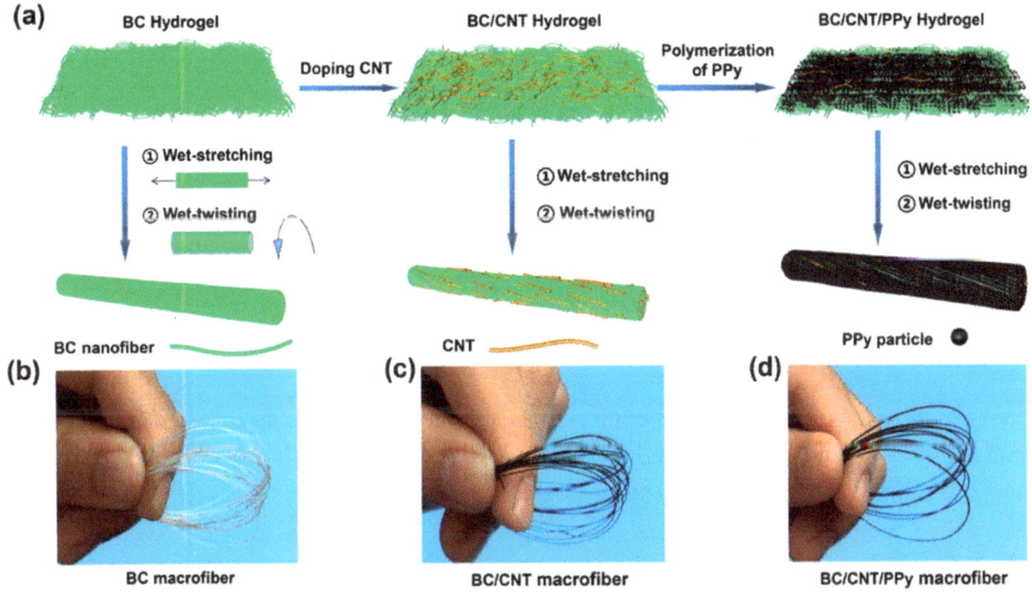

Figure 4. (**a**) Schematic view of the fabrication process of BC, BC/CNT/PPy macrofibers used in the fabric-based TENG developed by Hu et al. [155]. Images of (**b**) BC macrofibers, (**c**) BC/CNT macrofibers, and (**d**) BC/CNT/PPy macrofibers. Reprinted with permission from [155]. Copyright ©2022, Springer Nature.

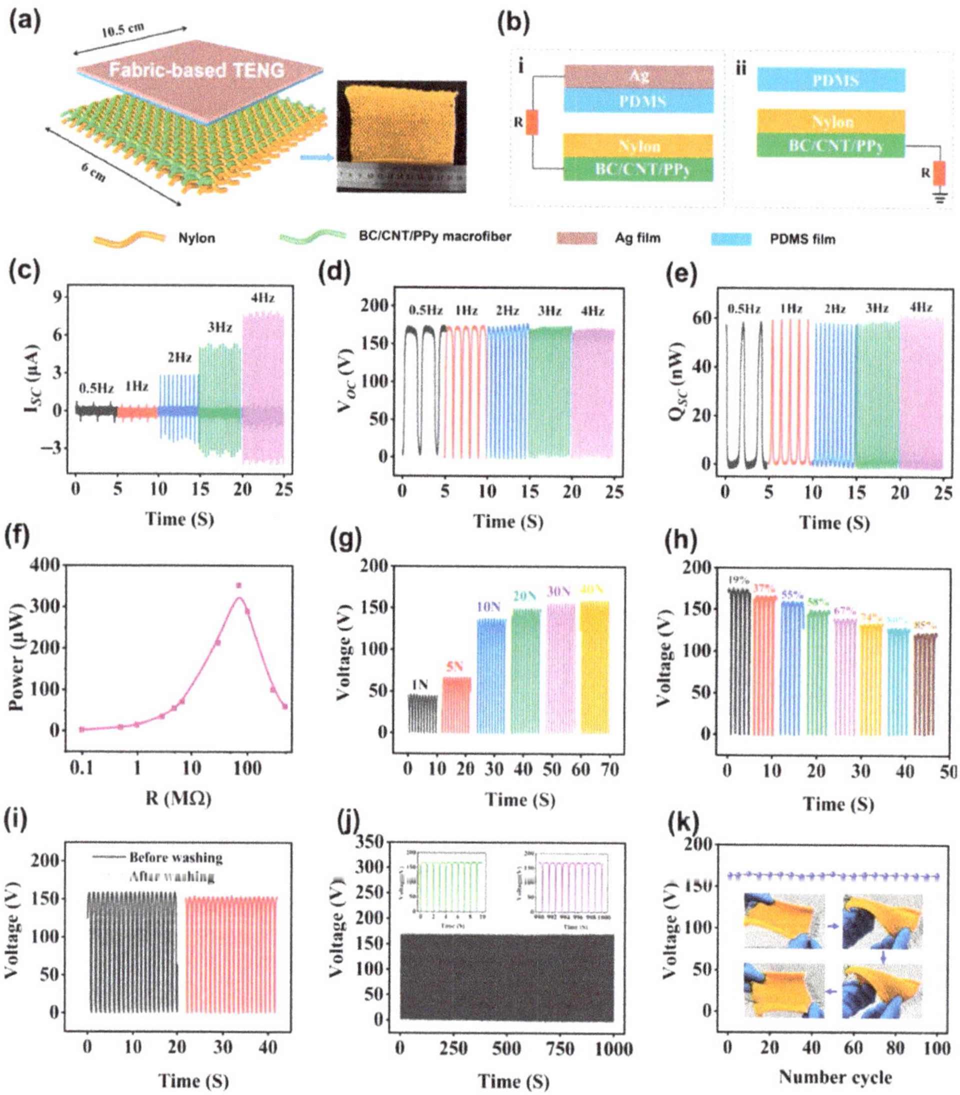

Figure 5. (**a**) Schematic view of the main components and materials of the fabric-based TENG designed by Hu et al. [155]; (**b**) two operating modes of the fabric-based TENG, (**i**) contact-separation mode and (**ii**) single electrode mode. Response of the (**c**) short-circuit current, (**d**) open-circuit voltage, and (**e**) transferred charges of the fabric-based TENG under different frequencies. (**f**) Results of the instantaneous power in relation to external load resistance, measurements of the output voltage of the fabric-based TENG considering (**g**) several impact forces at 1 Hz, (**h**) relative humidity variations, (**i**) before and after washing, (**j**) contact-separation mode with a frequency of 1 Hz during 100 s, and (**k**) mechanical strains with repetition of 100 cycles. Reprinted with permission from [155]. Copyright ©2022, Springer Nature.

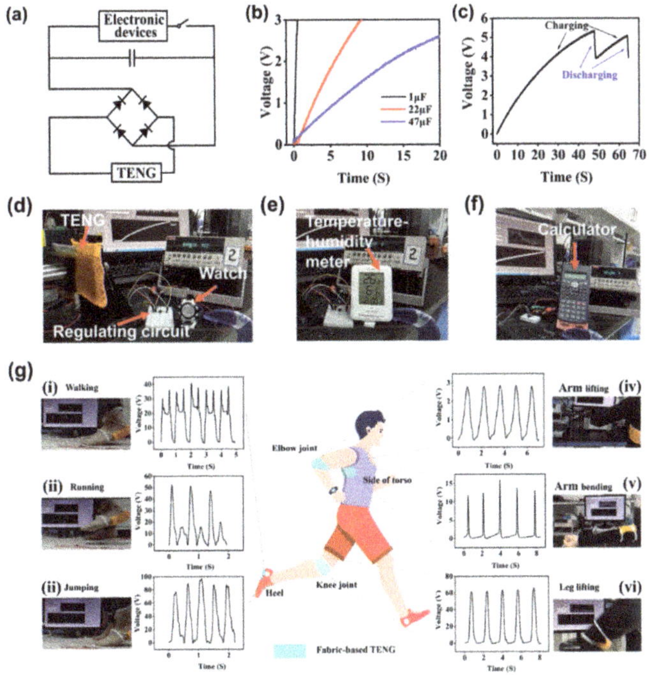

Figure 6. Applications of the fabric-based TENG were reported by Hu et al. [155]. (**a**) Diagram of the fabric-based TENG with rectifier bridge circuit for charging capacitors and powering electronic devices. (**b**) Response of the charging process of three commercial capacitors when the TENG is working in contact-separation mode with a frequency of 1 Hz. (**c**) Real-time measurements of the capacitor voltage, which is used for powering an electronic watch. (**d**) An electronic watch, (**e**) a temperature–humidity meter, and (**f**) a calculator powered using the fabric-based TENG with capacitors of 22 μF, 47 μF, and 100 μF, respectively. (**g**) Photographs and output voltages of the fabric-based TENG working as a self-powered device fixed to different sections of the human body for monitoring the body motion, (**i**) walking, (**ii**) running, (**iii**) jumping, (**iv**) arm lifting, (**v**) arm bending, and (**vi**) leg lifting. Reprinted with permission from [155]. Copyright ©2022, Springer Nature.

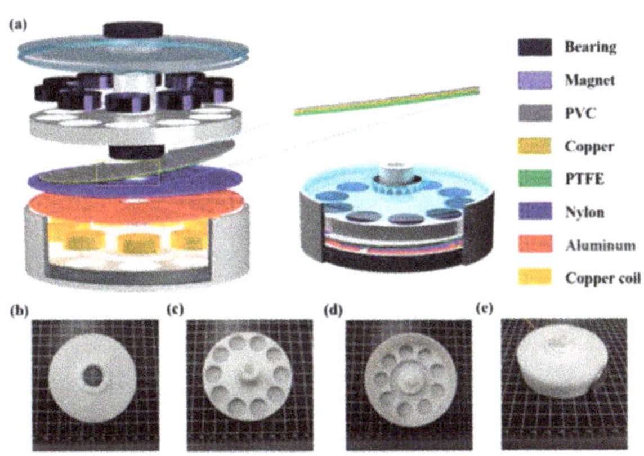

Figure 7. The TEHG structure developed by Zhao et al. [156]. (**a**) Schematic view of the main components and materials of the TEHG. (**b**) Image of the sealing cover of the cylindrical shell.

(**c**) Image of the rotor disk. (**d**) Image of the cylindrical shell. (**e**) Image of the assembled structure of the TEHG. Reprinted with permission from [156]. Copyright ©2021, John Wiley and Sons.

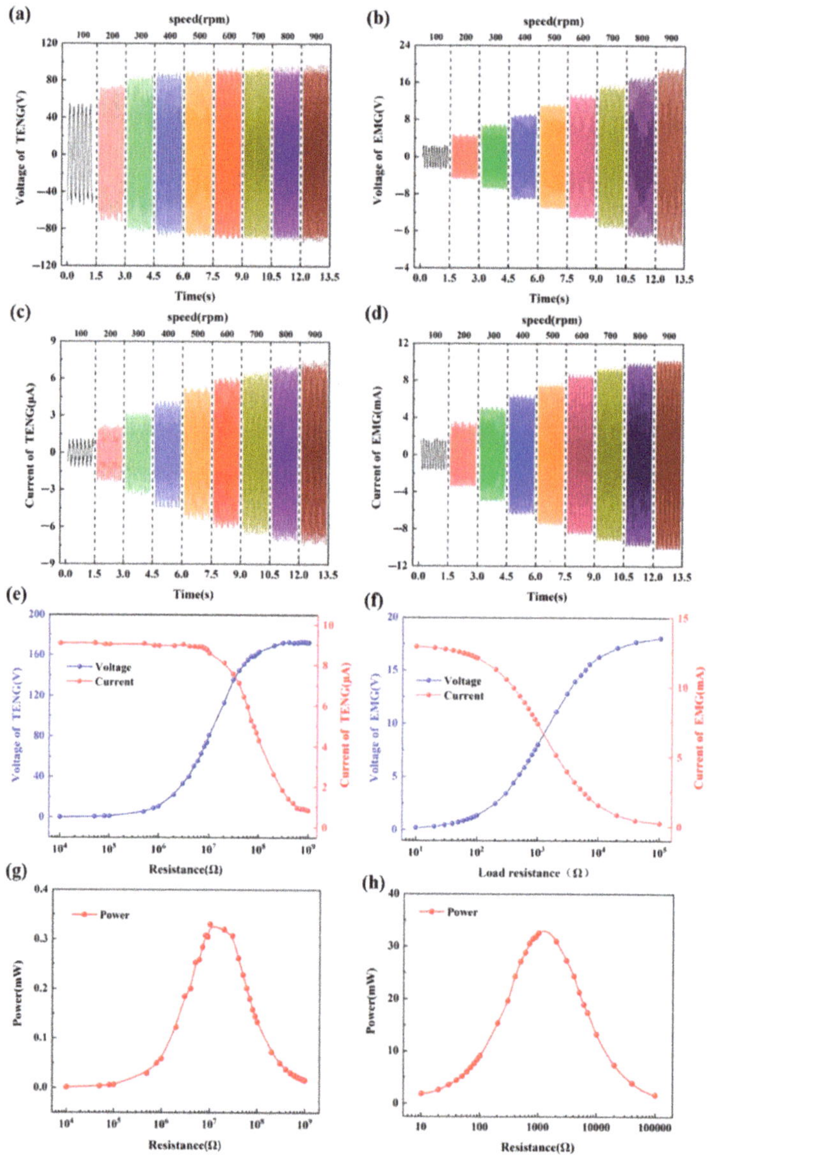

Figure 8. The output response of the TEHG was reported by Zhao et al. [156]. (**a**,**b**) The output open-circuit voltage of the TENG and EMG under several rotation speeds. (**c**,**d**) The output short-circuits current under several rotation speeds. (**e**,**f**) The output voltages and currents of the TENG and EMG as a function of external load resistance at a rotation speed of 400 rm. (**g**,**h**) The average output power of the TENG and EMG as a function of external load resistance at a rotation speed of 400 rpm. Reprinted with permission from [156]. Copyright ©2021, John Wiley and Sons.

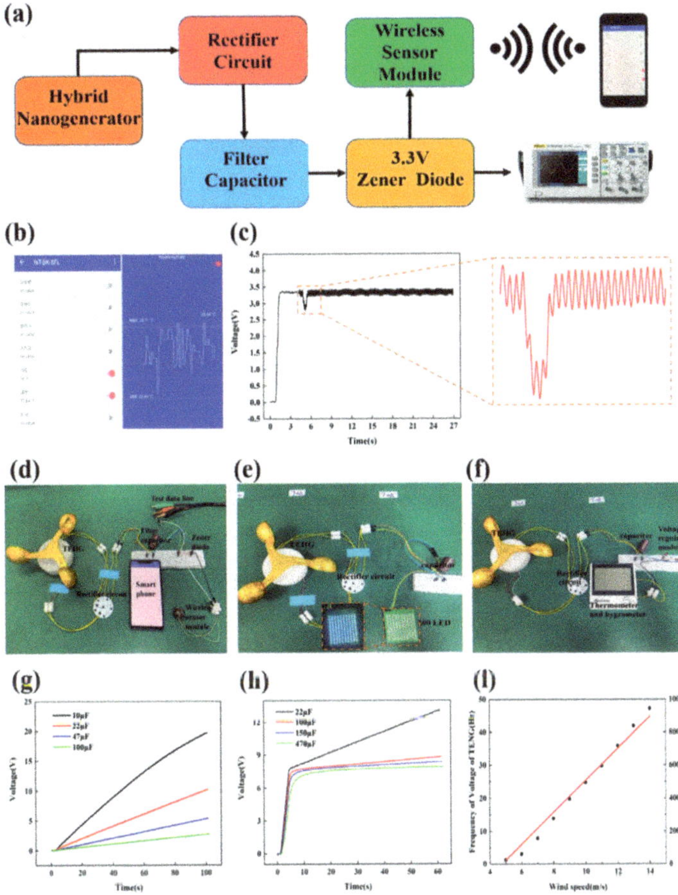

Figure 9. The application of the TEHG was proposed by Zhao et al. [156]. (**a**) Schematic view of the designed circuit for continuously supplying energy for the wireless sensor network node. (**b**) Mobile phone display to receive sensor data. (**c**) Mobile phone display to receive voltage data. (**d**) The test circuit of the TEHG to power the wireless sensor network node. (**e**) TEHG used to light up 200 LEDs. (**f**) TEHG supplies power to the thermometer and hygrometer device. (**g**) TENG is used to charge different capacitors. (**h**) TEHG is employed to charge several capacitors. (**i**) Relationship between the output voltage frequency, wind speed, and rotation speed. Reprinted with permission from [156]. Copyright ©2021, John Wiley and Sons.

Wu et al. [157] introduced a hybrid energy cell (Figure 10) that combines a TENG, an electrochemical cell (EC), and eight amorphous silicon-based solar cells (SCs) to gather wind, chemical, and solar energies from the environment simultaneously or independently. This hybrid energy cell might power low-power electronic devices such as wind speed sensors and temperature sensors. The key benefit of this technology is its capacity to scavenge three separate energy sources at the same time, which improves the usage of energy from the environment. A polytetrafluoroethylene (PTFE) film and an Al film are bonded to two acrylic tubes to form the TENG. The periodic contact/separation between the PTFE film and the Al film can generate charge transfer between the Al electrode and the ground by utilizing the coupling between the triboelectric effect and the electrostatic effect (Figure 11). First, both the Al and PTFE films are in an aligned position, where the two surfaces are completely in touch with one other. The two films have opposing charge polarities and are entirely balanced in this configuration, resulting in no electron flow across from Al film to PTFE film. This sequence is completed, and the mismatch between the two films is obtained. Since the relative rotation of both films continues, the PTFE film travels back to touch the Al film, creating electrons that flow from the Al electrode to the ground. This electrostatic induction action can lead the TENG's output signals to increase, indicating that the charges are entirely balanced. An alternating electric shape output is obtained during a TENG operation cycle. Figure 12 depicts the manufactured TENG's output signals. This TENG has an open-circuit voltage approaching 90 V, a short-circuit current density close to 0.5 mA/m^2, and a maximum power density of 16 mW/m^2, allowing it to power up 20 blue light-emitting diodes directly (LEDs). For charging a capacitor, the hybrid energy cell outperformed separate energy units significantly. The hybrid device's gathered energy can be stored in a Li-ion battery as a controlled power module for powering electronic equipment. Increasing the surface roughness and effective surface area of the triboelectric material induces a higher triboelectric charge density and improves TENG output performance.

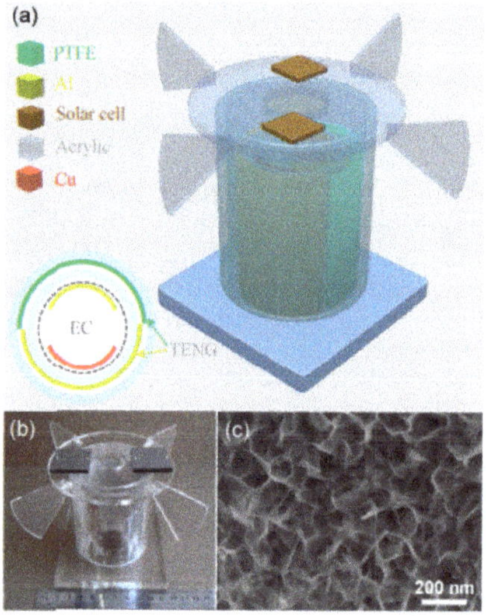

Figure 10. (**a**) Schematic view of the main components of the hybrid energy cell developed by Wu et al. [157]. (**b**) Image of the fabricated hybrid energy cell that includes solar cells and electrochemical cells placed on and in the TENG, respectively. (**c**) SEM image of Al film surface of the TENG, which was modified using nanostructures. Reprinted with permission from [157]. Copyright ©2014, Tsinghua University Press and Springer-Verlag Berlin Heidelberg.

The collagen fibrils of vegetables, fruits, and plants may be the responsible constituents of these natural materials' piezoelectricity. The piezoelectricity in collagen fibrils is caused by intermolecular hydrogen bonding, which results in a uniaxial orientation of the molecular dipoles [158]. Tomato peels (TPs), for example, include 16 different amino acids [159] and non-centrosymmetric properties due to their low symmetrical orthorhombic and monoclinic space groups, which might contribute to the piezoelectric effect [159]. Furthermore, the TPs feature structures with significant porosity, which causes additional displacement owing to applied external stresses, boosting the TPs' piezoelectricity [160,161]. Furthermore, the hydroxyl groups in TPs' lutein and zeaxanthin contribute to their piezoelectricity. Hydrogen bonding occurs in the hydroxyl group due to the extremely electropositive hydrogen and electronegative oxygen atoms [160]. According to the findings of these studies, TPs can be employed to generate piezoelectric energy.

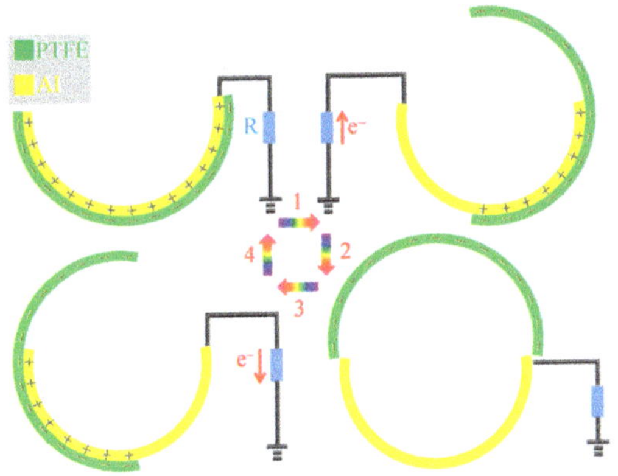

Figure 11. Stages of the operation principle of the TENG used in the hybrid energy cell fabricated by Wu et al. [157]. Reprinted with permission from [157]. Copyright ©2014, Tsinghua University Press and Springer-Verlag Berlin Heidelberg.

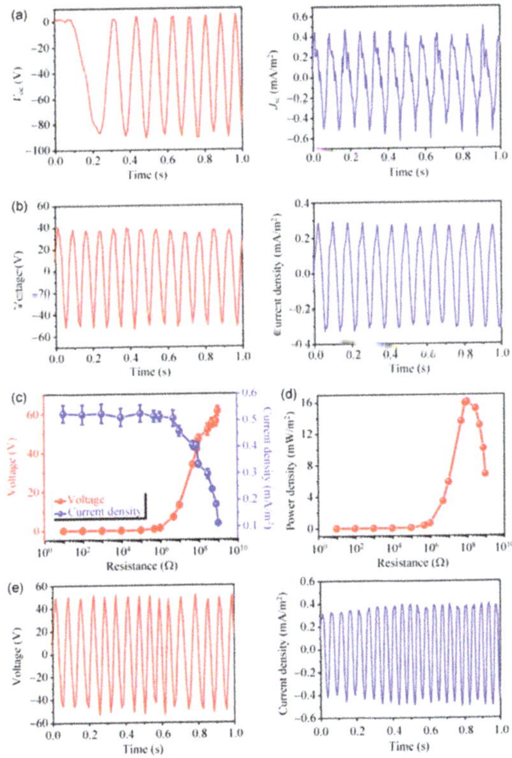

Figure 12. (**a**) Experimental results of the open-circuit voltage and short-circuit current density of the TENG using one strip unit. (**b**) Output voltage and current density of the TENG with one strip unit considering a load resistance of 100 MΩ. (**c**) Variation in the output voltage and current density of the TENG as a function of load resistance. (**d**) Response of the power density as a function of load resistance. (**e**) Output voltage and current density of the TENG with two strip units. Reprinted with permission from [157]. Copyright ©2014, Tsinghua University Press and Springer-Verlag Berlin Heidelberg.

Saqib et al. [160] examined the triboelectric and piezoelectric action of tomato peel (TP) in order to build a hybrid nanogenerator (TP-TPENG) with bio-organic nature materials for collecting green energy with potential applications in pollution-free and self-powered devices. The tomato's very porous structure boosts the TP-output TPENG's responsiveness. The open-circuit voltage, short-circuit current, and highest instantaneous power of a TP-based piezoelectric nanogenerator (TP-PENG) are 24.5 V, 2.5 μA, and 19.5 μW, respectively. The open-circuit voltage, short-circuit current, and highest instantaneous power of the TP-based triboelectric nanogenerator (TP-TENG), on the other hand, are 135 V, 81 μA, and 3750 μW, respectively. The combination of triboelectric and piezoelectric effects resulted in an enhanced TP-TPENG output response with a rectifier circuit. The rectified open circuit voltage, short circuit current, and maximum instantaneous power of this TP-TPENG are 150 V, 84 μA, and 5400 μW, respectively. Thus, TPs may be used to create unique non-toxic and eco-friendly hybrid nanogenerators based on their piezoelectric and triboelectric capabilities. This hybrid nanogenerator powered 141 commercial LEDs while also charging a 10 μF capacitor. Figure 13 depicts the primary components and materials of the hybrid nanogenerator. The TPs' hydroxyl and carbonyl groups contribute to their piezoelectric and triboelectric characteristics. Furthermore, TPs offer high flexibility and robustness. Figure 14 depicts the electrical output behavior of the TP-based nanogenerator when the piezoelectric and triboelectric effects are taken into account. The electrical output performance of the TP-TENG is superior to that of the TP-PENG. On the other hand, the combined responsiveness of TP's piezoelectric and triboelectric capabilities allows for a TP-TPENG with more superior electric output response than both TP-PENG and TP-TENG. The alternating output signal of the hybrid nanogenerator is rectified using two rectifier circuits. In addition, the hybrid nanogenerator was used to charge four different capacitors (0.22 μF, 10 μF, 50 μF, and 100 μF). The three nanogenerators were used to power several commercial LEDs. In addition, TP-TENG and TP-TPENG with rectifier circuits were used to power many commercial stopwatches (Figure 15).

Gokana et al. [91] developed a pyroelectric nanogenerator (PyNG) capable of producing electric energy from waste heat in the environment. As illustrated in Figure 16, this PyNG is made using a screen-printed serpentine electrode (SRE) that has been modified with cesium tungsten bronze ($Cs_{0.33}WO_3$). Furthermore, $Cs_{0.33}WO_3$ was applied to both the electrode and the PVDF sheets. With a load resistance of 20 ΩM, the PyNG with 7 wt% $Cs_{0.33}WO_3$ can reach a temperature of 121 °C and electrical output voltage, current, and power density of 4.36 V, 214 nA, and 23.38 $\mu W/m^2$, respectively. A liquid crystal display (LCD) and four LEDs were powered by the proposed PyNG. This PyNG is an alternate source for capturing solar energy and powering low-power electrical gadgets. Figure 16 depicts the various materials and components utilized in the fabrication of the PyNG, as well as the temperature values of its electrodes determined by IR thermographic imaging. The thermoelectric conversion behavior of the PyNG was evaluated by measuring its thermal and electrical output responses during heating and cooling at 8 mHz switching frequency. Under near-infrared (NIR) radiation, the PVDF PyNG with 7 wt% $Cs_{0.33}WO_3$ quickly registered 75 °C and reverted to 29 °C after 60 s of radiation removal. Using the same experimental conditions, PVDF PyNG without $Cs_{0.33}WO_3$ reached 58 °C and reverted to 29 °C. As a result, the rate of temperature changes of PVDF/$Cs_{0.33}WO_3$ PyNG was increased by over 27% when compared to pure PVDF PyNG. Furthermore, as compared to the pure PVDF PyNG, the electrical output voltage and current of the PVDF/ $Cs_{0.33}WO_3$ PyNG increased by around 26% and 16%, respectively (Figure 17). The PyNG's better performance might be attributed to its high photothermal conversion behavior and NIR light absorption. When NIR light is shone on the improved PyNG with a 10 μF capacitor, the PyNG may activate four LEDs and display an LCD, as illustrated in Figure 18.

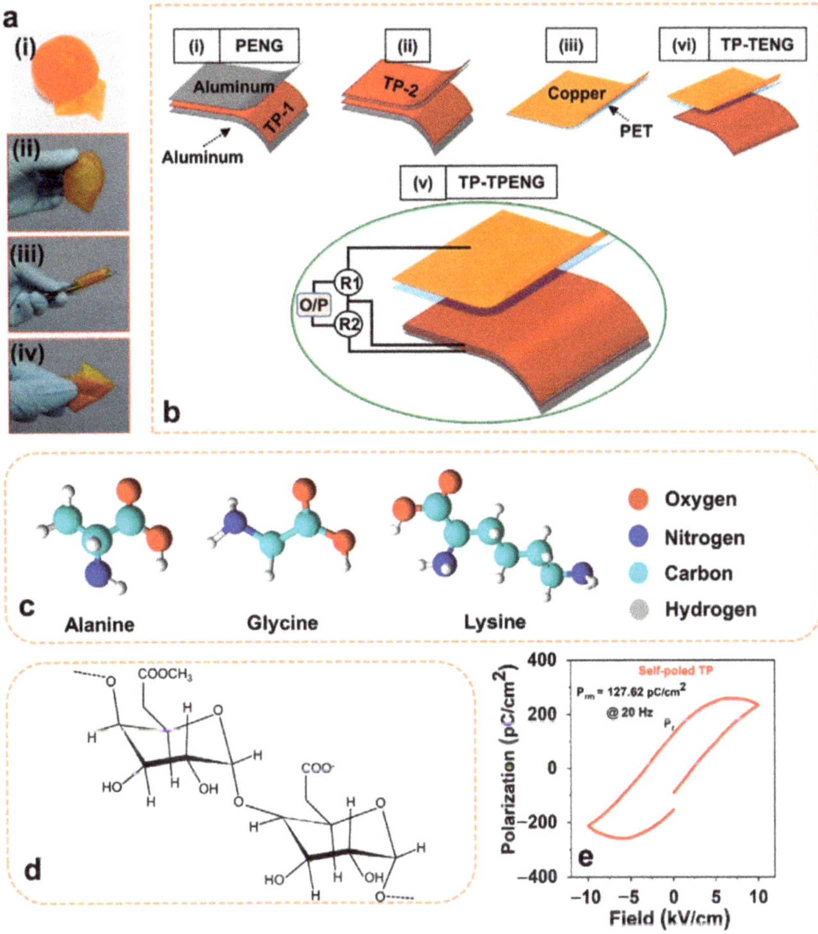

Figure 13. (**a**) Images of (**i**) peeling of a tomato, (**ii**) unfolded TP, (**iii**) folded TP, and (**iv**) rolled TP. (**b**) Schematic view of the main elements and materials used in the hybrid nanogenerator. (**c**) Structures of the three amino acids (alanine, glycine, and lysine) of the TP that allow the presence of C, O, N, and H. (**d**) Schematic view of the carbonyl and hydroxyl groups in the chain of the TP pectin structure. (**e**) Response of TP ferroelectric hysteresis considering a frequency of 20 Hz. Reprinted with permission from [160]. Copyright ©2021, John Wiley and Sons.

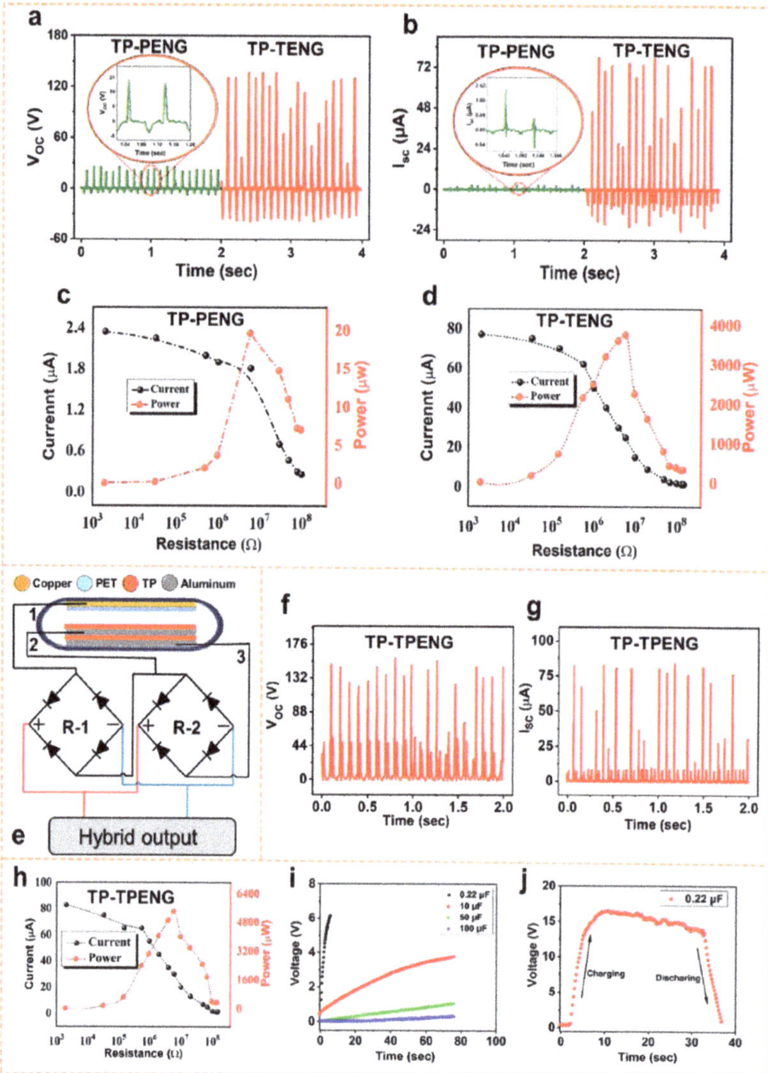

Figure 14. Electric output performance of the TP-PENG, TP-TENG, and TP-TPENG fabricated by Saqib et al. [160]. (**a**) Open circuit voltage and (**b**) short circuit current of the TP-based nanogenerator without considering the combination of both piezo and triboelectric effects. Output current and instantaneous power of the (**c**) TP-PENG and (**d**) TP-TENG as a function of the load resistance. (**e**) Schematic view of the main elements and materials of the hybrid nanogenerator. (**f**) Open circuit voltage and (**g**) short circuit current of the TP-TPENG. (**h**) Variation in the generated output current and instantaneous power of the TP-TPENG related with several external load resistances. (**i**) Charging curve of four different capacitors employing TP-TPENG. (**j**) The charging and discharging behavior of capacitor using hybrid nanogenerator. Reprinted with permission from [160]. Copyright ©2021, John Wiley and Sons.

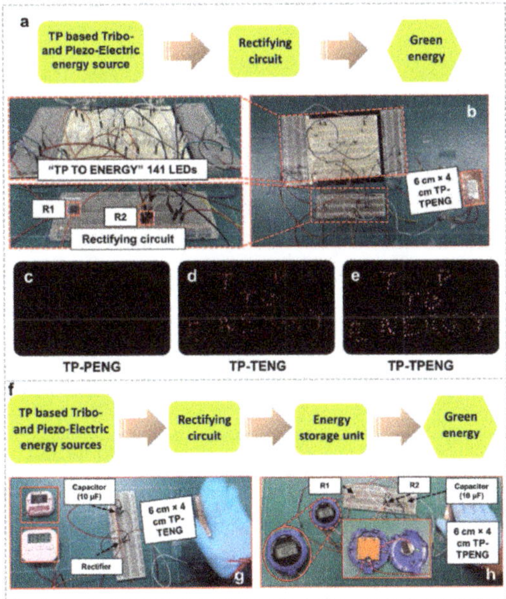

Figure 15. Application of the TP-based nanogenerators reported by Saqib et al. [160]. (**a**) Schematic diagram and (**b**) application of the TP-TPENG with rectifier circuit to light up LEDs. Lighted LEDs using (**c**) TP-PENG, (**d**) TP-TENG, and (**e**) TP-TPENG, respectively, under a simple hand pressing force. (**f**) Schematic diagram of the TP-TPENG with rectifier circuit and energy storage unit. Different stopwatches are powered using (**g**) TP-TENG and (**h**) TP-TPENG, respectively. Reprinted with permission from [160]. Copyright ©2021, John Wiley and Sons.

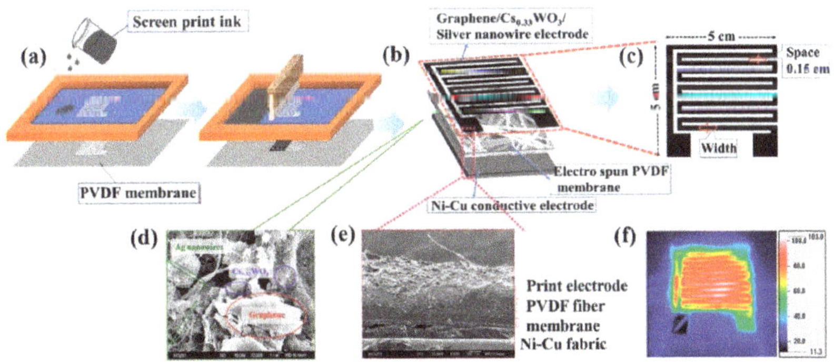

Figure 16. The main components and materials of the PyNG were reported by Gokana et al. [91]. (**a**) Schematic diagram of the elaboration of the serpentine electrode (SRE) using screen-printing. (**b**) Schematic diagram of structural design and materials of the SRE PyNG. (**c**) Dimensions of the SRE pattern. FESEM images of (**d**) surface and (**e**) cross-sectional view of the PyNG. (**f**) Temperature response of the PyNG using IR thermographic. Reprinted with permission from [91]. Copyright ©2022, Elsevier B.V.

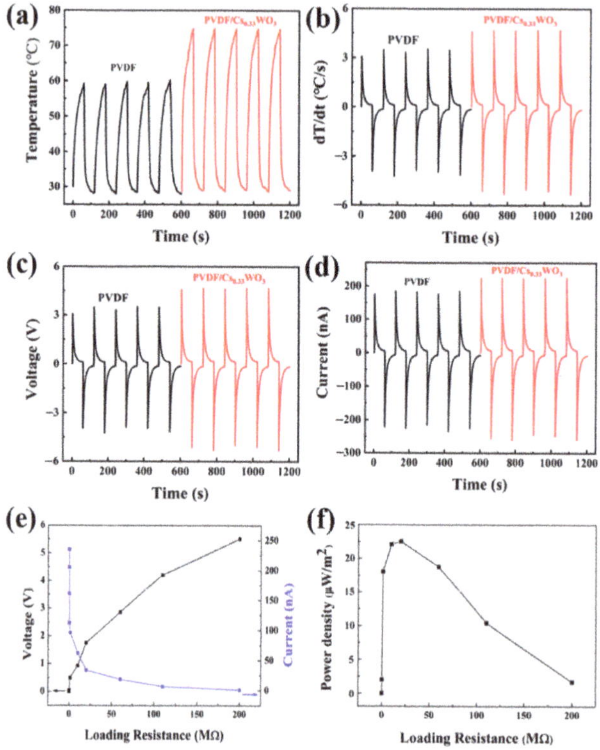

Figure 17. The thermal and electrical output performance of the PyNG developed by Gokana et al. [91]. (**a**) Temperature variation, (**b**) rate of temperature shift, (**c**) output voltage, and (**d**) output current of both pure PVDF PyNG and PVDF/$Cs_{0.33}WO_3$ PyNG. (**e**) Output voltage and current of PVDF/$Cs_{0.33}WO_3$ PyNG as a function of load resistance. (**f**) Output power density of PVDF/$Cs_{0.33}WO_3$ PyNG as a function of load resistance. Reprinted with permission from [91]. Copyright ©2022, Elsevier B.V.

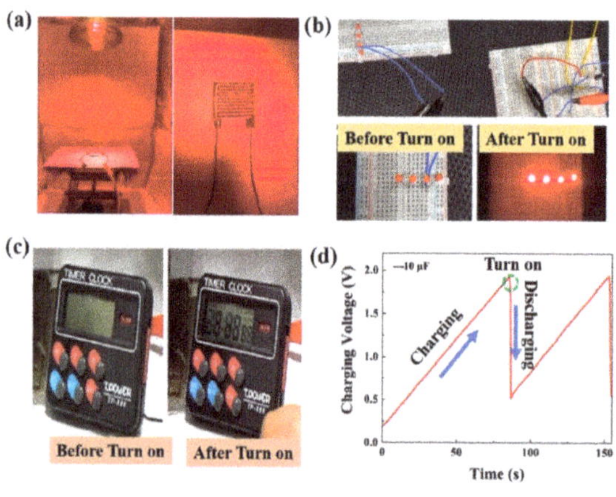

Figure 18. Application of the PVDF/$Cs_{0.33}WO_3$ PyNG designed by Gokana et al. [91]. (**a**) Experimental setup of the PyNG to turn on (**b**) four lighted LEDs and (**c**) display LCD. (**d**) Charging and discharging voltage of 10 μF capacitor using the PVDF/$Cs_{0.33}WO_3$ PyNG. Reprinted with permission from [91]. Copyright ©2022, Elsevier B.V.

Table 1 reported the comparison of the main characteristics of various nanogenerator types, including the transduction mechanism, energy source, materials, advantages, and weaknesses. For a single transduction mechanism, the triboelectric nanogenerators have important characteristics such as high electrical output performance, good stability, simple structure, and cost-efficient fabrication. In addition, this nanogenerator can be designed to convert different energy sources (e.g., biomechanical, water waves, environmental vibrations, and wind) into electrical energy. On the other hand, the hybrid nanogenerators present an enhanced electrical output performance in comparison with nanogenerators based on a single transduction mechanism. The hybrid nanogenerators that include the triboelectric effect have the best operation parameters such as compact and simple structure, good flexibility, stability, low-cost manufacturing, and high electrical output performance.

Table 1. Summary of the main characteristics of several nanogenerators used to harvest different green energy sources.

Transduction Mechanism	Energy Source and Main Materials	Advantages	Weaknesses	Potential Application	Reference
Piezoelectric	Biomechanical PVDF/Gly-MoS_2 composite film	High electro-mechanical performance, large-area compliant, long-time output signal stability, and low-power manufacturing	Performance depends on the Gly-MoS_2 nanosheet's content	Self-powered sensory systems, biomedical monitoring, and wearable electronics	[37]
Piezoelectric	Biomechanical PDMS/PPy composite film	Low-cost fabrication and flexible and robust devices	Performance depends on the PPy content	Self-powered biocompatible electronic devices	[38]
Piezoelectric	Biomechanical PVDF/coconut husk powder (CHP) composite film	Biowaste materials, simple fabrication process, and good electromechanical stability	Performance power depends on the CHP content	Biomedical devices and sustainable sensors	[39]
Piezoelectric	Biomechanical PVDF film	Flexible materials and low-cost fabrication	Electromechanical behavior requires more tests	Self-powered blood pressure sensors and wearable biomedical devices	[42]
Piezoelectric	Biomechanical 3D PPy/PVDF-poly-hexafluoropropylene (PHFP) composite film	Flexible materials and good performance stability	Performance depends on the 3D PPy content	Flexible biomedical devices	[44]
Triboelectric	Water waves and wind Cu electrodes, deionized water, and fluorinated ethylene–propylene tube	High output power density, simple structure, and easy fabrication process	Output power is highly dependent on the acceleration of water motion	Ship attitude sensors, multi-module devices, and ultra-sensitive sensor systems	[50]
Triboelectric	Water waves Stainless steel electrodes and PTFE film	High surface charge density, high output power, and low friction-induced loss	Wear of film by friction	Self-powered marine sensors, ocean buoys, and self-powered distributed energy for the marine IoT	[51]
Triboelectric	Water waves Cu electrodes and PTFE balls	Compact structure, ease of integration, and simple operation	Output performance depends on the wave direction and amplitude	Ocean buoys and cost-efficient beacon in night time marine operations	[52]
Triboelectric	Water waves Spring steel sheet and PFTE film	High electrical output performance for any direction of movement	Complex structure	Self-powered smart marine sensors and distributed power systems in oceans	[55]

Table 1. *Cont.*

Transduction Mechanism	Energy Source and Main Materials	Advantages	Weaknesses	Potential Application	Reference
Triboelectric	Biomechanical, environmental vibration and wind Ag electrodes, commercial VHB 4905 and Chitin films	Biodegradable materials, simple and cost-efficient fabrication, and high output performance	Performance depends on the chitin concentration	Self-powered flexible sensors, health monitoring of subtle pressures, non-contact sensing, and human–machine interfaces	[64]
Triboelectric	Biomechanical Cu electrodes, flourinated ethylene propylene (FEP) film	Simple operation and easy fabrication process	Wear of film by friction	Flexible electronic devices for real-time monitoring of human physiological states	[67]
Triboelectric	Biomechanical Cu and Ni fabrics and PDMS with $BaTiO_3$ nanoparticles	Stable electric behavior, ultra-flexibility, mechanical durability, and cyclic washing ability	Complex fabrication	Self-powered, wireless, and intelligent monitoring of human motions, portable power sources, and multifunctional human–machine interfaces	[71]
Thermoelectric	Human body heat rGO and PEDOT:PSS-coated textile fabric	Improved thermoelectric efficiency, high flexibility, breathable, washable, and bendable textile fabric	Complex manufacturing process and performance depends on the rGO concentration	Self-powered flexible devices and wearable e-textiles	[78]
Thermoelectric	Human body heat Cu, PET, $Bi_2(Te_{1-x}Se_x)_3$, and $(Bi_xSb_{1-x})_2Te_3$	Good flexibility and high output power density	Complex manufacturing process	Self-powered flexible and wearable sensors	[80]
Thermoelectric	Thermal ITO/PET, PEDOT:PSS, and MoS_2/graphene composite	Ultra-flexible and shape-adaptive	Complex manufacturing process and performance depends on the MoS_2/graphene content	Self-powered temperature sensors	[83]
Thermoelectric	Thermal and human body heat PEDOT:PSS/single-walled carbon nanotube (SWCNT) composite fibers	High flexibility and good bending durability	Complex manufacturing process and performance depends on the SWCNT content	Wearable electronic devices	[85]
Thermoelectric	Human body heat PDMS/boro nitride, n-Bi_2Te_3, and p-Sb_2Te_3 composite films	Portable and good flexibility	Complex manufacturing process	Self-powered wearable sensors for monitoring of human physiological signals and body motions	[90]
Pyroelectric	Near-infrared light PVDF/$Cs_{0.33}WO_3$ composite film	High output performance and high charge/discharge stability	Performance depends on the $Cs_{0.33}WO_3$ content	Implantable stimulator, high sensitivity sensors, and self-powered electronic devices	[91]
Pyroelectric	Thermal Au and ITO electrodes and ferroelectric antimony sulfoiodide (SbSI)-TiO_2 composite film	Simple fabrication process	Performance depends on the size and concentration of the TiO_2 nanoparticles	Pyroelectric sensors	[92]

Table 1. *Cont.*

Transduction Mechanism	Energy Source and Main Materials	Advantages	Weaknesses	Potential Application	Reference
Piezoelectric–pyroelectric	Biomechanical MWCNT doped PVDF nanofibers	High electrical output performance and high mechano-sensitivity	Complex manufacturing process and performance depends on the MWCNT content	Biomedical sensors integrated with IoT and remote care of infectious diseases	[96]
Piezoelectric–pyroelectric	Biomechanical Graphene oxide (GO), graphene (Gr), and halloysite (HNT) nanofillers and PVDF nanofibers	Improved electrical output performance and thermal stability	Complex manufacturing process and performance depends on the content of GO, Gr, and HNT	Wearable biomedical devices	[97]
Triboelectric–piezoelectric	Biomechanical Ag and Cu electrodes, PTFE, Nylon, PVDF films	Self-supported structure, high electrical output performance, low-cost and large-scale fabrication process, and high stability	For low-frequency and low amplitude mechanical vibrations	Self-powered flexible pressure sensors and electronic devices	[31]
Triboelectric–piezoelectric	Mechanical vibrations Al electrodes and polyvinylidene fluoridetrifluoroethylene (PVDF-TrFE), barium titanate (BTO), and PDMS composite	Large deformations, low-cost fabrication process, high electrical output performance, and stable electrical behavior	Performance depends on the concentration of PVDF-TrFE and BTO	Self-powered sensors for body motion monitoring, functional keyboards, and self-powered electronic devices placed in vehicles, bicycles, and pavements	[99]
Triboelectric–piezoelectric	Biomechanical BTO/silicon rubber (SR) composite film	Large deformation, stretchable, and high electrical output performance	Performance depends on the BTO content	Self-powered sensors for body motion monitoring, e-skin, and flexible wearable electronics	[101]
Triboelectric–piezoelectric	Mechanical vibrations Au electrodes, antimony selenoiodide (SbSeI) nanowires, and Kapton film	Simple and cost-effective fabrication process	Low-output power density	Low-power consumption electronic devices	[111]
Triboelectric–electromagnetic	Mechanical vibrations Cu foil electrode, polypropylene (PP) film, six magnets and nine coils	Improved electrical output performance	Large volume of magnets and coils	Self-powered flexible gas and motion monitoring, and charge smartphones	[120]
Triboelectric–electromagnetic	Wind Cu electrodes, FEP film, two magnets, and four Cu coils	High electrical output performance	Large volume of electromagnetic module	Self-powered electronics devices	[129]
Triboelectric–electromagnetic	Mechanical vibrations and biomechanical Al electrodes, BTO/PVDF film, five magnets, and five coils	High electrical output performance	Performance depends on the BTO content	Self-charging power systems for outdoor search and rescue, and electronic devices in the IoT	[27]
Triboelectric–piezoelectric–pyroelectric	Mechanical vibrations, wind, and thermal PVDF nanowires-PDMS composite film and ITO electrodes	Highly transparent and flexible	Complex fabrication process	Self-powered electronics	[134]

Table 1. *Cont.*

Transduction Mechanism	Energy Source and Main Materials	Advantages	Weaknesses	Potential Application	Reference
Triboelectric–piezoelectric–electromagnetic	Mechanical vibrations and biomechanical NdFeB magnet array structure, Cu coils, beryllium bronze electrodes, lead zirconate titanate (PZT) sheets, FEP films, and Cu electrodes	Small structure, ultra-low-frequency, multi-stable, portable, and high electrical output performance	Complex assembly of the three nanogenerators modules	Self-powered portable devices for body motion monitoring, sensors for detection of bridge motions, and construction safety monitoring	[142]
Triboelectric–piezoelectric–electromagnetic	Wind PVDF/PET film, PTFE film, PVDF film, PET sheet, Al electrodes, and eight NdFeB magnets and six Cu coils	Compact and small structure, and high electrical output performance	Complex assembly of the three nanogenerators modules	Self-powered wireless environmental monitoring system in subway tunnels	[146]

4. Challenges and Perspectives

This section discusses the main difficulties and prospects for nanogenerators in terms of design, materials, output performance, reliability, and prospective applications.

4.1. Design

The nanogenerator's design phase is critical to achieving the greatest performance for certain applications. This stage of design must take into account the various requirements and working circumstances of the prospective application of the nanogenerators. Thus, nanogenerator designers should investigate the required electric power, size and weight limitations, working time, and environmental conditions (relative humidity, temperature, pressure, radiation, wind, vibrations, dust, and so on), green energy acquisition mechanisms, materials more suitable for nanogenerator electromechanical behavior, fabrication process, packaging type, minimum electronic components, and so on. Analytical and numerical modeling may be utilized in the design of nanogenerators to find the best operation principle, materials, and electromechanical configuration that allow for the safety and reliability of the nanogenerators under various working conditions.

Furthermore, numerical simulation tools such as ANSYS, COMSOL, NASTRAN, ABAQUS, and MATLAB may be used to evaluate the performance of nanogenerators. As a result, the designers may create bidimensional or tridimensional models of the nanogenerators, complete with their major components, materials, and operating conditions. However, due to errors in the selection of the materials' characteristics, load values and analysis types, boundary conditions, and mesh size, the findings of the analytical and numerical simulation models of the nanogenerators might have a large error percentage in comparison to experimental results. To reduce this error percentage, designers should evaluate the important elements that determine the performance of the nanogenerators, such as genuine values of material characteristics, the right selection of load and boundary conditions, and the mesh quality and type of the suggested models.

4.2. Materials

The appropriate material selection for nanogenerators is critical for improving their performance and reliability for each prospective application. Each nanogenerator application necessitates unique performance characteristics in order to maximize the green energy collecting process from various natural sources and under varying climatic conditions. Thus, the design of a specific nanogenerator requires superior materials to meet the electrical signal requirements and performance stability for future applications. Nanogenerators for biomechanical applications, for example, may need stretchable, lightweight, and flexible materials to improve their output electrical responsiveness and mechanical behavior [162]. These materials must have the structural strength to minimize wear and mechanical failures in this application, as well as low density to reduce the weight of the nanogenerators. Furthermore, the operation of nanogenerators might be harmed by environmental dampness. The application of hydrophobic materials or materials to nanogenerator packing can be used to solve the humidity problem. For instance, nanogenerators for blue energy harvesting are built with specific packaging made of low-density materials that are resistant to corrosion and solar radiation. Nanogenerators that operate under mechanical vibrations, on the other hand, should be designed with materials that have adequate structural strength to reduce structural failures caused by cracks, wear, fatigue, or fracture. Recent nanogenerator research [39,63,64,160,163–169] has focused on organic or waste materials from the environment, such as tomato, chitin, eggshell, fish swim bladder, spider silk, peanut shell, sunflower husks, rice paper, garbage soda cans, silk fibroin, coconut husk, and so on. However, measuring the piezoelectric and triboelectric properties of these materials is extremely difficult. Another significant constraint is the analytical modeling of the piezoelectric and triboelectric response of organic materials. More research is needed to determine the piezoelectric and triboelectric effects of organic or waste materials employed in nanogenerators for green energy harvesting.

4.3. Energy Storage and Electrical Interfaces

The electrical output performance of nanogenerators can be impacted by changes in ambient circumstances and green energy stability, which can vary over time and exhibit erratic behavior. Due to these circumstances, the output electrical signals of nanogenerators might exhibit erratic behavior. In addition, most electronic equipment must be powered by DC voltage and current. The nanogenerators require rectifier circuits to convert their AC output electrical impulses into DC signals in order to power these devices. These DC signals must also be stored in capacitors or batteries in order to power electronic devices with controlled electrical signals. The development of effective energy storage devices is an intriguing research problem for nanogenerators.

Another important challenge of the nanogenerators is the development of electrical interfaces that achieve high efficiency with minimum power consumption [170–172]. For instance, these electrical interfaces could be self-powered and consider cold-start circuit architectures [173,174]. To reduce the size of the nanogenerators, the electrical interfaces should have a small footprint. It could be obtained using the Application-Specific Integrated Circuit (ASIC) implementation. Moreover, the electrical interfaces could be adaptive in order to maximize the harvested power, considering low-power maximum power point tracking (MPPT) algorithms [174–177].

4.4. Fabrication

To create nanogenerators for commercial uses, large-scale manufacture should be enabled. Alternatives to this difficulty include no-complex manufacturing procedures and new infrastructure with flexible phases for new nanogenerator designs and use. Furthermore, a low-cost production technique with few processing steps is critical for the market feasibility of nanogenerators. Another possibility to reduce nanogenerator manufacture costs is to re-use inorganic or organic materials from trash [63]. Furthermore, future nanogenerator manufacturing processes may contain biodegradable and environmentally benign materials.

4.5. Reliability

Future studies will focus on the stability and reliability of the electromechanical behavior of nanogenerators. For commercial applications, nanogenerators must provide output electrical signals that are stable throughout time. During the life of a nanogenerator, its electrical and structural components may have performance issues owing to abrasion, mechanical impact, crack development, fatigue, humidity, radiation, high temperature, environmental pollution, and other factors. To extend the life of nanogenerators, basic structural layouts with the fewest number of electrical and mechanical components should be considered. Electrical or mechanical failures of some of these components can affect the operation of nanogenerators with complicated structural designs that involve multiple components. To decrease the possibility of electrical and mechanical failures in the various nanogenerator components, the use of robust materials and appropriate packaging might increase the service life of nanogenerators.

5. Conclusions

The most recent advances in nanogenerators for green energy harvesting via various transduction processes were discussed. Triboelectric, piezoelectric, electromagnetic, and thermoelectric effects were all explored in these processes. The principles of operation and materials of several nanogenerators are reviewed. In addition, the behavior of the output electrical signals (voltage, current, and power) of multiple nanogenerators was considered, taking into account the combination of green energy acquisition processes. It was stated that nanogenerators were used to power several commercial electronic products. This review also discussed the problems and perspectives of nanogenerators in design, materials, energy storage, fabrication, and reliability.

Author Contributions: E.D.-A., E.A.E.-H., J.H.-H., J.H.-C., H.V.-L., J.M.-C., P.J.G.-R. and A.L.H.-M. wrote, coordinated, reviewed, and contributed to scientific aspects of the article. All authors have read and agreed to the published version of the manuscript.

Funding: This research received no external funding.

Institutional Review Board Statement: Not applicable.

Informed Consent Statement: Not applicable.

Data Availability Statement: Request the corresponding authors of this article.

Acknowledgments: Thanks to "Instituto Tecnológico Superior de Poza Rica" and "Universidad Veracruzana" for the support and facilities for the publication of this article.

Conflicts of Interest: The authors declare no conflict of interest.

References

1. Keshavarz, R.; Shariati, N. Highly sensitive and compact quad-band ambient RF energy harvester. *IEEE Trans. Ind. Electron.* **2022**, *69*, 3609–3621. [CrossRef]
2. Tao, K.; Chen, Z.; Yi, H.; Zhang, R.; Shen, Q.; Wu, J.; Tang, L.; Fan, K.; Fu, Y.; Miao, J.; et al. Hierarchical Honeycomb-Structured Electret/Triboelectric Nanogenerator for Biomechanical and Morphing Wing Energy Harvesting. *Nano-Micro Lett.* **2021**, *13*, 123. [CrossRef] [PubMed]
3. Zhao, T.; Xu, M.; Xiao, X.; Ma, Y.; Li, Z.; Wang, Z.L. Recent progress in blue energy harvesting for powering distributed sensors in ocean. *Nano Energy* **2021**, *88*, 106199. [CrossRef]
4. Tremmel, S.; Luo, X.; Rothammer, B.; Seynstahl, A.; Wang, B.; Rosenkranz, A.; Marian, M.; Zhu, L. Evaluation of DLC, MoS2, and Ti3C2Tx thin films for triboelectric nanogenerators. *Nano Energy* **2022**, *97*, 107185. [CrossRef]
5. Niu, Q.; Wei, H.; Hsiao, B.S.; Zhang, Y. Biodegradable silk fibroin-based bio-piezoelectric/triboelectric nanogenerators as self-powered electronic devices. *Nano Energy* **2022**, *96*, 107101. [CrossRef]
6. Lu, M.; Fu, G.; Osman, N.B.; Konbr, U. Green energy harvesting strategies on edge-based urban computing in sustainable internet of things. *Sustain. Cities Soc.* **2021**, *75*, 103349. [CrossRef]
7. Gao, Y.; Liu, G.; Bu, T.; Liu, Y.; Qi, Y.; Xie, Y.; Xu, S.; Deng, W.; Yang, W.; Zhang, C. MXene based mechanically and electrically enhanced film for triboelectric nanogenerator. *Nano Res.* **2021**, *14*, 4833–4840. [CrossRef]
8. Islam, E.; Abdullah, A.M.; Chowdhury, A.R.; Tasnim, F.; Martinez, M.; Olivares, C.; Lozano, K.; Uddin, M.J. Electromagnetic-triboelectric-hybrid energy tile for biomechanical green energy harvesting. *Nano Energy* **2020**, *77*, 105250. [CrossRef]
9. Shaukat, R.A.; Saqib, Q.M.; Khan, M.U.; Chougale, M.Y.; Bae, J. Bio-waste sunflower husks powder based recycled triboelectric nanogenerator for energy harvesting. *Energy Rep.* **2021**, *7*, 724–731. [CrossRef]
10. Niu, Z.; Cheng, W.; Cao, M.; Wang, D.; Wang, Q.; Han, J.; Long, Y.; Han, G. Recent advances in cellulose-based flexible triboelectric nanogeneratorss. *Nano Energy* **2021**, *87*, 106175. [CrossRef]
11. Nguyen, Q.-T.; Ahn, K.-K.K. Fluid-Based Triboelectric Nanogenerators: A Review of Current Status and Applications. *Int. J. Precis. Eng. Manuf. Green Technol.* **2021**, *8*, 1043–1060. [CrossRef]

12. Sanad, M.F.; Shalan, A.E.; Abdellatif, S.O.; Serea, E.S.A.; Adly, M.S.; Ahsan, M. Thermoelectric Energy Harvesters: A Review of Recent Developments in Materials and Devices for Different Potential Applications. *Top. Curr. Chem. (Z)* **2020**, *378*, 48. [CrossRef] [PubMed]
13. Karan, S.K.; Maiti, S.; Lee, J.H.; Mishra, Y.K.; Khatua, B.B.; Kim, J.K. Recent Advances in Self-Powered Tribo-/Piezoelectric Energy Harvesters: All-In-One Package for Future Smart Technologies. *Adv. Funct. Mater.* **2020**, *30*, 2004446. [CrossRef]
14. Nie, W. A sliding hybrid triboelectric-electromagnetic nanogenerator with staggered electrodes for human motion posture. *Energy Rep.* **2022**, *8*, 617–625. [CrossRef]
15. Sahu, M.; Hajra, S.; Panda, S.; Rajaitha, M.; Panigrahi, B.K.; Rubahn, H.-G.; Mishra, Y.K.; Kim, H.J. Waste textiles as the versatile triboelectric energy-harvesting platform for self-powered applications in sports and athletics. *Nano Energy* **2022**, *97*, 107208. [CrossRef]
16. Pang, Y.; Zhu, X.; Lee, C.; Liu, S. Triboelectric nanogenerator as next-generation self-powered sensor for cooperative vehicle-infrastructure system. *Nano Energy* **2022**, *97*, 107219. [CrossRef]
17. Bhatia, D.; Lee, K.-S.; Niazi, M.U.K.; Park, H.-S. Triboelectric nanogenerator integrated origami gravity support device for shoulder rehabilitation using exercise gaming. *Nano Energy* **2022**, *97*, 107179. [CrossRef]
18. Fang, L.; Zheng, Q.; Hou, W.; Zheng, L.; Li, H. A self-powered vibration sensor based on the coupling of triboelectric nanogenerator and electromagnetic generator. *Nano Energy* **2022**, *97*, 107164. [CrossRef]
19. Shi, S.; Jiang, Y.; Xu, Q.; Zhang, J.; Zhang, Y.; Li, J.; Xie, Y.; Cao, Z.-P. A self-powered triboelectric multi-information motion monitoring sensor and its application in wireless real-time control. *Nano Energy* **2022**, *97*, 107150. [CrossRef]
20. Miskovsky, N.M.; Cutler, P.H.; Mayer, A.; Weiss, B.L.; Willis, B.; Sullivan, T.E.; Lerner, P.B. Nanoscale devices for rectification of high frequency Radiation from the infrared through the visible: A new approach. *J. Nanotechnol.* **2012**, *2012*, 512379. [CrossRef]
21. Donchev, E.; Pang, J.S.; Gammon, P.M.; Centeno, A.; Xie, F.; Petrov, P.K.; Breeze, J.D.; Ryan, M.P.; Riley, D.J.; Alford, N. The rectenna device: From theory to practice (a review). *MRS Energy Sustain.* **2014**, *1*, E1. [CrossRef]
22. Mupparapu, R.; Cunha, J.; Tantussi, F.; Jacassi, A.; Summerer, L.; Patrini, M.; Giugni, A.; Maserati, L.; Alabastri, A.; Garoli, D.; et al. High-Frequency Light Rectification by Nanoscale Plasmonic Conical Antenna in Point-Contact-Insulator-Metal Architecture. *Adv. Energy Mater.* **2022**, *12*, 2103785. [CrossRef]
23. Shanawani, M.; Masotti, D.; Costanzo, A. THz Rectennas and Their Design Rules. *Electronics* **2017**, *6*, 99. [CrossRef]
24. Byrnes, S.J.; Blanchard, R.; Capasso, F. Harvesting renewable energy from Earth's mid-infrared emissions. *Proc. Natl. Acad. Sci. USA* **2014**, *111*, 3927–3932. [CrossRef] [PubMed]
25. Joshi, S.; Moddel, G. Rectennas at optical frequencies: How to analyze the response. *J. Applied Phys.* **2015**, *118*, 084503. [CrossRef]
26. Zhao, J.; Cong, Z.; Hu, J.; Lu, H.; Wang, L.; Wang, H.; Malyi, O.I.; Pu, X.; Zhang, Y.; Shao, H.; et al. Regulating zinc electroplating chemistry to achieve high energy coaxial fiber Zn ion supercapacitor for self-powered textile-based monitoring system. *Nano Energy* **2022**, *93*, 106893. [CrossRef]
27. Zhang, Y.; Gao, X.; Zhang, Y.; Gui, J.; Sun, C.; Zheng, H.; Guo, S. High-efficiency self-charging power systems based on performance-enhanced hybrid nanogenerators and asymmetric supercapacitors for outdoor search and rescue. *Nano Energy* **2022**, *92*, 106788. [CrossRef]
28. Yu, Z.; Zhang, Y.; Wang, Y.; Zheng, J.; Fu, Y.; Chen, D.; Wang, G.; Cui, J.; Yu, S.; Zheng, L.; et al. Integrated piezo-tribo hybrid acoustic-driven nanogenerator based on porous MWCNTs/PVDF-TrFE aerogel bulk with embedded PDMS tympanum structure for broadband sound energy harvesting. *Nano Energy* **2022**, *97*, 107205. [CrossRef]
29. Li, R.; Wei, X.; Shi, Y.; Yuan, Z.; Wang, B.; Xu, J.; Wang, L.; Wu, Z.; Wang, Z.L. Low-grade heat energy harvesting system based on the shape memory effect and hybrid triboelectric-electromagnetic nanogenerator. *Nano Energy* **2022**, *96*, 107106. [CrossRef]
30. Banerjee, S.; Bairagi, S.; Ali, S.W. A lead-free flexible piezoelectric-triboelectric hybrid nanogenerator composed of uniquely designed PVDF/KNN-ZS nanofibrous web. *Energy* **2022**, *244*, 123102. [CrossRef]
31. Song, C.; Xia, K.; Xu, Z. A self-supported structure hybrid triboelectric/piezoelectric nanogenerator for biomechanical energy harvesting and pressure sensing. *Microelectron. Eng.* **2022**, *256*, 111723. [CrossRef]
32. Zhao, Z.; Dai, Y.; Dou, S.X.; Liang, J. Flexible nanogenerators for wearable electronic applications based on piezoelectric materials. *Mater. Today Energy* **2021**, *20*, 100690. [CrossRef]
33. Hinchet, R.; Seung, W.; Kim, S.W. Recent Progress on Flexible Triboelectric Nanogenerators for SelfPowered Electronics. *ChemSusChem* **2015**, *8*, 2327–2344. [CrossRef]
34. Quelen, A.; Morel, A.; Gasnier, P.; Grézaud, R.; Monfray, S.; Pillonnet, G. A 30nA quiescent 80nW-to-14mW power-range shock-optimized SECE-based piezoelectric harvesting interface with 420% harvested-energy improvement. In Proceedings of the 2018 IEEE International Solid-State Circuits Conference—(ISSCC), San Francisco, CA, USA, 11–15 February 2018; pp. 150–152. [CrossRef]
35. Hehn, T.; Hagedorn, F.; Maurath, D.; Marinkovic, D.; Kuehne, I.; Frey, A.; Manoli, Y. A Fully Autonomous Integrated Interface Circuit for Piezoelectric Harvesters. *IEEE J. Solid-State Circuits* **2012**, *47*, 2185–2198. [CrossRef]
36. Mahapatra, A.; Ajimsha, R.S.; Misra, P. Oxygen annealing induced enhancement in output characteristics of ZnO based flexible piezoelectric nanogenerators. *J. Alloys Compd.* **2022**, *913*, 165277. [CrossRef]
37. Huang, X.; Wang, Y.; Zhang, X. Ultrarobust, hierarchically anisotropic structured piezoelectric nanogenerators for self-powered sensing. *Nano Energy* **2022**, *99*, 107379. [CrossRef]

38. Veeralingam, S.; Bharti, D.K.; Badhulika, S. Lead-free PDMS/PPy based low-cost wearable piezoelectric nanogenerator for self-powered pulse pressure sensor application. *Mater. Res. Bull.* **2022**, *151*, 111815. [CrossRef]
39. Sahu, M.; Hajra, S.; Jadhav, S.; Panigrahi, B.K.; Dubal, D.; Kim, H.J. Bio-waste composites for cost-effective self-powered breathing patterns monitoring: An insight into energy harvesting and storage properties. *Sustain. Mater. Technol.* **2022**, *32*, e00396. [CrossRef]
40. Sarkar, L.; Sushma, M.V.; Yalagala, B.P.; Rengan, A.K.; Singh, S.G.; Vanjari, S.R.K. ZnO nanoparticles embedded silk fibroin-a piezoelectric composite for nanogenerator applications. *Nanotechnology* **2022**, *33*, 265403. [CrossRef]
41. Pusty, M.; Shirage, P.M. Insights and perspectives on graphene-PVDF based nanocomposite materials for harvesting mechanical energy. *J. Alloys Compd.* **2022**, *904*, 164060. [CrossRef]
42. Tan, P.; Xi, Y.; Chao, S.; Jiang, D.; Liu, Z.; Fan, Y.; Li, Z. An Artificial Intelligence-Enhanced Blood Pressure Monitor Wristband Based on Piezoelectric Nanogenerator. *Biosensors* **2022**, *12*, 234. [CrossRef] [PubMed]
43. Mahmud, M.A.P.; Adhikary, P.; Zolfagharian, A.; Adams, S.; Kaynak, A.; Kouzani, A.Z. Advanced design, fabrication, and applications of 3D-printable piezoelectric nanogenerators. *Electron. Mater. Lett.* **2022**, *18*, 129–144. [CrossRef]
44. Shi, S.; Pan, Z.; Cheng, Y.; Zhai, Y.; Zhang, Y.; Ding, X.; Liu, J.; Zhai, J.; Xu, J. Three-dimensional polypyrrole induced high-performance flexible piezoelectric nanogenerators for mechanical energy harvesting. *Comp. Sci. Technol.* **2022**, *219*, 109260. [CrossRef]
45. Pan, Q.; Wang, B.; Zhang, L.; Li, Z.; Yang, Z. Whisk-inspired Motion Converter for Ocean Wave Energy Harvesting. *IEEE/ASME Trans. Mechatron.* **2022**, *27*, 1808–1811. [CrossRef]
46. Wang, J.; Pan, L.; Guo, H.; Zhang, B.; Zhang, R.; Wu, Z.; Wu, C.; Yang, L.; Liao, R.; Wang, Z.L. Rational Structure Optimized Hybrid Nanogenerator for Highly Efficient Water Wave Energy Harvesting. *Adv. Energy Mater.* **2019**, *9*, 1802892. [CrossRef]
47. Wu, Y.; Zeng, Q.; Tang, Q.; Liu, W.; Liu, G.; Zhang, Y.; Wu, J.; Hu, C.; Wang, X. A teeterboard-like hybrid nanogenerator for efficient harvesting of low-frequency ocean wave energy. *Nano Energy* **2020**, *67*, 104205. [CrossRef]
48. Zhang, X.; Yang, Q.; Ji, P.; Wu, Z.; Li, Q.; Yang, H.; Li, X.; Zheng, G.; Xi, Y.; Wang, Z.L. Modeling of liquid-solid hydrodynamic water wave energy harvesting system based on triboelectric nanogenerator. *Nano Energy* **2022**, *99*, 107362. [CrossRef]
49. Zaw, N.Y.W.; Yun, J.; Goh, T.S.; Kim, I.; Kim, Y.; Lee, J.S.; Kim, D. All-polymer waterproof triboelectric nanogenerator towards blue energy harvesting and self-powered human motion detection. *Energy* **2022**, *247*, 123422. [CrossRef]
50. Zhou, H.; Dong, J.; Liu, H.; Zhu, L.; Xu, C.; He, X.; Zhang, S.; Song, Q. The coordination of displacement and conduction currents to boost the instantaneous power output of a water-tube triboelectric nanogenerator. *Nano Energy* **2022**, *95*, 107050. [CrossRef]
51. Li, W.; Wan, L.; Lin, Y.; Liu, G.; Qu, H.; Wen, H.; Ding, J.; Ning, H.; Yao, H. Synchronous nanogenerator with intermittent sliding friction self-excitation for water wave energy harvesting. *Nano Energy* **2022**, *95*, 106994. [CrossRef]
52. Zhang, Z.; Hu, Z.; Wang, Y.; Wang, Y.; Zhang, Q.; Liu, D.; Wang, H.; Xu, M. Multi-Tunnel Triboelectric Nanogenerator for Scavenging Mechanical Energy in Marine Floating Bodies. *J. Mar. Sci. Eng.* **2022**, *10*, 455. [CrossRef]
53. Wang, X.; Shi, Y.; Yang, P.; Tao, X.; Li, S.; Lei, R.; Liu, Z.; Wang, Z.L.; Chen, X. Fish-Wearable Data Snooping Platform for Underwater Energy Harvesting and Fish Behavior Monitoring. *Small* **2022**, *18*, 2107232. [CrossRef] [PubMed]
54. Xu, L.; Jiang, T.; Lin, P.; Shao, J.J.; He, C.; Zhong, W.; Chen, X.Y.; Wang, Z.L. Coupled Triboelectric Nanogenerator Networks for Efficient Water Wave Energy Harvesting. *ACS Nano* **2018**, *12*, 1849–1858. [CrossRef]
55. Wen, H.; Yang, P.; Liu, G.; Xu, S.; Yao, H.; Li, W.; Qu, H.; Ding, J.; Li, J.; Wan, L. Flower-like triboelectric nanogenerator for blue energy harvesting with six degrees of freedom. *Nano Energy* **2022**, *93*, 106796. [CrossRef]
56. Wang, A.; Chen, J.; Wang, L.; Han, J.; Su, W.; Li, A.; Liu, P.; Duan, L.; Xu, C.; Zeng, Z. Numerical analysis and experimental study of an ocean wave tetrahedral triboelectric nanogenerator. *Appl. Energy* **2022**, *307*, 118174. [CrossRef]
57. Qu, Z.; Huang, M.; Chen, C.; An, Y.; Liu, H.; Zhang, Q.; Wang, X.; Liu, Y.; Yin, W.; Li, X. Spherical triboelectric nanogenerator based on eccentric structure for omnidirectional low frequency water wave energy harvesting. *Adv. Funct. Mater.* **2022**, *32*, 2202048. [CrossRef]
58. Xu, S.; Liu, G.; Wang, J.; Wen, H.; Cao, S.; Yao, H.; Wan, L.; Wang, Z.L. Interaction between water wave and geometrical structures of floating triboelectric nanogenerators. *Adv. Energy Mater.* **2021**, *12*, 2103408. [CrossRef]
59. Feng, Y.; Han, J.; Xu, M.; Liang, X.; Jiang, T.; Li, H.; Wang, Z.L. Blue energy for green hydrogen fuel: A self-powered electrochemical conversion driven by triboelectric nanogenerators. *Adv. Energy Mater.* **2021**, *10*, 1902467. [CrossRef]
60. Zheng, F.; Sun, Y.; Wei, X.; Chen, J.; Yuan, Z.; Jin, X.; Tao, L.; Wu, Z. A hybridized water wave energy harvester with a swing magnetic structure toward intelligent fishing ground. *Nano Energy* **2021**, *90*, 106631. [CrossRef]
61. Torres, F.G.; Troncoso, O.P.; De-la-Torre, G.E. Hydrogel-based triboelectric nanogenerators: Properties, performance, and applications. *Int. J. Energy Res.* **2022**, *46*, 5603–5624. [CrossRef]
62. Chen, B.; Wang, Z.L. Toward a new era of sustainable energy: Advanced triboelectric nanogenerator for harvesting high entropy energy. *Small* 2022, in press. [CrossRef] [PubMed]
63. Bukhari, M.U.; Khan, A.; Maqbool, K.Q.; Arshad, A.; Riaz, K.; Bermak, A. Waste to energy: Facile, low-cost and environment-friendly triboelectric nanogenerators using recycled plastic and electronic wastes for self-powered portable electronics. *Energy Rep.* **2022**, *8*, 1687–1695. [CrossRef]
64. Zhang, J.; Hu, Y.; Lin, X.; Qian, X.; Zhang, L.; Zhou, J.; Lu, A. High-performance triboelectric nanogenerator based on chitin for mechanical-energy harvesting and self-powered sensing. *Carbohydr. Polym.* **2022**, *291*, 119586. [CrossRef] [PubMed]
65. Gravesen, J.; Willatzen, M.; Shao, J.; Wang, Z.L. Energy optimization of a mirror-symmetric spherical triboelectric nanogenerator. *Adv. Funct. Mater.* **2022**, *32*, 2110516. [CrossRef]

66. Wu, H.; Wang, J.; Wu, Z.; Kang, S.; Wei, X.; Wang, H.; Luo, H.; Yang, L.; Liao, R.; Wang, Z.L. Multi-parameter optimized triboelectric nanogenerator based self-powered sensor network for broadband aeolian vibration online-monitoring of transmission lines. *Adv. Energy Mater.* **2022**, *12*, 2103654. [CrossRef]
67. Wang, J.; Jiang, Z.; Sun, W.; Xu, X.; Han, Q.; Chu, F. Yoyo-ball inspired triboelectric nanogenerators for harvesting biomechanical energy. *Appl. Energy* **2022**, *308*, 118322. [CrossRef]
68. Mathew, A.A.; Vivekanandan, S. Design and Simulation of Single-Electrode Mode Triboelectric Nanogenerator-Based Pulse Sensor for Healthcare Applications Using COMSOL Multiphysics. *Energy Technol.* **2022**, *10*, 2101130. [CrossRef]
69. Zheng, Z.; Xia, J.; Wang, B.; Guo, Y. Hierarchically designed nanocomposites for triboelectric nanogenerator toward biomechanical energy harvester and smart home system. *Nano Energy* **2022**, *95*, 107047. [CrossRef]
70. Sun, Z.; Yang, W.; Chen, P.; Zhang, Y.; Wang, X.; Hu, Y. Effects of PDMS Base/Agent Ratios and Texture Sizes on the Electrical Performance of Triboelectric Nanogenerators. *Adv. Mater. Interfaces* **2022**, *9*, 2102139. [CrossRef]
71. Jiang, C.; Lai, C.L.; Xu, B.; So, M.Y.; Li, Z. Fabric-rebound triboelectric nanogenerators with loops and layered structures for energy harvesting and intelligent wireless monitoring of human motions. *Nano Energy* **2022**, *93*, 106807. [CrossRef]
72. Wang, Z.L. Triboelectric nanogenerators as new energy technology and self-powered sensors-principles, problems and perspectives. *Faraday Discuss.* **2014**, *176*, 447. [CrossRef] [PubMed]
73. Wang, Z. Nanopiezotronics. *Adv. Mater.* **2007**, *19*, 889–892. [CrossRef]
74. Wang, Z.L. Towards Self-Powered Nanosystems: From Nanogenerators to Nanopiezotronics. *Adv. Funct. Mater.* **2008**, *18*, 3553–3567. [CrossRef]
75. Wang, Z.L. The new field of nanopiezotronics. *Mater. Today* **2007**, *10*, 20–28. [CrossRef]
76. Yang, Y. Pyroelectric and Thermoelectric Nanogenerators. In *Hybridized and Coupled Nanogenerators*; Yang, Y., Ed.; Wiley-VCH GmbH: Weinheim, Germany, 2020. [CrossRef]
77. Klochko, N.P.; Klepikova, K.S.; Kopach, V.R.; Tyukhov, I.I.; Starikov, V.V.; Sofronov, D.S.; Khrypunova, I.V.; Zhadan, D.O.; Petrushenko, S.I.; Dukarov, S.V.; et al. Development of semi-transparent ZnO/FTO solar thermoelectric nanogenerator for energy efficient glazing. *Sol. Energy* **2019**, *184*, 230–239. [CrossRef]
78. Khoso, N.A.; Xu, G.; Xie, J.; Sun, T.; Wang, J. The fabrication of a graphene and conductive polymer nanocomposite-coated highly flexible and washable woven thermoelectric nanogenerator. *Mater. Adv.* **2021**, *2*, 3695–3704. [CrossRef]
79. Khoso, N.A.; Ahmed, A.; Deb, H.; Tian, S.; Jiao, X.; Gong, X.Y.; Wang, J. Controlled template-free in-situ polymerization of PEDOT for enhanced thermoelectric performance on textile substrate. *Org. Electron.* **2019**, *75*, 105368. [CrossRef]
80. Feng, R.; Tang, F.; Zhang, N.; Wang, X. Flexible, High-Power Density, Wearable Thermoelectric Nanogenerator and Self-Powered Temperature Sensor. *ACS Appl. Mater. Interfaces* **2019**, *11*, 42. [CrossRef]
81. He, M.; Lin, Y.-J.; Chiu, C.M.; Yang, W.; Zhang, B.; Yun, D.; Xie, Y.; Lin, Z.H. A flexible photo-thermoelectric nanogenerator based on MoS2/PU photothermal layer for infrared light harvesting. *Nano Energy* **2018**, *49*, 588–595. [CrossRef]
82. Fan, Z.; Zhang, Y.; Pan, L.; Ouyang, J.; Zhang, Q. Recent developments in flexible thermoelectrics: From materials to devices. *Renew. Sust. Energ. Rev.* **2021**, *137*, 110448. [CrossRef]
83. Xie, Y.; Chou, T.M.; Yang, W.; He, M.; Zhao, Y.; Li, N.; Lin, Z.-H. Flexible thermoelectric nanogenerator based on the MoS_2/graphene nanocomposite and its application for a self-powered temperature sensor. *Semicond. Sci. Technol.* **2017**, *32*, 044003. [CrossRef]
84. Huang, X.-L.; Ao, D.-W.; Chen, T.-B.; Chen, Y.-X.; Li, F.; Chen, S.; Liang, G.-X.; Zhang, X.-H.; Zheng, Z.-H.; Fan, P. High performance copper selenide thermoelectric thin films for flexible thermoelectric application. *Mater. Today Energy* **2021**, *21*, 100743. [CrossRef]
85. Xu, C.; Yang, S.; Li, P.; Wang, H.; Li, H.; Liu, Z. Wet-spun PEDOT:PSS/CNT composite fibers for wearable thermoelectric energy harvesting. *Compos. Commun.* **2022**, *32*, 101179. [CrossRef]
86. Palaporn, D.; Mongkolthanaruk, W.; Tanusilp, S.-A.; Kurosaki, K.; Pinitsoontorn, S. A simple method for fabricating flexible thermoelectric nanocomposites based on bacterial cellulose nanofiber and Ag_2Se. *Appl. Phys. Lett.* **2022**, *120*, 073901. [CrossRef]
87. Zhang, X.; Shiu, B.-C.; Li, T.T.; Liu, X.; Ren, H.-T.; Wang, Y.; Lou, C.-W.; Lin, J.-H. Synergistic work of photo-thermoelectric and hydroelectric effects of hierarchical structure photo-thermoelectric textile for solar energy harvesting and solar steam generation simultaneously. *Chem. Eng. J.* **2021**, *426*, 131923. [CrossRef]
88. Wang, Y.; Zhu, W.; Deng, Y.; Zhu, P.; Yu, Y.; Hu, S.; Zhang, R. High-sensitivity self-powered temperature/pressure sensor based on flexible Bi-Te thermoelectric film and porous microconed elastomer. *J. Mater. Sci. Technol.* **2022**, *103*, 1–7. [CrossRef]
89. Yuan, J.F.; Zhu, R.; Li, G.Z. Self-Powered Electronic Skin with Multisensory Functions Based on Thermoelectric Conversion. *Adv. Mater. Technol.* **2020**, *5*, 2000419. [CrossRef]
90. Wang, Y.; Zhu, W.; Deng, Y.; Fu, B.; Zhu, P.; Yu, Y.; Li, J.; Guo, J. Self-powered wearable pressure sensing system for continuous healthcare monitoring enabled by flexible thin-film thermoelectric generator. *Nano Energy* **2020**, *73*, 104773. [CrossRef]
91. Gokana, M.R.; Wu, C.-M.; Motora, K.G.; Qi, J.Y.; Yen, W.-T. Effects of patterned electrode on near infrared light-triggered cesium tungsten bronze/poly(vinylidene)fluoride nanocomposite-based pyroelectric nanogenerator for energy harvesting. *J. Power Sources* **2022**, *536*, 231524. [CrossRef]
92. Mistewicz, K. Pyroelectric Nanogenerator Based on an SbSI–TiO_2 Nanocomposite. *Sensors* **2022**, *22*, 69. [CrossRef]
93. Feng, Y.; Zhang, Y.; Wang, Y.; Wang, Z. Frequency response characteristics of pyroelectric effect in p-n junction UV detectors. *Nano Energy* **2018**, *54*, 429–436. [CrossRef]

94. Korkmaz, S.; Kariper, İ.A. Pyroelectric nanogenerators (PyNGs) in converting thermal energy into electrical energy: Fundamentals and current status. *Nano Energy* **2021**, *84*, 105888. [CrossRef]
95. Ali, I.; Hassan, G.; Shuja, A. Fabrication of self-healing hybrid nanogenerators based on polyurethane and ZnO for harvesting wind energy. *J Mater. Sci. Mater. Electron.* **2022**, *33*, 3982–3993. [CrossRef]
96. Mahanty, B.; Ghosh, S.K.; Maity, K.; Roy, K.; Sarkar, S.; Mandal, D. All-fiber pyro- and piezo-electric nanogenerator for IoT based self-powered health-care monitoring. *Mater. Adv.* **2021**, *2*, 4370–4379. [CrossRef]
97. Abbasipour, M.; Khajavi, R.; Yousefi, A.A.; Yazdanshenas, M.E.; Razaghian, F.; Akbarzadeh, A. Improving piezoelectric and pyroelectric properties of electrospun PVDF nanofibers using nanofillers for energy harvesting application. *Polym. Adv. Technol.* **2019**, *30*, 279–291. [CrossRef]
98. Zhu, L.; Lin, P.; Chen, B.; Wang, L.; Chen, L.; Li, D.; Wang, Z.L. Piezo-phototronic and pyro-phototronic effects to enhance Cu(In, Ga)Se2 thin film solar cells. *Nano Res.* **2018**, *11*, 3877–3885. [CrossRef]
99. Liu, X.; Liu, Y.; Cheng, T.; Gao, Y.; Yang, Z. A high performing piezoelectric and triboelectric nanogenerator based on a large deformation of the novel lantern-shaped structure. *Nano Energy* **2022**, *92*, 106699. [CrossRef]
100. Wang, Z.; Liu, Z.; Zhao, G.; Zhang, Z.; Zhao, X.; Wan, X.; Zhang, Y.; Wang, Z.L.; Li, L. Stretchable Unsymmetrical Piezoelectric BaTiO3 Composite Hydrogel for Triboelectric Nanogenerators and Multimodal Sensors. *ACS Nano* **2022**, *16*, 1661–1670. [CrossRef]
101. Hou, X.; Zhong, J.; Yang, C.; Yang, Y.; He, J.; Mu, J.; Geng, W.; Chou, X. A high-performance, single-electrode and stretchable piezo-triboelectric hybrid patch for omnidirectional biomechanical energy harvesting and motion monitoring. *J. Mater.* 2022, *in press*. [CrossRef]
102. Joo, H.; Lee, K.Y.; Lee, J. Piezo/Triboelectric Effect Driven Self-Powered Gas Sensor for Environmental Sensor Networks. *Energy Technol.* **2022**, *10*, 2200113. [CrossRef]
103. Negedu, S.D.; Tromer, R.; Gowda, C.C.; Woellner, C.F.; Olu, F.E.; Roy, A.K.; Pandey, P.; Galvao, D.S.; Ajayan, P.M.; Kumbhakar, P.; et al. Two-dimensional cobalt telluride as a piezo-tribogenerator. *Nanoscale* **2022**, *14*, 7788–7797. [CrossRef] [PubMed]
104. Zhang, J.-H.; Zhou, Z.; Li, J.; Shen, B.; Zhu, T.; Gao, X.; Tao, R.; Guo, X.; Hu, X.; Shi, Y.; et al. Coupling Enhanced Performance of Triboelectric–Piezoelectric Hybrid Nanogenerator Based on Nanoporous Film of Poly(vinylidene fluoride)/BaTiO3 Composite Electrospun Fibers. *ACS Mater. Lett.* **2022**, *4*, 847–852. [CrossRef]
105. Nazar, A.M.; Egbe, K.I.; Jiao, P. Hybrid Piezoelectric and Triboelectric Nanogenerators for Energy Harvesting and Walking Sensing. *Energy Technol.* **2022**, *10*, 2200063. [CrossRef]
106. Du, M.; Cao, Y.; Qu, X.; Xue, J.; Zhang, W.; Pu, X.; Shi, B.; Li, Z. Hybrid Nanogenerator for Biomechanical Energy Harvesting, Motion State Detection, and Pulse Sensing. *Adv. Mater. Technol.* **2022**, *7*, 2101332. [CrossRef]
107. Kao, F.-C.; Ho, H.-H.; Chiu, P.-Y.; Hsieh, M.-K.; Liao, J.-C.; Lai, P.-L.; Huang, Y.-F.; Dong, M.-Y.; Tsai, T.T.; Lin, Z.-H. Self-assisted wound healing using piezoelectric and triboelectric nanogenerators. *Sci. Technol. Adv. Mater.* **2022**, *23*, 1–16. [CrossRef]
108. Manchi, P.; Graham, S.A.; Patnam, H.; Paranjape, M.V.; Yu, J.S. rGO-ZnSnO$_3$ Nanostructure-Embedded Triboelectric Polymer-Based Hybridized Nanogenerators. *Adv. Mater. Technol.* 2022, in press. [CrossRef]
109. Zhu, Y.; Sun, F.; Jia, C.; Zhao, T.; Mao, Y. A Stretchable and Self-Healing Hybrid Nano-Generator for Human Motion Monitoring. *Nanomaterials* **2022**, *12*, 104. [CrossRef]
110. García-Casas, X.; Ghaffarinejad, A.; Aparicio, F.J.; Castillo-Seoane, J.; López-Santos, C.; Espinós, J.P.; Cotrino, J.; Sánchez-Valencia, J.R.; Barranco, A.; Borrás, A. Plasma engineering of microstructured piezo—Triboelectric hybrid nanogenerators for wide bandwidth vibration energy harvesting. *Nano Energy* **2022**, *91*, 106673. [CrossRef]
111. Toroń, B.; Mistewicz, K.; Jesionek, M.; Kozioł, M.; Zubko, M.; Stróż, D. A new hybrid piezo/triboelectric SbSeI nanogenerator. *Energy* **2022**, *238C*, 122048. [CrossRef]
112. Li, X.; Ji, D.; Yu, B.; Ghosh, R.; He, J.; Qin, X.; Ramakrishna, S. Boosting piezoelectric and triboelectric effects of PVDF nanofiber through carbon-coated piezoelectric nanoparticles for highly sensitive wearable sensors. *Chem. Eng. J.* **2021**, *426*, 130345. [CrossRef]
113. Yang, Z.; Zhu, Z.; Chen, Z.; Liu, M.; Zhao, B.; Liu, Y.; Cheng, Z.; Wang, S.; Yang, W.; Yu, T. Recent Advances in Self-Powered Piezoelectric and Triboelectric Sensors: From Material and Structure Design to Frontier Applications of Artificial Intelligence. *Sensors* **2021**, *21*, 8422. [CrossRef] [PubMed]
114. Yang, X.; Li, P.; Wu, B.; Li, H.; Zhou, G. A flexible piezoelectric-triboelectric hybrid nanogenerator in one structure with dual doping enhancement effects. *Curr. Appl. Phys.* **2021**, *32*, 50–58. [CrossRef]
115. Lee, T.; Kim, I.; Kim, D. Flexible Hybrid Nanogenerator for Self-Powered Weather and Healthcare Monitoring Sensor. *Adv. Electron. Mater.* **2021**, *7*, 2100785. [CrossRef]
116. Singh, H.H.; Khare, N. A ferroelectric nanocomposite-film-based device for harvesting energy from water droplets using both piezoelectric and triboelectric effects. *Nanotechnology* **2021**, *32*, 485406. [CrossRef]
117. Hajra, S.; Padhan, A.M.; Sahu, M.; Alagarsamy, P.; Lee, K.; Kim, H.J. Lead-free flexible Bismuth Titanate-PDMS composites: A multifunctional colossal dielectric material for hybrid piezo-triboelectric nanogenerator to sustainably power portable electronics. *Nano Energy* **2021**, *89A*, 106316. [CrossRef]
118. Gai, Y.; Bai, Y.; Cao, Y.; Wang, E.; Xue, J.; Qu, X.; Liu, Z.; Luo, D.; Li, Z. A Gyroscope Nanogenerator with Frequency Up-Conversion Effect for Fitness and Energy Harvesting. *Small* **2022**, *18*, 2108091. [CrossRef]

119. Mu, J.; He, H.; Song, J.; He, J.; Hou, X.; Han, X.; Feng, C.; Zou, J.; Yu, J.; Chou, X. Functional structure enhanced synergistic sensing from triboelectric–electromagnetic hybrid nanogenerator for self-powered rotating speed monitoring. *Energy Rep.* **2022**, *8*, 5272–5283. [CrossRef]
120. Wang, D.; Zhang, D.; Tang, M.; Zhang, H.; Chen, F.; Wang, T.; Li, Z.; Zhao, P. Rotating triboelectric-electromagnetic nanogenerator driven by tires for self-powered MXene-based flexible wearable electronics. *Chem. Eng. J.* **2022**, *446*, 136914. [CrossRef]
121. Yu, D.; Sun, C.; Wang, K.; Yin, S.; Sun, L.; Chen, H.; Kong, F. A novel direct-driven triboelectric–electromagnetic hybridized wave energy converter for buoy power supply. *Appl. Nanosci.* **2022**, *12*, 1697–1711. [CrossRef]
122. Cho, H.; Kim, I.; Park, J.; Kim, D. A waterwheel hybrid generator with disk triboelectric nanogenerator and electromagnetic generator as a power source for an electrocoagulation system. *Nano Energy* **2022**, *95*, 107048. [CrossRef]
123. He, J.; Fan, X.; Zhao, D.; Cui, M.; Han, B.; Hou, X.; Chou, X. A high-efficient triboelectric-electromagnetic hybrid nanogenerator for vibration energy harvesting and wireless monitoring. *Sci. China Inf. Sci.* **2022**, *65*, 142401. [CrossRef]
124. Hu, Y.; Wang, X.; Qin, Y.; Li, Z.; Wang, C.; Wu, H. A robust hybrid generator for harvesting vehicle suspension vibration energy from random road excitation. *Appl. Energy* **2022**, *309*, 118506. [CrossRef]
125. Chen, Y.; Jie, Y.; Zhu, J.; Lu, Q.; Cheng, Y.; Cao, X.; Wang, Z.L. Hybridized triboelectric-electromagnetic nanogenerators and solar cell for energy harvesting and wireless power transmission. *Nano Res.* **2022**, *15*, 2069–2076. [CrossRef]
126. Mu, J.; Zou, J.; Song, J.; He, J.; Hou, X.; Yu, J.; Han, X.; Feng, C.; He, H.; Chou, X. Hybrid enhancement effect of structural and material properties of the triboelectric generator on its performance in integrated energy harvester. *Energy Convers. Manag.* **2022**, *254*, 115151. [CrossRef]
127. Kong, F.; Yin, S.; Sun, C.; Yang, C.; Chen, H.; Liu, H. Design and optimization of a maglev electromagnetic–triboelectric hybrid energy converter for supplying power to intelligent sensing equipment. *Sustain. Energy Fuels* **2022**, *6*, 800–814. [CrossRef]
128. Hong, H.; Yang, X.; Cui, H.; Zheng, D.; Wen, H.; Huang, R.; Liu, L.; Duan, J.; Tang, Q. Self-powered seesaw structured spherical buoys based on a hybrid triboelectric–electromagnetic nanogenerator for sea surface wireless positioning. *Energy Environ. Sci.* **2022**, *15*, 621–632. [CrossRef]
129. Li, X.; Gao, Q.; Cao, Y.; Yang, Y.; Liu, S.; Wang, Z.L. Tinghai Cheng. Optimization strategy of wind energy harvesting via triboelectric-electromagnetic flexible cooperation. *Appl. Energy* **2022**, *307*, 118311. [CrossRef]
130. Xin, C.; Guo, H.; Shen, F.; Peng, Y.; Xie, S.; Li, Z.; Zhang, Q. A Hybrid Generator with Electromagnetic Transduction for Improving the Power Density of Triboelectric Nanogenerators and Scavenging Wind Energy. *Adv. Mater. Technol.* 2022, in press. [CrossRef]
131. Zhao, T.; Niu, B.; Xie, G.; Hu, C.; Liu, B.; Xu, M.; Ma, Y. A High Output Triboelectric–Electromagnetic Hybrid Generator Based on In-Phase Parallel Connection. *Adv. Mater. Technol.* **2022**, *7*, 2101485. [CrossRef]
132. Zhang, B.; Zhang, S.; Li, W.; Gao, Q.; Zhao, D.; Wang, Z.L.; Cheng, T. Self-Powered Sensing for Smart Agriculture by Electromagnetic–Triboelectric Hybrid Generator. *ACS Nano* **2021**, *15*, 20278–20286. [CrossRef]
133. Askari, H.; Xu, N.; Barbosa, B.H.G.; Huang, Y.; Chen, L.; Khajepour, A.; Chen, H.; Wang, Z.L. Intelligent systems using triboelectric, piezoelectric, and pyroelectric nanogenerators. *Mater. Today* **2022**, *52*, 188–206. [CrossRef]
134. Wang, S.; Wang, Z.L.; Yang, Y. A One-Structure-Based Hybridized Nanogenerator for Scavenging Mechanical and Thermal Energies by Triboelectric–Piezoelectric Pyroelectric Effects. *Adv. Mater.* **2016**, *28*, 2881–2887. [CrossRef] [PubMed]
135. Zhang, H.; Zhang, S.; Yao, G.; Huang, Z.; Xie, Y.; Su, Y.; Yang, W.; Zheng, C.; Lin, Y. Simultaneously Harvesting Thermal and Mechanical Energies based on Flexible Hybrid Nanogenerator for Self Powered Cathodic Protection. *ACS Appl. Mater. Interfaces* **2015**, *7*, 28142–28147. [CrossRef] [PubMed]
136. Zi, Y.; Lin, L.; Wang, J.; Wang, S.; Chen, J.; Fan, X.; Yang, P.-K.; Yi, F.; Wang, Z.L. Triboelectric–Pyroelectric–Piezoelectric Hybrid Cell for High-Efficiency Energy-Harvesting and Self-Powered Sensing. *Adv. Mater.* **2015**, *27*, 2340–2347. [CrossRef] [PubMed]
137. Khan, A.A.; Saritas, R.; Rana, M.M.; Tanguy, N.; Zhu, W.; Mei, N.; Kokilathasan, S.; Rassel, S.; Leonenko, Z.; Yan, N.; et al. Performance-Improved Highly Integrated Uniaxial Tristate Hybrid Nanogenerator for Sustainable Mechanical Energy Harvesting. *ACS Appl. Mater. Interfaces* **2022**, *14*, 4119–4131. [CrossRef]
138. Tang, G.; Wang, Z.; Hu, X.; Wu, S.; Xu, B.; Li, Z.; Yan, X.; Xu, F.; Yuan, D.; Li, P.; et al. A Non-Resonant Piezoelectric–Electromagnetic–Triboelectric Hybrid Energy Harvester for Low-Frequency Human Motions. *Nanomaterials* **2022**, *12*, 1168. [CrossRef]
139. Zhao, L.-C.; Zou, H.-X.; Zhao, Y.-J.; Wu, Z.-Y.; Liu, F.-R.; Wei, K.-X.; Zhang, W.-M. Hybrid energy harvesting for self-powered rotor condition monitoring using maximal utilization strategy in structural space and operation process. *Appl. Energy* **2022**, *314*, 118983. [CrossRef]
140. Zhang, C.; Yuan, W.; Zhang, B.; Yang, O.; Liu, Y.; He, L.; Wang, J.; Wang, Z.L. High Space Efficiency Hybrid Nanogenerators for Effective Water Wave Energy Harvesting. *Adv. Funct. Mater.* **2022**, *32*, 2111775. [CrossRef]
141. Xue, F.; Chen, L.; Li, C.; Ren, J.; Yu, J.; Hou, X.; Geng, W.; Mu, J.; He, J.; Chou, X. A static-dynamic energy harvester for a self-powered ocean environment monitoring application. *Sci. China Technol. Sci.* **2022**, *65*, 893–902. [CrossRef]
142. Wang, C.; Lai, S.-K.; Wang, J.-M.; Feng, J.-J.; Ni, Y.-Q. An ultra-low-frequency, broadband and multi-stable tri-hybrid energy harvester for enabling the next-generation sustainable power. *Appl. Energy* **2021**, *291*, 116825. [CrossRef]
143. Xue, X.; Zhang, Z.; Wu, B.; He, S.; Wang, Q.; Zhang, W.; Bi, R.; Cui, J.; Zheng, Y.; Xue, C. Coil-levitated hybrid generator for mechanical energy harvesting and wireless temperature and vibration monitoring. *Sci. China Technol. Sci.* **2021**, *64*, 1325–1334. [CrossRef]

144. Ma, T.; Gao, Q.; Li, Y.; Wang, Z.; Lu, X.; Cheng, T. An Integrated Triboelectric–Electromagnetic–Piezoelectric Hybrid Energy Harvester Induced by a Multifunction Magnet for Rotational Motion. *Adv. Eng. Mater.* **2020**, *22*, 1900872. [CrossRef]
145. Rodrigues, C.; Gomes, A.; Ghosh, A.; Pereira, A.; Ventura, J. Power-generating footwear based on a triboelectric-electromagnetic-piezoelectric hybrid nanogenerator. *Nano Energy* **2019**, *62*, 660–666. [CrossRef]
146. Rahman, M.T.; Salauddin, M.; Maharjan, P.; Rasel, M.S.; Cho, H.; Park, J.Y. Natural wind-driven ultra-compact and highly efficient hybridized nanogenerator for self-sustained wireless environmental monitoring system. *Nano Energy* **2019**, *57*, 256–268. [CrossRef]
147. Du, X.; Zhao, S.; Xing, Y.; Li, N.; Wang, J.; Zhang, X.; Cao, R.; Liu, Y.; Yuan, Z.; Yin, Y.; et al. Hybridized Nanogenerators for Harvesting Vibrational Energy by Triboelectric–Piezoelectric–Electromagnetic Effects. *Adv. Mater. Technol.* **2018**, *3*, 1800019. [CrossRef]
148. He, J.; Wen, T.; Qian, S.; Zhang, Z.; Tian, Z.; Zhu, J.; Mu, J.; Hou, X.; Geng, W.; Cho, J.; et al. Triboelectric-piezoelectric-electromagnetic hybrid nanogenerator for high-efficient vibration energy harvesting and self-powered wireless monitoring system. *Nano Energy* **2018**, *43*, 326–339. [CrossRef]
149. Liu, Y.; Sun, N.; Liu, J.; Wen, Z.; Sun, X.; Lee, S.-T.; Sun, B. Integrating a Silicon Solar Cell with a Triboelectric Nanogenerator via a Mutual Electrode for Harvesting Energy from Sunlight and Raindrops. *ACS Nano* **2018**, *12*, 2893–2899. [CrossRef]
150. Im, B.; Lee, S.-K.; Kang, G.; Moon, J.; Byun, D.; Cho, D.-H. Electrohydrodynamic jet printed silver-grid electrode for transparent raindrop energy-based triboelectric nanogenerator. *Nano Energy* **2022**, *95*, 107049. [CrossRef]
151. Zheng, Y.; Liu, T.; Wu, J.; Xu, T.; Wang, X.; Han, X.; Cui, H.; Xu, X.; Pan, C.; Li, X. Energy Conversion Analysis of Multilayered Triboelectric Nanogenerators for Synergistic Rain and Solar Energy Harvesting. *Adv. Mater.* **2022**, *34*, 2202238. [CrossRef]
152. Wu, Y.; Qu, J.; Chu, P.K.; Shin, D.-M.; Luo, Y.; Feng, S.-P. Hybrid photovoltaic-triboelectric nanogenerators for simultaneously harvesting solar and mechanical energies. *Nano Energy* **2021**, *89*, 106376. [CrossRef]
153. Sivasubramanian, R.; Vaithilingam, C.A.; Indira, S.S.; Paiman, S.; Misron, N.; Abubakar, S. A review on photovoltaic and nanogenerator hybrid system. *Mater. Today Energy* **2021**, *20*, 100772. [CrossRef]
154. Yang, D.; Ni, Y.; Su, H.; Shi, Y.; Liu, Q.; Chen, X.; He, D. Hybrid energy system based on solar cell and self-healing/self-cleaning triboelectric nanogenerator. *Nano Energy* **2021**, *79*, 105394. [CrossRef]
155. Hu, S.; Han, J.; Shi, Z.; Chen, K.; Xu, N.; Wang, Y.; Zheng, R.; Tao, Y.; Sun, Q.; Wang, Z.L.; et al. Biodegradable, Super-Strong, and Conductive Cellulose Macrofibers for Fabric-Based Triboelectric Nanogenerator. *Nano-Micro Lett.* **2022**, *14*, 115. [CrossRef]
156. Zhao, J.; Mu, J.; Cui, H.; He, W.; Zhang, L.; He, J.; Gao, X.; Li, Z.; Hou, X.; Chou, X. Hybridized Triboelectric-Electromagnetic Nanogenerator for Wind Energy Harvesting to Realize Real-Time Power Supply of Sensor Nodes. *Adv. Mater. Technol.* **2021**, *6*, 2001022. [CrossRef]
157. Wu, Y.; Zhong, X.; Wang, X.; Yang, Y.; Wang, Z.L. Hybrid energy cell for simultaneously harvesting wind, solar, and chemical energies. *Nano Res.* **2014**, *7*, 1631–1639. [CrossRef]
158. Denning, D.; Kilpatrick, J.I.; Fukada, E.; Zhang, N.; Habelitz, S.; Fertala, A.; Gilchrist, M.D.; Zhang, Y.; Tofail, S.A.M.; Rodriguez, B.J. Piezoelectric Tensor of Collagen Fibrils Determined at the Nanoscale. *ACS Biomater. Sci. Eng.* **2017**, *3*, 929–935. [CrossRef]
159. Knoblich, M.; Anderson, B.; Latshaw, D. Analyses of tomato peel and seed byproducts and their use as a source of carotenoids. *J. Sci. Food Agric.* **2005**, *85*, 1166–1170. [CrossRef]
160. Saqib, Q.M.; Khan, M.U.; Song, H.; Chougale, M.Y.; Shaukat, R.A.; Kim, J.; Bae, J.; Choi, M.J.; Kim, S.C.; Kwon, O.; et al. Natural Hierarchically Structured Highly Porous Tomato Peel Based Tribo- and Piezo-Electric Nanogenerator for Efficient Energy Harvesting. *Adv. Sustain. Syst.* **2021**, *5*, 2100066. [CrossRef]
161. Mao, Y.; Zhao, P.; McConohy, G.; Yang, H.; Tong, Y.; Wang, X. Sponge-Like Piezoelectric Polymer Films for Scalable and Integratable Nanogenerators and Self-Powered Electronic Systems. *Adv. Energy Mater.* **2014**, *4*, 1301624. [CrossRef]
162. Zhu, Y.; Xia, Y.; Wu, M.; Guo, W.; Jia, C.; Wang, X. Wearable, freezing-tolerant, and self-powered electroluminescence system for long-term cold-resistant displays. *Nano Energy* **2022**, *98*, 107309. [CrossRef]
163. Karan, S.K.; Maiti, S.; Paria, S.; Maitra, A.; Si, S.K.; Kim, J.K.; Khatua, B.B. A new insight towards eggshell membrane as high energy conversion efficient bio-piezoelectric energy harvester. *Mater. Today Energy* **2018**, *9*, 114–125. [CrossRef]
164. Zhang, M.; Du, H.; Liu, K.; Nie, S.; Xu, T.; Zhang, X.; Si, C. Fabrication and applications of cellulose-based nanogenerators. *Adv. Compos. Hybrid Mater.* **2021**, *4*, 865–884. [CrossRef]
165. Ghosh, S.K.; Mandal, D. Efficient natural piezoelectric nanogenerator: Electricity generation from fish swim bladder. *Nano Energy* **2016**, *28*, 356–365. [CrossRef]
166. Karan, S.K.; Maiti, S.; Kwon, O.; Paria, S.; Maitra, A.; Si, S.K.; Kim, Y.; Kim, J.K.; Khatua, B.B. Nature driven spider silk as high energy conversion efficient bio-piezoelectric nanogenerator. *Nano Energy* **2018**, *49*, 655–666. [CrossRef]
167. Saqib, Q.M.; Shaukat, R.A.; Khan, M.U.; Chougale, M.; Bae, J. Biowaste Peanut Shell Powder-Based Triboelectric Nanogenerator for Biomechanical Energy Scavenging and Sustainably Powering Electronic Supplies. *ACS Appl. Electron. Mater.* **2020**, *2*, 3953–3963. [CrossRef]
168. Chougale, M.Y.; Saqib, Q.M.; Khan, M.U.; Shaukat, R.A.; Kim, J.; Bae, J. Novel Recycled Triboelectric Nanogenerator Based on Polymer-Coated Trash Soda Can for Clean Energy Harvesting. *Adv. Sustain. Syst.* **2021**, *5*, 2100161. [CrossRef]
169. Jiang, W.; Li, H.; Liu, Z.; Li, Z.; Tian, J.J.; Shi, B.J.; Zou, Y.; Ouyang, H.; Zhao, C.C.; Zhao, L.M.; et al. Fully Bioabsorbable Natural-Materials-Based Triboelectric Nanogenerators. *Adv. Mater.* **2018**, *30*, 1801895. [CrossRef]

170. Morel, A.; Brenes, A.; Gibus, D.; Lefeuvre, E.; Gasnier, P.; Pillonnet, G.; Badel, A. A comparative study of electrical interfaces for tunable piezoelectric vibration energy harvesting. *Smart Mater. Struct.* **2022**, *31*, 045016. [CrossRef]
171. Rincón-Mora, G.A.; Yang, S. Tiny piezoelectric harvesters: Principles, constraints, and power conversion. *IEEE Trans. Circuits Syst. I Regul. Pap.* **2016**, *63*, 639–649. [CrossRef]
172. Ghaffarinejad, A.; Hasani, J.Y.; Hinchet, R.; Lu, Y.; Zhang, H.; Karami, A.; Galayko, D.; Kim, S.-W.; Basset, P. Aconditioning circuit with exponential enhancement of output energy for triboelectric nanogenerator. *Nano Energy* **2018**, *51*, 173–184. [CrossRef]
173. Coustans, M.; Krummenacher, F.; Kayal, M. A Fully Integrated 60 mV cold-start circuit for single coil DC–DC boost converter for thermoelectric energy harvesting. *IEEE Trans. Circuits Syst. II Express Briefs* **2019**, *66*, 1668–1672. [CrossRef]
174. Long, Z.; Li, P.; Chen, J.; Chung, H.S.-H.; Yang, Z. Self-Powered Single-Inductor Rectifier-Less SSHI Array Interface with the MPPT Technique for Piezoelectric Energy Harvesting. *IEEE Trans. Ind. Electron.* **2022**, *69*, 10172–10181. [CrossRef]
175. Wang, J.; Chen, Z.; Li, Z.; Jiang, J.; Liang, J.; Zeng, X. Piezoelectric energy harvesters: An overview on design strategies and topologies. *IEEE Trans. Circuits Syst. II Express Briefs* **2022**, *69*, 3057–3063. [CrossRef]
176. Chandrarathna, S.C.; Graham, S.A.; Ali, M.; Yu, J.S.; Lee, J.-W. An efficient power management system using dynamically configured multiple triboelectric nanogenerators and dual-parameter maximum power point tracking. *Adv. Energy Mater.* **2022**, *12*, 2103249. [CrossRef]
177. Wang, X.; Xia, Y.; Shi, G.; Xia, H.; Chen, M.; Chen, Z.; Ye, Y.; Qian, L. A novel MPPT technique based on the envelope extraction implemented with passive components for piezoelectric energy harvesting. *IEEE Trans. Power Electron.* **2021**, *36*, 12685–12693. [CrossRef]

nanomaterials

Article

Self-Powered Resistance-Switching Properties of $Pr_{0.7}Ca_{0.3}MnO_3$ Film Driven by Triboelectric Nanogenerator

Yanzi Huang [1], Lingyu Wan [1,*], Jiang Jiang [1], Liuyan Li [1] and Junyi Zhai [2,3,*]

1 Center on Nano-Energy Research, Guangxi Key Laboratory for Relativistic Astrophysics, School of Physical Science and Technology, Guangxi University, Nanning 530004, China; 1907301031@st.gxu.edu.cn (Y.H.); 2007301061@st.gxu.edu.cn (J.J.); 2007301071@st.gxu.edu.cn (L.L.)
2 CAS Center for Excellence in Nanoscience, Beijing Key Laboratory of Micro-Nano Energy and Sensor, Beijing Institute of Nanoenergy and Nanosystems, Chinese Academy of Sciences, Beijing 100083, China
3 College of Nanoscience and Technology, University of Chinese Academy of Sciences, Beijing 100049, China
* Correspondence: lyw2017@gxu.edu.cn (L.W.); jyzhai@binn.cas.cn (J.Z.); Tel.: +8613878156436 (L.W.); +86-13911845281 (J.Z.)

Abstract: As one of the promising non-volatile memories (NVMs), resistive random access memory (RRAM) has attracted extensive attention. Conventional RRAM is deeply dependent on external power to induce resistance-switching, which restricts its applications. In this work, we have developed a self-powered RRAM that consists of a $Pr_{0.7}Ca_{0.3}MnO_3$ (PCMO) film and a triboelectric nanogenerator (TENG). With a traditional power supply, the resistance switch ratio achieves the highest switching ratio reported so far, 9×10^7. By converting the mechanical energy harvested by a TENG into electrical energy to power the PCMO film, we demonstrate self-powered resistance-switching induced by mechanical movement. The prepared PCMO shows excellent performance of resistance switching driven by the TENG, and the resistance switch ratio is up to 2×10^5, which is higher than the ones ever reported. In addition, it can monitor real-time mechanical changes and has a good response to the electrical signals of different waveforms. This self-powered resistance switching can be induced by random movements based on the TENG. It has potential applications in the fields of self-powered sensors and human-machine interaction.

Keywords: self-powered; resistance-switching; triboelectric nanogenerator

Citation: Huang, Y.; Wan, L.; Jiang, J.; Li, L.; Zhai, J. Self-Powered Resistance-Switching Properties of $Pr_{0.7}Ca_{0.3}MnO_3$ Film Driven by Triboelectric Nanogenerator. *Nanomaterials* **2022**, *12*, 2199. https://doi.org/10.3390/nano12132199

Academic Editor: Alessia Irrera

Received: 12 June 2022
Accepted: 24 June 2022
Published: 27 June 2022

1. Introduction

With the advantages of fast switching speed, high storage density, and low power consumption, resistive random access memory (RRAM) has been considered a promising candidate for the next generation of non-volatile memories (NVMs) storage technology [1–4]. The RRAM cell has a capacitor-like structure composed of insulating or semiconducting materials sandwiched between two metal electrodes. In 1962, Hickmott first reported hysteretic current-voltage (I–V) characteristics in the metal-insulator-metal (MIM) structure of $Al/Al_2O_3/Al$ [5], indicating that resistive switching occurs as a result of applied electric fields, i.e., the resistive switching phenomenon. RRAM switches back and forth between a high resistance state (HRS) and a low resistance state (LRS) by the action of an applied electric field. A recent study has shown that the switching speed was faster than 5 ns [6]. In addition, I.G. Baek et al. showed that the metal oxide-based RRAM reached up to 10^{12} read cycles [7], proving that oxide-based RRAM has excellent memory performance. Meanwhile, due to the simplicity of the RRAM structure, the highest-density integrated array could be realized. The RRAM-based intersection structure reported by ShuKai Duan et al. demonstrated the feasibility and effectiveness of this scheme [8,9]. Besides, T. Kawauchi et al. reported the resistive switch-based artificial synaptic devices [10–12], indicating the great potential of RRAM for the development of highly networked and neuromorphic circuits. Various studies have demonstrated that RRAM has great prospects in the application of NVMs. Currently, with the intensification of the energy crisis, reducing energy consumption has become a consensus, and developing self-powered devices has become a research front. RRAM driven by traditional external power sources will consume huge energy in large-scale applications. Therefore, self-powered non-volatile memory devices are highly desirable.

Recently, the triboelectric nanogenerator (TENG) has attracted extensive interest as a novel energy collector. It converts mechanical energy into electrical energy based on the coupling between triboelectrification and electrostatic induction [13,14]. With the advantages of simple structure and high output in low frequency and low cost, TENG has been widely used in the development of self-powered sensors. Lots of self-powered sensors have been successfully demonstrated, such as pressure sensors [15], displacement sensors [16], acceleration sensors [17], vibration and acoustic sensors [18,19], multi-function sensors [20,21], and so on. The TENG that works in mechanical motion mode also supplies a new concept to prepare a self-powered RRAM. The TENG collects external mechanical signals and converts them into electrical signals for driving resistive switching, writing, and storing data without an external power supply. It can also monitor in real-time mechanical changes in the environment. Bo-Yun Kim reported a resistive switching memory Integrated with a nanogenerator, the switching ratio of $(Na_{0.5}K_{0.5})NbO_3$ is greater than two orders of magnitude [22]. Zhengchun Yang designed a self-powered RRAM based on Ga_2O_3 with a switching ratio of about 10^2 [23]. Although very few self-powered RRAMs have been reported, their performance is relatively low. The property of self-powered RRAM is crucial and can be further improved.

In this work, we report a self-powered RRAM and its self-driven resistive switching (RS) performance based on the TENG. We have prepared a sandwiched structure of Pt/PCMO/Ag, in which the PCMO film acts as a resistive switching layer. The TENG harvests the kinetic energy of movement and converts it into electrical energy to induce resistive switching in PCMO films. With a traditional power supply, the resistance switch ratio achieves the highest switching ratio reported so far, 9×10^7. The self-powered RRAM demonstrates a switching ratio of up to 2×10^5 in self-powered mode, which effectively converts external stimuli into electrical signals to realize resistive switching. The performance is 1000 times higher than the existing reported values. Meanwhile, this device exhibits a good response to electrical signals of different waveforms. The constructed device exhibits excellent sensitivity to mechanical movement and records each mechanical change in real-time. The self-powered RRAM shows an excellent response and good stability over multiple switching cycles. Besides the application in non-volatile memory, this self-powered RRAM has potential applications in the fields of self-powered sensors and human-machine interaction.

2. Materials and Methods

Preparation of RRAM: First, the $Pt/SiO_2/Si$ substrate is cleaned with acetone and ethanol, followed by ultrasonic treatment in pure water. The PCMO thin films are deposited on a $Pt/SiO_2/Si$ substrate by means of radio-frequency (RF) magnetron sputtering. The growth temperature is 383.15 K, and the base pressure of the sputtering chamber is 3×10^{-4} Pa. The working pressure is 0.43 Pa and is maintained by a gas mixture of oxygen and argon with a flow of 1 sccm and 24 sccm, respectively. After the fabrication, X-ray diffraction (XRD) and scanning electron microscope (SEM) are used to assess the crystalline quality of PCMO films. In addition, a 30-nm-thick Ag top electrode is deposited on the PCMO by vacuum evaporation with a shadow mask. The resistance switching performances of Pt/PCMO/Ag are measured by Keithley 2612B Source Meter at room temperature.

Fabrication of TENG: The TENG consists of a conductive sponge attached to a polymethyl methacrylate (PMMA) substrate and a Polytetrafluoroethylene (PTFE) membrane with a conductive tape attached to another PMMA substrate. The conductive sponge and the conductive tape are purchased from Guangzhou Beilong Electronics Co., Ltd., Guangzhou, China. The size of the TENG is 4 cm by 4 cm. It can work in both contact-separation and sliding modes. The output voltage and current of TENG are measured by an electrometer (Keithley 6514). The TENG is driven by a mechanical linear motor to work.

3. Results and Discussion

Figure 1a shows the schematic of the self-powered RRAM. The RRAM has a vertical sandwiched structure of Pt/PCMO/Ag. The two electrodes of the RRAM are connected with those of the TENG to form a current loop. Meanwhile, the TENG harvests the kinetic energy of motions to induce the resistive switching (RS) of RRAM.

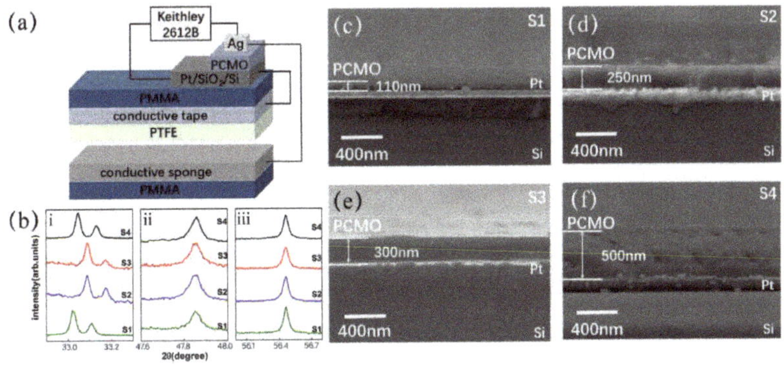

Figure 1. (**a**) Schematic illustration of self-powered RRAM; (**b**) XRD patterns of PCMO films deposited on Pt/SiO_2/Si substrates at room temperature, and three main diffraction peaks are shown in (i), (ii), and (iii); (**c–f**) Cross-sectional SEM images of the four samples with PCMO of different thicknesses, (**c**) 110 nm, (**d**) 250 nm, (**e**) 300 nm.

Firstly, the properties of PCMO thin films are characterized by using the XRD, and the XRD patterns of four samples are shown in Figure 1b. Three main diffraction peaks can be clearly observed at about 33.1°, 47.8°, and 56.4° in the XRD pattern, respectively, corresponding to (121), (202), and (113) diffraction peaks of the face-centered cubic structure of PCMO [24]. In order to measure the thicknesses of PCMO films prepared under different sputtering times, the cross-sectional images of four samples are observed by SEM, as shown in Figure 1c–f. Samples with a sputtering time of 0.5 h, 1 h, 1.5 h, and 2 h are labeled as S1, S2, S3, and S4, respectively. The thickness of the PCMO layer in each sample is designed as 110 nm, 250 nm, 300 nm, and 500 nm, respectively.

Figure 2 shows the RS characteristics of the PCMO memory cells measured through a Keithley 2612B Source Meter with a traditional power supply. The inset shows typical I–V curves of a PCMO memory device. In order to avoid the dielectric breakdown of the device, the limiting current is set to 100 mA during the measurements. Applying a dc voltage cycle from positive to negative (0 V → 2.5 V → 0 V → −2.5 V → 0 V) to the PCMO memory device, four samples exhibit similar I–V features. When a positive dc voltage scan is applied to the PCMO memory devices, samples S1, S3, and S4 are in a low resistance state (LRS), and sample S2 is in a high resistance state (HRS). When a negative dc voltage scan is applied to the PCMO memory device, the samples are in the HRS at first and then produce a SET-process-induced resistance shift from the HRS to the LRS. As the applied reverse bias increases to a certain value (SET voltage), the device switches from an HRS to an LRS. There is an abrupt increase in the current at the SET voltage. The SET voltages of the four samples are −2.2 V, −0.6 V, −2.1 V, and −1.7 V, respectively. The SET voltages of the samples are relatively low, which is conducive to reducing the requirement of the TENG output and making them easier to realize self-powered RRAM. In Figure 2, the I–V curves of the four samples show obvious hysteresis characteristics, and a large hysteresis window indicates that there is a large switching ratio of HRS/LRS, namely, a high resistance-switching ratio. The switching ratio of the four samples is 9×10^7, 2.5×10^3, 1.2×10^4, and 1×10^6, respectively. Among them, sample S1 achieves the highest switching ratio reported so far.

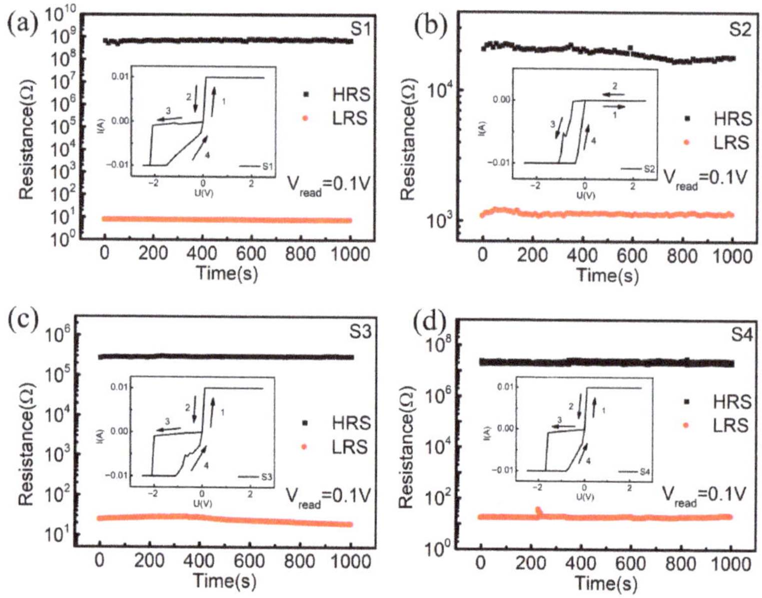

Figure 2. The resistive switching characteristics of (**a**) S1, (**b**) S2, (**c**) S3 and (**d**) S4. The insets show the I–V curves of each sample.

Retention performance is an important factor in evaluating the RS behavior for an RRAM device. Thus, we measured the retention performance of the device in the HRS and the LRS at a reading voltage of 0.1 V. As shown in Figure 2, the PCMO memory unit had excellent retention characteristics during the 10^3 s testing. In the absence of other electrical stimulation, the device will remain in its original state. No significant change is observed in the HRS and LRS; the resistance states of the resistors are also not reversed. The results indicate that the Ag/PCMO/Pt resistive storage structure has good retention characteristics. Furthermore, we investigate self-powered resistance properties driven by the TENG.

To evaluate the feasibility of TENG inducing the RS, its electrical output properties are measured. Among the four working modes of TENG, the vertical contact-separation mode and the sliding mode can generate higher output. The TENG working in these two modes is chosen to apply in four RRAM samples. The TENG is operated by a mechanical linear motor to contact and separate periodically, and its electrical outputs are measured using an electrometer (Keithley 6514). The typical open-circuit voltage (V_{oc}) and short-circuit current (I_{sc}) of TENG working in the vertical contact-separation mode are shown in Figure 3, which are alternating pulse outputs. As the movement amplitudes are 10 mm, 20 mm, 30 mm, and 40 mm, the peak values of V_{oc} are 15.2 V, 16.1 V, 18.0 V, and 21.1 V, respectively, as shown in Figure 3a. The peak values of I_{sc} are 380 nA, 403 nA, 458 nA, and 469 nA, as shown in Figure 3b. The outputs of V_{oc} and I_{sc} of the TENG are increased with the variation of amplitudes.

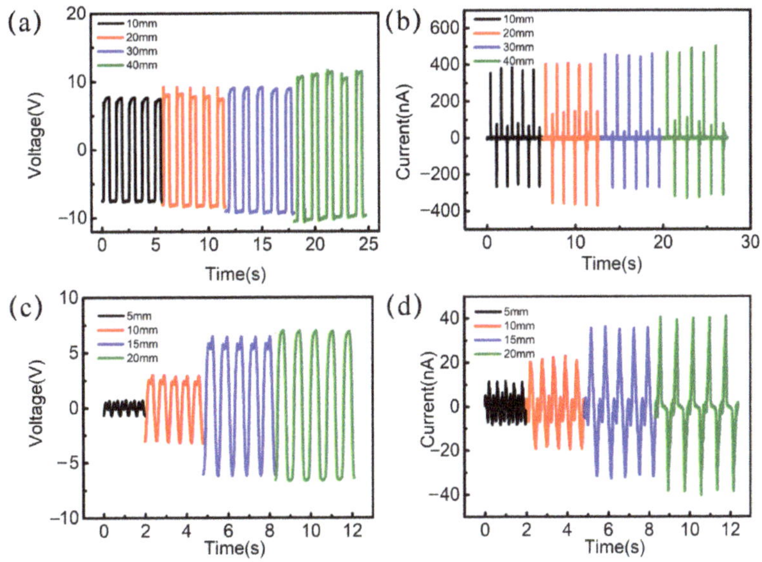

Figure 3. Output performance of the TENG. (**a**) Open-circuit voltage and (**b**) short-circuit current with a cyclic contact force of different amplitudes based on contact mode. (**c**) Open-circuit voltage and (**d**) short-circuit current of the TENG based on lateral sliding mode.

Figure 3c,d shows the V_{oc} and I_{sc} of TENG working in lateral sliding mode. When the top layer slides away about 5 mm, there is a 1.17 V potential difference between the two electrodes of the TENG. With an increase in the sliding amplitudes from 5 mm to 20 mm, the V_{oc} is gradually increased from 1.17 V to 13.4 V, as shown in Figure 3c. In addition, the I_{sc} of the TENG are 11 nA, 20.4 nA, 35.7 nA, and 40.5 nA, as shown in Figure 3d. The outputs of the TENG increase with the sliding amplitudes.

It is noted that whether in the vertical contact-separation mode or the lateral sliding mode, the electrical outputs of the TENG are alternating cyclic signals with a change of 0 V $\rightarrow$ positive peak Voltage $\rightarrow$ 0 V $\rightarrow$ negative peak voltage $\rightarrow$ 0 V, which are exactly the required stimulation signal of RS. Such a pulse alternating electric signal is very suitable for the application of RS. Due to the sensitive response to external forces, the RRAM with TENG has great potential in the application of self-powered sensing and human-machine interaction. Moreover, TENG is a high efficient energy harvester in low frequency by converting mechanical energy to electricity. It is an excellent candidate for developing a self-powered RRAM device.

By utilizing a TENG working in the vertical contact-separation mode to our four PCMO samples, as shown in Figure 4a, the RS properties are measured. With an amplitude of 13 mm of linear motor motion, the TENG successfully induces resistance switching in four PCMO films. It is observed that the RS of PCMO synchronously changes with the movement state of the TENG. While the TENG is being separated and outputs a positive electrical pulse, the PCMO exhibits the initial HRS. With the TENG moving from the separation state to the contact state, the positive electrical pulse becomes a negative pulse, and the RS of PCMO correspondingly shifts from the HRS to the LRS. In turn, as the TENG moves from the contact state to the separation state, the PCMO returns to the HRS. The RS of PCMO successfully switches with the movement state of TENG. The switching ratios of four samples, S1, S2, S3, and S4, are 2×10^5, 588, 9.6×10^3, and 1.8×10^5, respectively. Compared with the switching ratios induced by the electric field powered by a Source Meter, sample S4 shows the same switching ratio, while the switching ratios of samples S1, S2, and S3 are slightly lower. The HRS of sample S1 is more affected by TENG transient output in the switching resistance state. Samples S2, S3, and S4 are relatively stable in HRS.

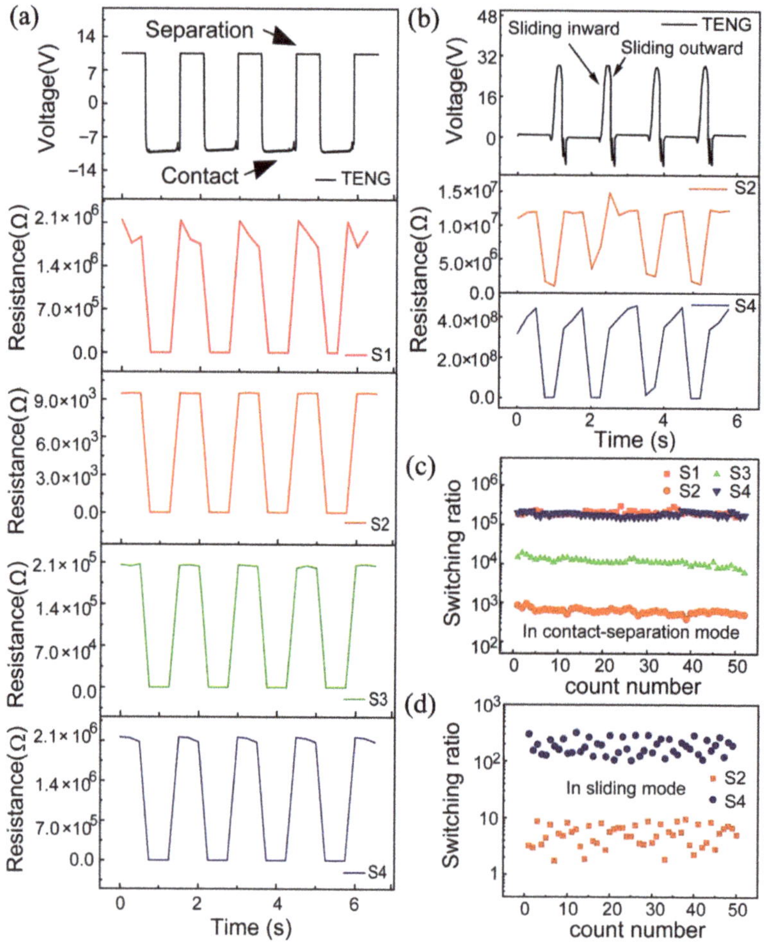

Figure 4. RS in an Ag/PCMO/Pt cell at room temperature. By applying voltages from TENG, the resistance of the cell changes reversibly between HRS and LRS. (**a**) RS characteristics of four samples based on contact mode TENG. (**b**) RS characteristics curves of samples S2 and S4 based on sliding mode TENG. (**c**) Endurance of switching behavior in the contact-separation mode. (**d**) Endurance of switching behavior in the sliding friction mode.

As the output of the TENG in the sliding mode is lower than that in the contact-separation mode, only samples S2 and S4 with lower set voltages are observed. The induced resistance switching effect, when the sliding amplitude of the TENG is 30 mm, is shown in Figure 4b. While the upper friction layer slides outward, a positive peak voltage of 28 V is generated, and the resistance value of the PCMO decreases and switches to the LRS. When the upper friction layer slides inward, a reversely increasing voltage is generated to make the PCMO return to the HRS. The switching ratios of the samples S2 and S4 are in the order of 10 and 2.4×10^2, respectively. The switching ratios are the same as those measured by a Source Meter at that time.

Figure 4c shows the stability of the switching ratio test over 50-cycles for the device as the TENG is working in contact-separation mode. During the test, the switching ratios of S2 and S3 slightly decrease. The switching ratios of samples S2 and S3 are about 844 and 1.8×10^4 at the beginning. Then they decrease slightly and stabilize at about 500 and 1×10^4. The switching ratio of the sample S1 fluctuates between 3×10^5 and 1.6×10^5.

Sample S4 is similar to S1. All the samples demonstrate good stability in the retention test, which means the self-powered RRAM devices all have the perfect capability to store data and keep the data quite stable. This approach shows good reliability since it can ensure the correct switching of the device in each cycle.

Figure 4d shows the stability of the switching ratio test over 50-cycles for the device as the TENG is working in the sliding mode, exhibiting a slight change in the switching ratio. In this mode, the pulse signal with narrow width can also excite RS, while the HRS and LRS fluctuate within a certain range due to the fluctuation of TENG output. The TENG working in the contact-separation mode is more likely to stably drive RS, while the TENG working in the sliding-friction mode is not easy to drive the stable RS phenomenon due to the asymmetric output voltage.

As shown in Figure 5, the possible mechanism is a summary to explain the effect of TENG on the RS behaviors. In p-type oxide semiconductors, it is generally considered that the migration of oxygen vacancies induces resistance switching [25,26]. In the contact-separation mode, when the conductive foam and PTFE approach, the negative charges on the PTFE films induce the positive charges in the conductive tape, which produces the current from the conductive tape to the conductive foam through RRAM, leading to the migration of positively charged oxygen vacancy and forming conductive filaments. Therefore, the PCMO switches from the HRS to the LRS. As the conductive foam and the PTFE are being separated, the negative charges on the PTFE film appeal to a number of positive charges in the conductive tape, which produces a current flowing from the conductive foam to the conductive tape through RRAM. The filaments are ruptured, and the PCMO is back to the HRS.

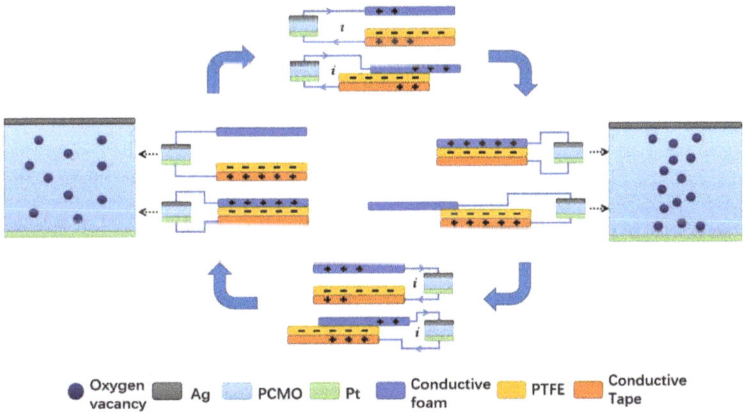

Figure 5. Schematic illustration of the underlying mechanism of the self-powered RRAM.

In sliding mode, when the conductive foam is slid backward, the redundant transferred charges on the electrodes will flow from the conductive tape to the conductive foam with an increase in the contact area. Therefore, the PCMO switches from the HRS to the LRS. When the conductive foam starts to slide outward, triboelectric charges are not fully compensated at the mismatched areas, resulting in the potential difference across the two electrodes, which produces a current flowing from the conductive foam to the conductive tape and forms conductive filaments. The filaments are ruptured, and the PCMO is back to the HRS.

4. Conclusions

In summary, we demonstrate a self-powered RRAM based on the PCMO film and the TENG. The TENG collects external mechanical energy and converts it into electrical energy to provide an electric field in PCMO, and the electric field successfully induces the RS effect. The prepared PCMO exhibits excellent self-powered RS performances, and the largest switching ratio is up to 2 × 10^5, which is the highest self-powered performance of RRAM reported at present. The self-powered RRAM effectively responds to external mechanical motions to achieve resistance switching. The constructed device exhibits excellent sensitivity to mechanical movement and records each mechanical change in real-time. We find that the TENG working in the contact-separation mode is more likely to drive the RS effect. In addition, the switching ratios maintain good stability in multiple switching cycles. The results are of significance for applications of RRAM in fields such as self-powered sensors and human-machine interaction. All authors have read and agreed to the published version of the manuscript.

Author Contributions: Conceptualization, L.W.; Data curation, Y.H., J.J. and L.L.; Investigation, Y.H. and L.L.; Project administration, L.W.; Supervision, L.W. and J.Z.; Validation, Y.H. and J.J.; Writing—original draft, Y.H.; Writing—review and editing, L.W. and J.Z. All authors have read and agreed to the published version of the manuscript.

Funding: This research was funded by the National Key R & D Project from the Minister of Science and Technology (2021YFA1201603).

Institutional Review Board Statement: Not applicable.

Data Availability Statement: Not applicable.

Conflicts of Interest: The authors declare no conflict of interest.

References

1. Waser, R.; Dittmann, R.; Staikov, G.; Szot, K. Redox-Based Resistive Switching Memories—Nanoionic Mechanisms, Prospects, and Challenges. *Adv. Mater.* **2009**, *21*, 2632–2663.
2. Waser, R.; Aono, M. Nanoionics-based resistive switching memories. *Nat. Mater.* **2007**, *6*, 833–840.
3. Gupta, R.; Jash, P.; Mondal, P.C. Nanoscale molecular layers for memory devices: Challenges and opportunities for commercialization. *J. Mater. Chem. C* **2021**, *9*, 11497–11516.
4. Shen, Z.; Zhao, C.; Qi, Y.; Xu, W.; Liu, Y.; Mitrovic, I.Z.; Yang, L.; Zhao, C. Advances of RRAM Devices: Resistive Switching Mechanisms, Materials and Bionic Synaptic Application. *Nanomaterials* **2020**, *10*, 1437.
5. Hickmott, T.W. Low Frequency Negative Resistance in Thin Anodic Oxide Films. *J. Appl. Phys.* **1962**, *33*, 2669–2682.
6. Yoshida, C.; Tsunoda, K.; Noshiro, H.; Sugiyama, Y. High speed resistive switching in Pt/TiO_2/TiN film for nonvolatile memory application. *Appl. Phys. Lett.* **2007**, *9*, 223510.
7. Baek, I.G.; Lee, M.S.; Seo, S.; Lee, M.J.; Seo, D.H.; Suh, D.S.; Park, J.C.; Park, S.O.; Kim, H.S.; Yoo, K.I.; et al. Highly Scalable Non-volatile Resistive Memory using Simple Binary Oxide Driven by Asymmetric Unipolar Voltage Pulses. *IEEE Xplore* **2004**, *4*, 587–590.
8. Duan, S.K.; Hu, X.F.; Wang, L.D.; Li, C.D. Analog memristive memory with applications in audio signal processing. *Sci. China Inf. Sci.* **2014**, *57*, 1–15.
9. Chen, Y.C.; Hu, S.T.; Lin, C.Y.; Fowler, B.; Huang, H.C.; Lin, C.C.; Kim, S.J.; Chang, Y.F.; Lee, J.C. Graphite-based selectorless RRAM: Improvable intrinsic nonlinearity for array applications. *Nanoscale* **2018**, *10*, 15608–15614.
10. Kawauchi, T.; Kano, S.; Fujii, M. Electrically Stimulated Synaptic Resistive Switch in Solution Processed Silicon Nanocrystal Thin Film: Formation Mechanism of Oxygen Vacancy Filament for Synaptic Function. *ACS Appl. Electron. Mater.* **2019**, *1*, 2664–2670.
11. Cho, H.; Kim, S. Emulation of Biological Synapse Characteristics from Cu/AlN/TiN Conductive Bridge Random Access Memory. *Nanomaterials* **2020**, *10*, 1709.
12. Cho, H.; Kim, S. Short-Term Memory Dynamics of TiN/Ti/TiO_2/SiOx/Si Resistive Random Access Memory. *Nanomaterials* **2020**, *10*, 1821. [CrossRef]
13. Wang, S.; Lin, L.; Wang, Z.L. Triboelectric nanogenerators as self-powered active sensors. *Nano Energy* **2015**, *11*, 436–462.
14. Si, J.; Duan, R.; Zhang, M.; Liu, X. Recent Progress Regarding Materials and Structures of Triboelectric Nanogenerators for AR and VR. *Nanomaterials* **2022**, *12*, 1385. [CrossRef]
15. Li, X.; Lin, Z.H.; Cheng, G.; Wen, X.; Liu, Y.; Niu, S.; Wang, Z.L. 3D Fiber-Based Hybrid Nanogenerator for Energy Harvesting and as a Self-Powered Pressure Sensor. *ACS Nano* **2014**, *8*, 10674–10681. [CrossRef]
16. Yang, Y.; Zhang, H.; Chen, J.; Jing, Q.; Zhou, Y.S.; Wen, X.; Wang, Z.L. Single-Electrode-Based Sliding Triboelectric Nanogenerator for Self-Powered Displacement Vector Sensor System. *ACS Nano* **2013**, *7*, 7342–7351.
17. Bian, J.; Wang, N.; Ma, J.M.; Jie, Y.; Zou, J.D.; Cao, X. Stretchable 3D Polymer for Simultaneously Mechanical Energy Harvesting and Biomimetic Force Sensing. *Nano Energy* **2018**, *47*, 442–450. [CrossRef]
18. Chen, J.; Zhu, G.; Yang, W.Q.; Jing, Q.S.; Bai, P.; Yang, Y.; Hou, T.C.; Wang, Z.L. Harmonic-Resonator-Based Triboelectric Nanogenerator as a Sustainable Power Source and a Self-Powered Active Vibration Sensor. *Adv. Mater.* **2013**, *25*, 42–47.

19. Liu, J.; Cui, N.; Gu, L.; Chen, X.; Bai, S.; Zheng, Y.; Hu, C.; Qin, Y. A three-dimensional integrated nanogenerator for effectively harvesting sound energy from the environment. *Nanoscale* **2016**, *8*, 4938–4944.
20. Xia, K.Q.; Du, C.L.; Zhu, Z.Y.; Wang, R.J.; Zhang, H.Z.; Xu, Z.W. Sliding-mode triboelectric nanogenerator based on paper and as a self-powered velocity and force sensor. *Appl. Mater. Today* **2018**, *13*, 190–197. [CrossRef]
21. Du, T.; Ge, B.; Mtui, A.E.; Zhao, C.; Dong, F.; Zou, Y.; Wang, H.; Sun, P.; Xu, M. A Robust Silicone Rubber Strip-Based Triboelectric Nanogenerator for Vibration Energy Harvesting and Multi-Functional Self-Powered Sensing. *Nanomaterials* **2022**, *12*, 1248. [CrossRef] [PubMed]
22. Kim, B.Y.; Lee, W.H.; Hwang, H.G.; Kim, D.H.; Kim, J.H.; Lee, S.H.; Nahm, S. Resistive Switching Memory Integrated with Nanogenerator for Self-Powered Bioimplantable Devices. *Adv. Funct. Mater.* **2016**, *26*, 5211–5221. [CrossRef]
23. Yang, Z.; Wu, J.; Li, P.; Chen, Y.; Yan, Y.; Zhu, B.; Hwang, C.S.; Mi, W.; Zhao, J.; Zhang, K.; et al. Resistive random access memory based on gallium oxide thin films for self-powered pressure sensor systems. *Ceram. Int.* **2020**, *46*, 21141–21148. [CrossRef]
24. Panwar, N.; Kumbhare, P.; Singh, A.K.; Venkataramani, N.; Ganguly, U. Effect of Morphological Change on Unipolar and Bipolar Switching Characteristics in $Pr_{0.7}Ca_{0.3}MnO_3$ Based RRAM. *Online Proc. Library Arch.* **2015**, *1729*, 47–52. [CrossRef]
25. Sawa, A. Resistive switching in transition metal oxides. *Mater. Today* **2008**, *11*, 28–36. [CrossRef]
26. Baek, K.; Park, S.; Park, J.; Kim, Y.M.; Hwang, H.; Oh, S.H. In situ TEM observation on the interface-type resistive switching by electrochemical redox reactions at a TiN/PCMO interface. *Nanoscale* **2017**, *9*, 582–593. [CrossRef]

nanomaterials

MDPI

Review

Advanced Implantable Biomedical Devices Enabled by Triboelectric Nanogenerators

Chan Wang [1,2,3,†], Qiongfeng Shi [1,2,3,†] and Chengkuo Lee [1,2,3,4,*]

1 Department of Electrical and Computer Engineering, National University of Singapore, 4 Engineering Drive 3, Singapore 117576, Singapore; chanwang@nus.edu.sg (C.W.); eleshiq@nus.edu.sg (Q.S.)
2 Center for Intelligent Sensors and MEMS, National University of Singapore, 5 Engineering Drive 1, Singapore 117608, Singapore
3 NUS Suzhou Research Institute (NUSRI), Suzhou Industrial Park, Suzhou 215123, China
4 NUS Graduate School-Integrative Sciences and Engineering Program (ISEP), National University of Singapore, Singapore 119077, Singapore
* Correspondence: elelc@nus.edu.sg
† These authors contributed equally to this work.

Abstract: Implantable biomedical devices (IMDs) play essential roles in healthcare. Subject to the limited battery life, IMDs cannot achieve long-term in situ monitoring, diagnosis, and treatment. The proposal and rapid development of triboelectric nanogenerators free IMDs from the shackles of batteries and spawn a self-powered healthcare system. This review aims to overview the development of IMDs based on triboelectric nanogenerators, divided into self-powered biosensors, in vivo energy harvesting devices, and direct electrical stimulation therapy devices. Meanwhile, future challenges and opportunities are discussed according to the development requirements of current-level self-powered IMDs to enhance output performance, develop advanced triboelectric nanogenerators with multifunctional materials, and self-driven close-looped diagnosis and treatment systems.

Keywords: implantable biomedical devices; triboelectric nanogenerator; energy harvester; self-powered biosensor; nerve stimulation; muscle stimulation

Citation: Wang, C.; Shi, Q.; Lee, C. Advanced Implantable Biomedical Devices Enabled by Triboelectric Nanogenerators. *Nanomaterials* **2022**, *12*, 1366. https://doi.org/10.3390/nano12081366

Academic Editor: Jyh-Ming Wu

Received: 3 March 2022
Accepted: 11 April 2022
Published: 15 April 2022

1. Introduction

Implantable biomedical devices (IMDs) have received rapid development in recent decades, effectively improving patients' life quality and prolonging lifespan [1–3]. Numerous IMDs have been developed in medical monitoring and treatment tools such as pacemakers [4,5], in situ blood pressure sensors [6], cardio-aid defibrillators, cochlear prosthesis, and deep brain electrical stimulators, as well as that of peripheral nerve and muscles, etc. [7–12]. The IMDs have contributed significantly to deepening the reorganization of the human body's biological processes, including the complex mechanisms of neural communication and the formation of memory [13–15]. Since the birth of the first implanted pacemaker in 1958, the fabrication and application of IMDs have made significant progress but still face many challenges [16–18]. On the one hand, IMDs need to be more miniaturized and lightweight to weaken the impact on daily life and improve comfort [19,20]. On the other hand, the battery life is typically 3 to 5 years, and then it must be surgically removed to replace the battery, which further increases the risk and cost of patients [21,22].

Harvesting energy from human activities and physiological environments to power the operation of IMDs provides a feasible solution to miniaturize implantable biomedical systems [23–33]. Researchers have demonstrated that various bioenergy, from a temperature gradient, glucose oxidation, muscle contraction, et al., can be harvested and converted into electricity [34–46]. Among all bioenergy sources, biomechanical energy from human movements, including blood circulation, and the contraction and relaxation of the heart/lung, is considered the most abundant and effective [47].

Numerous efforts have been devoted to pushing forward the development of biomechanical energy harvesting technologies, mainly including piezoelectric nanogenerator (PENG) and triboelectric nanogenerator (TENG) [26,28,30,40–42,44,48]. Researchers have reported a piezoelectric ZnO nanowire that successfully converted the periodic vibrational energy from respiration and heartbeat into electricity, relying on the mechanical-electrical coupling effect [49]. The first in vivo biomechanical energy harvester using a TENG was proposed in 2014, which outputted the power density of 8.44 mW/m^2 from a rat's normal respiration. Among them, the TENGs stand out as the biomechanical energy harvesters with the most potential for better biological compliance, high energy conversion efficiency, larger material sources and easier fabrication [50,51].

The triboelectric generator was first proposed by Zhong Lin Wang in 2012 [52], based on the principles of electrostatic induction and charge coupling, and was carried forward in the field of energy harvesting [53–60], wearable electronic devices [61–69], the Internet of Things [70–80], human-computer interaction [71,81–84] and security monitoring [71,85–88]. For instance, a self-powered and self-functional sock enabled by TENGs exhibited outstanding energy harvesting efficiency and multiple physiological signals monitoring performance (including gait, contact force, sweat level, et al.) [89]. The triboelectric self-powered wearable, flexible patch made applications as the 3D motion control interface in the robotic manipulator [90]. Notably, a series of implantable TENGs which demonstrated its potential for in vivo physiological signal sensors and biomechanical energy harvesting from heart rate, blood pressure, and respiration were developed [91–99]. The biomimetic membrane sensors (BMS) based on TENGs could identify the high-frequency vocal vibrations and the low-frequency fluctuations from human pulse [100,101].

In this review, we introduced the research progress of implantable medical devices enabled by advanced triboelectric nanogenerators incorporating self-powered biosensors, in vivo energy harvesters, and electrical stimulation, as shown in Figure 1. In Section 2, we introduced the TENG's working principle and four working modes. Section 3 summarized the self-powered biosensors and their applications in the real-time monitoring of blood pressure, heart rhythm, and bladder pressure. Section 4 reviewed the biomechanical energy harvesters and in vivo ultrasound-driven energy harvesting devices based on TENGs. Section 5 introduced the research about directly electrical stimulators with TENGs as current sources, including electrical nerve and muscle stimulation. Section 6 summarized the advanced TENGs with biodegradable and self-healing properties. In the last section, we talked about the challenges and opportunities of the self-powered IMDs mainly from three aspects: improving output performance, developing advanced materials with multifunction, and a self-powered close-looped IMDs system.

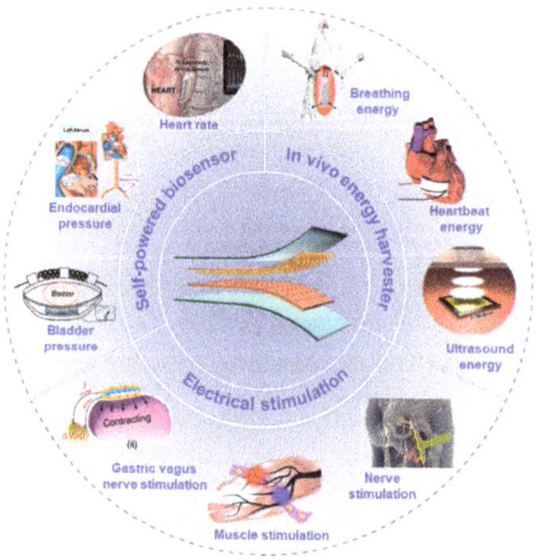

Figure 1. Overview of the implantable biomedical devices based on triboelectric nanogenerators

(TENGs) incorporating self-powered biosensors, in vivo energy harvesters, and electrical stimulation. Reprinted with permission from Ref. [102]. Copyright 2016 American Chemical Society; Reprinted with permission from Ref. [103]. Copyright 2018 John Wiley & Sons; Reprinted with permission from Ref. [104]. Copyright 2018 American Chemical Society; Reprinted with permission from Ref. [105]. Copyright 2018 American Chemical Society; reprinted with permission from Ref. [3]. Copyright 2016 American Chemical Society; reprinted with permission from Ref. [106]. Copyright 2019 American Association for the Advancement of Science; reprinted with permission from Ref. [107]. Copyright 2017 Elsevier.; Reprinted with permission from Ref. [108]. Copyright 2019 Published by Elsevier.; Reprinted from Ref. [109]. Copyright 2018 Springer Nature.

2. Triboelectric Nanogenerators

Triboelectric nanogenerators are based on triboelectric and electrostatic coupling effects [110–117]. In general, contact electrification occurs between two materials in different polarities. As shown in Figure 2a, during the contact process, the electrons transferred from the atom of material A to that of material B with their electron cloud are forced to overlap [118]. After separation, these two materials are equipped with equal positive charges and negative charges, respectively. At the same time, induced charges are generated on the electrodes under the electrostatic force from the triboelectric charges [119]. When the relative position of the two friction layers changes under the external force, the induced charges flow out and into the electrodes according to the corresponding change in the electric field [120–123]. By connecting with an electrical loading between two electrodes, the electrical pulses could be recorded, related to the change rate of induced charges, and satisfy the equation of Maxwell's displacement current [124].

$$J_D = \frac{\partial D}{\partial t} = \varepsilon \frac{\partial E}{\partial t} + \frac{\partial P_s}{\partial t}$$

where J_D is the density of free conduction current density in space due to charge flow, D is the electric displacement vector, t is time, E is the electric field, P_s is the polarization created by the electrostatic surface charges.

Triboelectrification or contact electrification exists widely in our lives, enabling an extensive selection of friction materials [125–128]. The structure of TENG could be designed according to the application scenario requirements [70,71,129–133]. As shown in Figure 2b–e, there are mainly four working modes, including a vertical contact-separation (CS) mode, single electrode (SE) mode, lateral sliding (LS) mode, and freestanding (FS) mode [124,134–136].

2.1. Vertical Contact-Separation Mode

The CS mode is first reported as the most basic TENG mode, which usually consists of two dielectric materials as the friction layers and the metal deposited on their back surface as electrodes (Figure 2a). Under external force, the two friction layers continuously contact and separate from each other. Meanwhile, the induced charges in the electrodes flow across the external circuit driven by the electrical field. The TENGs based on CS mode possessing the advantages of simple structure, easy preparation, and good output performance have substantial applications in wearable sensors. In particular, it has unique advantages as implantable devices attributed to easy, full-encapsulation, robustness, and high energy conversion efficiency in vivo.

2.2. Lateral Sliding Mode

The LS mode TENGs have the same structure as that of CS mode TENG but triggered triboelectric charges by the relative lateral sliding of the two friction layers. During periodic sliding separation and contact, the electrostatic field drives electrons to flow between two electrodes horizontally to realize a periodic alternating output. Compared to the CS mode, the TENGs based on LS mode possess higher charge transfer efficiency due to the sufficient contact of friction layers, but poor durability. In particular, when combined with the disc rotation or cylindrical rotation, the TENGs based on LS mode can realize up to several thousand volts [135,136], which shows significant potential in wind and ocean energy harvesting [134,137].

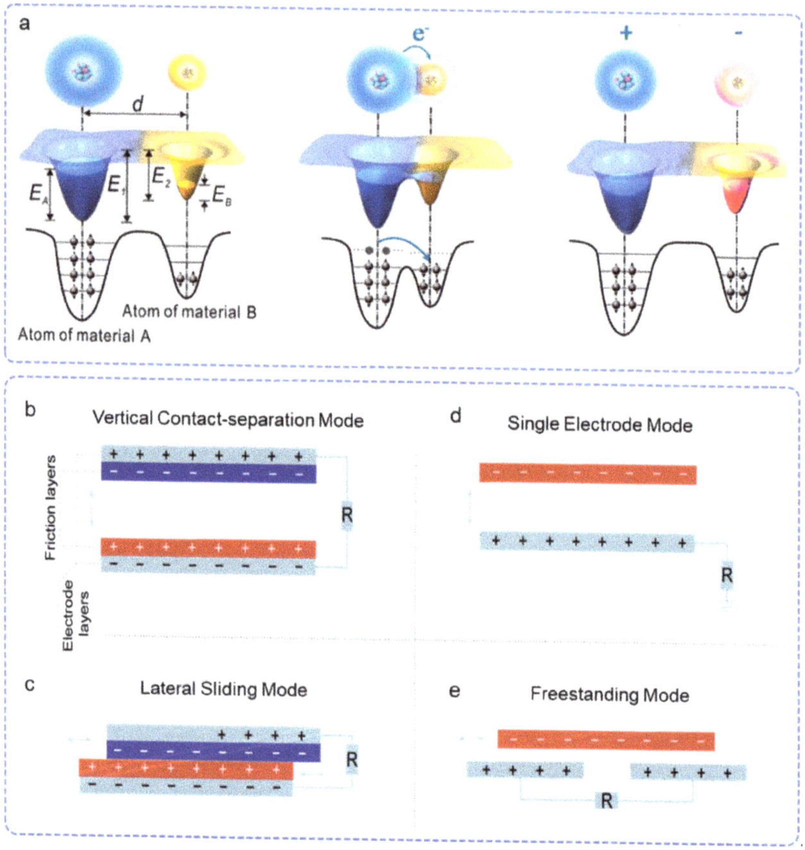

Figure 2. The working principle of triboelectric nanogenerators (TENGs). (**a**) Contact-electrification mechanism explained by electron cloud potential well model. Reprinted with permission from Reference [118]. Copyright 2018 John Wiley & Sons. Four modes of TENGs, including (**b**) vertical contact-separation mode, (**c**) single electrode mode, (**d**) lateral sliding mode, and (**e**) freestanding mode.

2.3. Single Electrode Mode

There are two electrodes connected by an external load of both CS and LS mode TENGs, while SE mode has only one friction layer and a single electrode, which is grounded. The proposal of the SE mode greatly broadens the application potential of TENG, for some moving objects operated directly as the second frictional layer, like the energy harvester driven by raindrops [138]. In addition, one of the disadvantages of the SE mode TENG is the relatively small output due to the reduced efficiency of charges transferred between two triboelectric materials.

2.4. Freestanding Mode

TENGs based on the FS mode consist of a charged moving object and two fixed triboelectric layers. The electrodes near triboelectric layers are connected with external loading. The reciprocating moving of the charged object between the two friction layers will cause a potential change between two electrodes, which drives the electrons to flow back and forth in the external circuit loop. It is considered as the most effective mode because the FS-TENGs can generate maximum power output and the highest energy conversion efficiency compared with other modes [139]. Meanwhile, the TENGs based on the noncontact FS mode exhibit the highest robustness, since the moving charged objects do not need to contact the fixed friction layers directly. The material abrasion can be greatly reduced, which significantly improves the durability of devices.

3. Self-Powered Biosensors

Accurate identification and long-term monitoring of the health status of various organs in the body is the most important and basic function for in vivo medical devices, which provide original support for subsequent targeted treatment plans. Organs and tissue dysfunction often accompanied by abnormal tissue morphology and respiratory action, like cardiovascular disease related to blood pressure that is too high or too low, can be judged by monitoring physiological differences. Self-powered sensors based on TENGs show high sensitivity to slight force, pressure, and mechanical deformation in thousandth or millionth magnitude, making them ideal for in vivo health monitoring [140,141]. At the same time, the cross fusion of functional materials and advanced manufacturing technology with TENGs enables active biosensors to have better accuracy, durability and biocompatibility, and flexibility [142,143].

For instance, Ma et al. proposed a multifunctional biomedical sensor (iTEAS) for providing real-time, accurate, and continuous monitoring of multiple physiological and pathological signs (Figure 3a) [102]. The iTEAS was designed in a core-shell structure with a 50 μm surface-nanostructured polytetrafluoroethylene film service as one triboelectric layer, and the 100 μm aluminum foil as both the other triboelectric layer and electrode. When driven by the small animals' breath and breathing, the iTEAS guarantees output of 10 V (Voc) and ~4 μA (Isc). Meanwhile, the events of atrial fibrillation and premature ventricular contractions can be accurately recognized in real-time through analyzing changes in iTEAS's output peak, which achieved a high measurement accuracy of 99% in large animals. The iTEAS can also estimate blood pressure and calculate blood flow velocity with the help of an arterial catheter. The test two weeks after the implanting operation claims good durability and bidirectional biocompatibility of the iTEAS in vivo.

Endocardial pressure status monitoring has important clinical significance for heart failure patients with impaired cardiac function. Liu et al. reported a transcatheter self-powered endocardial pressure sensor (SEPS) integrated with a surgical catheter for minimally invasive implantation [103]. As shown in Figure 3b, the oversize of SEPS was $1 \times 1.5 \times 0.1$ cm^3 with nano-PTFE and Al film as frictional layers. After the surface nanostructure and corona discharge process of one frictional layer, the open-circuit voltage of SEPS was enhanced from 1.2 V to 6.2 V. In the in vitro testing system, the SEPS demonstrated outstanding linearity ($R^2 = 0.997$) with a sensitivity of 1.195 mV mm/Hg at a wide pressure range from 0 to ≈350 mmHg. In an adult Yorkshire pig model, SEPS was implanted into the left ventricle for monitoring intracardiac pressure and diagnosing clinical conditions such as arrhythmias, ventricular fibrillation and premature ventricular contractions. On this basis, Ouyang et al. reported a completely bioabsorbable triboelectric sensor (BTS), which could directly convert blood pressure changes into electrical signals (Figure 3c) [93]. The BTS demonstrated excellent sensitivity (11 mV/mmHg), linearity ($R^2 = 0.993$), and good durability (450,000 cycles) with an air-gap structure. In large animals (dogs), BTS successfully identified abnormal vascular occlusion events and could be completely degraded and absorbed after maintaining stable output for five days.

Another group, Hassani et al., developed the self-control system for neurogenic underactive bladder (UAB) by integrating a TENG sensor with a bistable micro-actuator [104]. As shown in Figure 3d, the TENG sensor was designed with PET, copper, PDMS, and wet sponge, assessed layer by layer. With the 1 mm-thick wet sponge layer as an effective spacer, the TENG sensor well-identified 0 to 6.86 N force with output increased from

35.6 mV to 114 mV, covering the bladder pressure range. The TENG behaved as the sensing component in the self-control system for detecting the fullness status of the bladder for the consequent activation of the actuator. In the in vivo experiment, the constant output of the sensor when the bladder was almost fulfilled activated the actuator to give an extrusion force on bladder and sensor. Then, the voltage-gated shape bistable actuator empties 78% volume of an anesthetized rat within 20 s relay on memory alloy components integrated on biocompatible polyvinyl chloride sheets. The as-proposed self-control bistable actuator system helps improve the quality of life and reduce pain and severe complications for UAB patients.

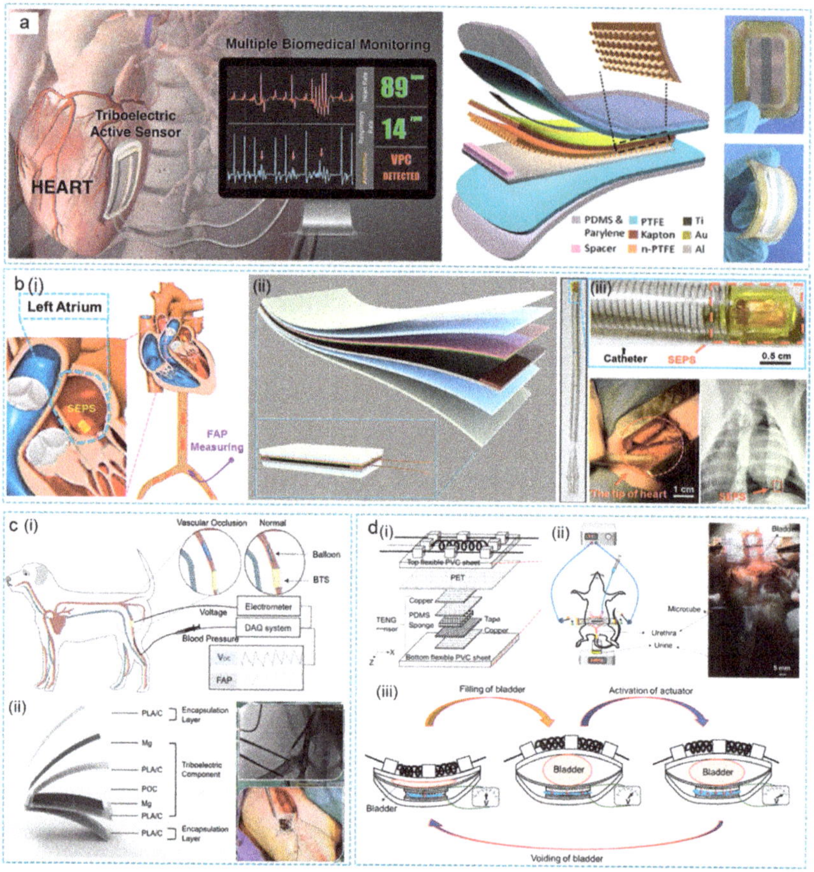

Figure 3. TENGs for implantable active sensors. (**a**) TENG as implantable self-powered cardiac sensors (iTES). The working principle (left) and structure (right) of the iTEAS. Reprinted with permission from Ref. [102]. Copyright 2016 American Chemical Society. (**b**) TENG as an implantable self-powered endocardial pressure sensor (SEPS). (i,iii) The SEPS implanted into an Adult Yorkshire swine's heart. (ii) Schematic structure of the SEPS. Reprinted with permission from Ref. [103]. Copyright 2018 John Wiley & Sons. (**c**) TENG as a blood pressure sensor (BTS). The schematic diagram of BTS used for abnormal cardiac event identification (i) in a large animal and its structure (ii). Reprinted with permission from Ref. [93]. Copyright 2021 John Wiley & Sons. (**d**) TENG as a bladder pressure sensor. The schematic diagram about the structure (i) and working principle (iii) of the bladder pressure sensor. (ii) The bladder pressure sensor in vivo. Reprinted with permission from Ref. [104]. Copyright 2018 American Chemical Society.

The implantation sites of these advanced self-powered biosensors were mainly concentrated inside/outside the heart, blood vessels, bladder and other places with large mechanical energy, as shown in Table 1. Adult Yorkshire pigs are ideal implantation animals for their similar characteristics of the cardiovascular system and weight to humans, enabling the larger electrical output. Some tissues or organs with less mechanical deformation that is still significant, such as vocal cords, tendons, ligaments, etc., should also be paid attention to for their in-situ monitoring and health status analysis. This requires IMDs to put forward higher requirements in terms of output performance, packaging technology, implantation technology and signal acquisition. At present, the long-term monitoring of most implantable self-powered biosensors relies on indwelling conductive wires, which easily leads to tissue inflammation and instability of implants. Equipping a wireless transmission module into the integration with the sensing unit and dates analysis module should be taken into consideration. Furthermore, the waveform, phase, amplitude, and change rates of the sensing dates contain rich biological health information. The in-depth analysis of these sensing dates is very important, which could be potentially realized by the rapidly advancing deep learning technologies.

Table 1. The summary of the self-powered biosensors.

Implant Site	Animal Model	Output In Vitro	Size	Application	Implantation Time	Ref.
Heart (out wall)	Adult pig	14 V, 5 μW	$2.5 \times 1 \times 0.15$ cm^3	Heart rate,	72 h	[3]
Heart (out wall)	Adult pig	10 V, 4 μA	$3 \times 2 \times 0.1$ cm^3	Heart/respiratory rate, atrial fibrillation, ventricular premature contraction	2 weeks	[102]
Heart (LV) [a]	Adult pig	6.2 V	$10 \times 5 \times 1$ mm^3	Endocardial pressure	—	[103]
Bladder	Adult rat	114 mV	1.4×2.2 cm^2	Bladder pressure	—	[104]

[a] LV: left ventricular.

4. In Vivo Energy Harvesters

Their direct role as bioenergy harvesters to power other implantable electronic medical devices is another important branch of the implantable TENGs. Although the self-powered biosensors are free from the constraints of bulky batteries, other components, such as the dates transmission unit, still require power sources. TENGs have been developed to harvest bioenergy from various human activities. The potential power from the shoulder movements is 2.2 W, and the corresponding power from the knee joint and ankle joint movements can reach 36.4 W and 66.8 W. Other biomechanical energy, involving heartbeat (0.93 W) and respiratory muscle (0.41 W) movement during breathing, is also potentially harvestable [38].

4.1. Bioenergy Harvester

In recent years, TENGs for bioenergy harvesting have sprung up. In 2014, Zheng et al. first demonstrated an implanted TENG (iTENG) in a living rat to harvest bioenergy from its periodic breathing [144]. As shown in Figure 4a, the iTENG was designed to be 2×2 cm^2 for the small rat thorax implantation. In one breath, the magnitude of the voltage and current signals was about 3.73 V and 0.14 μA, respectively. Then, the electrical energy was stored in a capacitor to power the pacemaker prototype. According to theoretical calculations, the energy obtained from five breaths can generate a pulse that regulates rat heart rate, which is a breakthrough in the research on self-driving implantable medical systems. The output of iTENG was further promoted by Zheng et al. by introducing a memory alloy as "keel structure", which facilitates the contact and separation of frictional layers in complex living environments (Figure 4b) [3]. Driven by the heartbeat of an adult Yorkshire pig, the V_{oc} and I_{sc} reach 14 V and 5 μA, which are 3.5-fold and 25-fold higher than the previous one, respectively. After 72 h of implantation, the iTENG exhibited excellent stability. By connecting the iTENG within an implantable wireless transmitter, a self-powered data transmission system was fabricated for a real-time remote cardiac monitoring system.

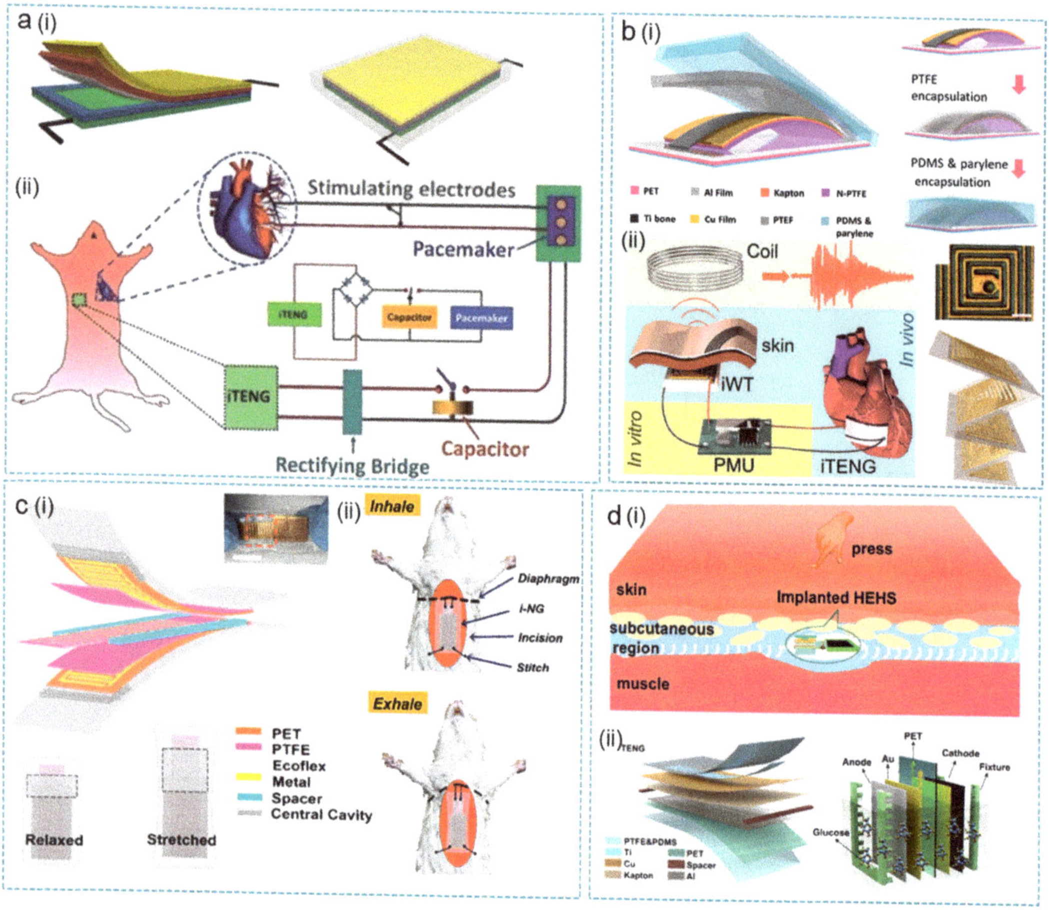

Figure 4. TENGs as in vivo energy harvesters. (**a**) In vivo power source for pacemaker driven by breathing. (i) The structure diagram of the implanted TENG (iTENG). (ii) Schematic diagram of iTENG for powering a pacemaker. Reprinted with permission from Ref. [144]. Copyright 2014 John Wiley & Sons. (**b**) In vivo self-powered wireless cardiac monitoring driven by heartbeat. (i) Structure diagram of the implanted TENG (iTENG). (ii) The self-powered wireless transmission system based on the iTENG. Reprinted with permission from Ref. [3]. Copyright 2016 American Chemical Society. (**c**) In vivo breath energy harvester. (i) Structure diagram of the implantable generator (i-NG). (ii) The i-NG for in vivo breath energy harvesting. Reprinted with permission from Ref. [105]. Copyright 2018 American Chemical Society. (**d**) TENG integrated with a hybrid biofuel cell (EHES) for bioenergy harvesting. (i) Conception graph of an implanted HEHS for biomechanical and biochemical energy in body. (ii) Structure diagram of the TENG and biofuel cell. Reprinted from Ref. [51]. Copyright 2020 Springer.

The low frequency (<5 Hz) of human activities, such as breathing and heartbeat, leads to the discrete output of implanted TENGs, which is impractical to use directly as a power source. As shown in Figure 4c, Li et al. designed a sliding mode implantable NG (i-NG) based on grating electrodes to convert low-frequency biomechanical motions into continuous electrical signals [105]. With 200 μm of each finger in the grating-electrode, the i-NG generated seven peaks within one envelope of 0.1 s, corresponding to a frequency of 70 Hz. In an adult SD rat, the i-NG generated 7–8 voltage peaks with V_{pp} of 0.8 V (two electrode-units) during one respiration cycle. Connected with the rectifier and capacitor, i-NG could stably power a green LED without blinking. Li et al. developed a hybrid energy-harvesting system (HEHS) for both biomechanical energy and biochemical energy simultaneously collecting with the integration of an implantable triboelectric nanogenerator (TENG) and glucose fuel cell (GFC) [51]. As shown in Figure 4d, driven by pressing and hand flapping, TENG realized 22 V (V_{oc}) and 0.24 μA (I_{sc}), respectively. Meanwhile, GFC generated a stable output voltage of 0.6 V and 10.5 μA at three times glucose loading. After integration in parallel, HEHS obtained a superimposed current of 1.2 μA and voltage of 21.7 V, stating a higher efficiency hybrid bioenergy harvesting system.

4.2. Ultrasound-Driven Energy Harvester

The above demonstrations show that directly harvesting energy from in vivo organ movements has been achieved with highly sensitive and compliant TENGs [145–149]. Though in vivo organs such as the heart and lung can exert regular vibrations as the internal energy sources, they generally have low frequency, low acceleration, and low amplitude that are difficult to be harvested. Especially for small animal models, the generated energy from the implantable TENGs may not be sufficient to power medical implants. On the other hand, wireless energy delivery using external sources to transfer energy to implantable energy harvesters offers an alternative approach that is more stable, adaptable, and provides power-on-demand [150]. The acoustic wave has been widely used in medical applications, including monitoring, diagnosis, and in vivo imaging. It can also penetrate deep tissue with low attenuation, and thus it is a safe and good medium for wireless energy delivery [151]. In the past few years, several implantable piezoelectric energy harvesters, as well as TENGs, have been proposed to study the performance and efficiency of acoustic energy transfer (AET) [106,132,152–154].

In 2019, Hinchet et al. developed a vibrating and implantable triboelectric generator (VI-TEG) to harvest the wirelessly delivered ultrasound energy from a controllable external source (Figure 5a) [106]. The VI-TEG mainly consists of a thin perfluoroalkoxy (PFA) membrane (~50 μm) and an Au/Cu electrode with an airgap of 80 μm to collect the charges when the membrane vibrates under ultrasound pressure. This working mechanism triggered by the incoming ultrasound eliminates the influence of implanted site, in vivo environment, and usage scenarios. The whole device is sealed with melt adhesive and further integrated with a power management circuit and a Li-ion battery on the backside. Operated in a single-electrode mode, the VI-TEG can produce a peak output voltage of 25 V (at 40 MΩ impedance) and current of 1.3 mA (at 1 Ω impedance), when it is tested in a grounded water medium 5 mm away from a 3-cm-diameter ultrasound probe (20 kHz and 3 W/cm^2). In an ex vivo AET experiment (20 kHz and 1 W/cm^2) through a 0.5 cm porcine tissue, the generated outputs are more than 2.4 V and 156 μA, which are high enough as power supplies for small medical implants.

Later in 2020, Chen et al. reported a micro-structured triboelectric ultrasonic device (μTUD) based on the technology fusion of TENG and microelectromechanical systems (MEMS) [155]. As indicated in Figure 5b, the μTUD is composed of an array of micro capacitors that has a suspended membrane over a vacuum cavity. Silicon oxide is adopted as the friction layer, while a highly conductive doped silicon layer is used as the electrode and another friction layer. When ultrasound pressure is applied to the μTUD, the suspended membranes vibrate and contact the oxide layer periodically, generating triboelectric outputs in a contact-separation mode. Ex vivo AET measurements are then performed through an oil medium (30 mm distance) without and with a sound-attenuation medium (porcine tissue). In an incident acoustic wave of 63 kPa at1 MHz, the generated output voltage reaches 16.8 mV and 12.7 mV, respectively. Even though the output power of the demonstrated μTUD is currently low, theoretical analysis by the authors shows that the device's output power could be enhanced by 10,000 times through optimizations in material selection, geometry structure, management circuit, etc.

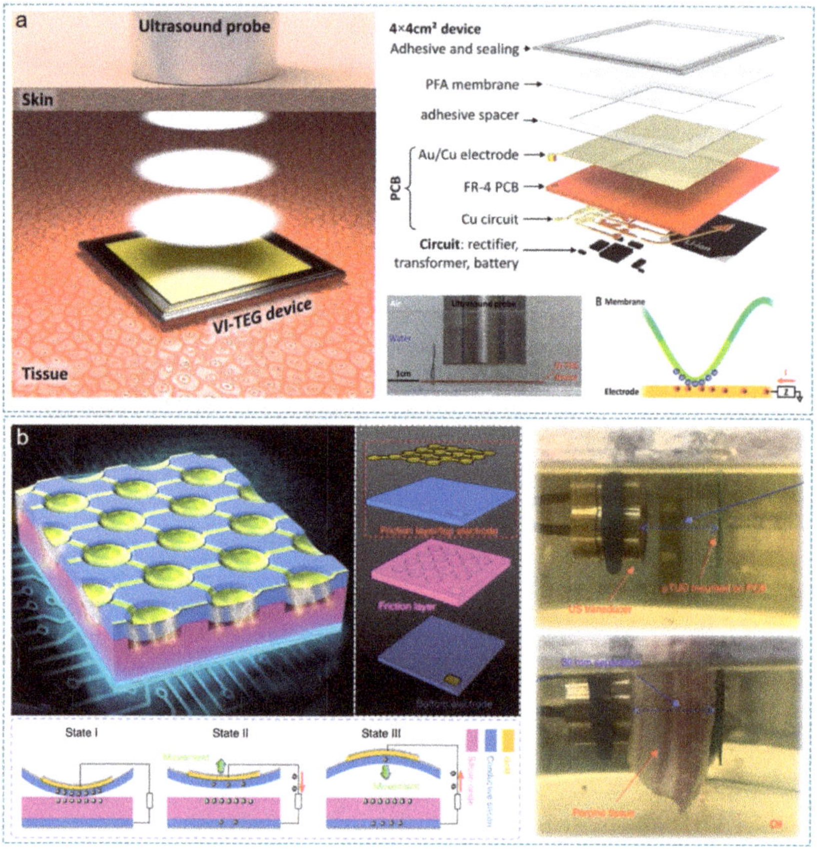

Figure 5. Ultrasound-driven energy harvester. (**a**) Ultrasound energy harvesting under skin based on triboelectric effects. Illustration of US energy harvesting under skin using the vibrating and implantable triboelectric generator (VI-TEG, left). The structure diagram and working principle of the VI-TEG (right). Reprinted with permission from Ref. [106]. Copyright 2019 American Association for the Advancement of Science. (**b**) Micro triboelectric ultrasonic device for acoustic energy transfer. Reprinted from Ref. [155]. Copyright 2020 Springer Nature.

From direct biomechanical energy harvesting to ultrasonic-driven in vivo energy harvesting based on the triboelectric effect, IMDs have been put forward to better output performance and longer operation time, but there is still a long way to go to fully realize self-sustained and real-time IMDs. Many significant factors should be taken into consideration, and we will give a summary and discussion in Section 7.1.

5. Electrical Stimulation Therapy

5.1. Nerve Stimulation

The rapid development of flexible implantable bioelectronics has opened up a new way to enable electroceuticals to more sensitively and accurately record and modulate the biosignals [155–158]. In this regard, neuromodulation has been widely adopted as a reversible and non-destructive therapeutic strategy to manipulate various bodily functions by stimulating or interfering with the neurophysiological signals in the neural network [159,160]. More specifically, neural signals transmitted in the nerve fibers can be tuned via stimulation on the targeted fibers for therapeutic intervention, such as restoring the sensory/motor function of neuro-prosthesis, controlling bodily functions, treating diseases, etc. [133,161–168].

Benefited by the self-generated characteristics and wearable/implantable compatibility, TENGs have been integrated with neural interfaces and explored for nerve stimulation, using the directly generated outputs from mechanical actuation [18,81,107,109,169–172].

As shown in Figure 6a, Zhang et al. first applied a high-performance TENG for a dissected frog leg [169]. The TENG is operated in the contact-separation mode, with Al and PDMS (produced with micro/nano structure and treated with fluorocarbon plasma) as the two triboelectric materials. Actuated by a 5 Hz external force, it can generate outputs of 265 V and 18.3 $\mu A/cm^2$ in terms of voltage and current density, respectively. The outputs are then applied to the sciatic nerve of a frog leg via a 3D microneedle electrode array (MEA) toward the realization of the self-powered neural prosthesis. The two electrodes of the MEA are directly connected to the outputs from the TENG without the use of any external circuits. With force applied to the TENG, the sciatic nerve is successfully stimulated by the instantaneous outputs from the TENG, leading to the muscle actuation of the frog leg.

To obtain effective obesity treatment, Yao et al. proposed an implanted system that can respond to stomach movement and perform vagus nerve stimulation to regulate food intake (Figure 6b) [109]. The implanted system only consists of a flexible and biocompatible TENG, with its body is attached to the stomach surface while the output electrodes are directly connected to the vagus nerve. With food intake, the peristalsis motion of the stomach induces cyclic contact and separation on the TENG, which then generates biphasic electrical signals (~0.05 to 0.12 V recorded on a 1 MΩ load) to stimulate the vagal afferent fibers for food intake reduction and weight control. This strategy is then verified using rat models, and the results show that the average body weight of the stimulation group is 38% less than the control groups, indicating the effectiveness of nerve stimulation in weight control and obesity treatment. On the basis of frog leg stimulation, Lee et al. took one step forward and investigated the feasibility of using TENG for sciatic nerve and common peroneal nerve stimulation in a rat model (Figure 6c) [107]. The adopted TENG has a stacked-layer structure, with five contact-separation TENGs (micro-pyramid PDMS and Cu as the triboelectric layers) connected in parallel to enhance the output performance. After optimizing the TENG's confinement height, it can generate a peak-to-peak voltage of 160 V and a current of 6.7 μA. The outputs are then directly applied for nerve stimulation through a flexible sling electrode as the designed neural interface. The experimental results show that direct stimulation of the rat's sciatic nerve and the common peroneal nerve is achieved, as proven by the corresponding EMG signals being successfully recorded.

To further investigate the efficacy, selectivity, and controllability of direct stimulation on peripheral nerves, Lee et al. developed a water–air TENG array with optimized force sensitivity and sensing range, as shown in Figure 6d [170]. The suspended thin PDMS in the middle can greatly improve the generated outputs of the liquid involved TENG under pressure, i.e., from the mV range to tens of volts. Meanwhile, the adopted water-filled sponge enables a much higher sensing range compared to the hydrogel. Using the multiple pixels in the TENG array and flexible neural clip interfaces, selective stimulation of plantar flexion (PF) and ankle dorsiflexor (DF) via tibial and common peroneal nerve branches are realized on rats, with controllable muscle activation by the pressing force. The exponentially decreasing waveform from TENGs is also compared with the traditionally used biphasic square waveform, demonstrating the high effectiveness of using TENG for direct nerve stimulation. In addition to muscle control, nerve stimulation can also be used for controlling bladder function. Lee et al. from the same group investigated the direct stimulation of the autonomic pelvic nerve by a stacked-layer TENG and a flexible neural clip interface for bladder function control (Figure 6e) [171]. The study of stimulation dependence on frequency and number of pulses reveals that only two pulses (~29.5 nC per pulse) of 0.83 Hz from the TENG are sufficient to induce the micturition response in rats. These in vivo results are also compared with the stimulation results by a commercial stimulator, showing once again that the direct stimulation waveforms from TENG are more effective than the biphasic pulses.

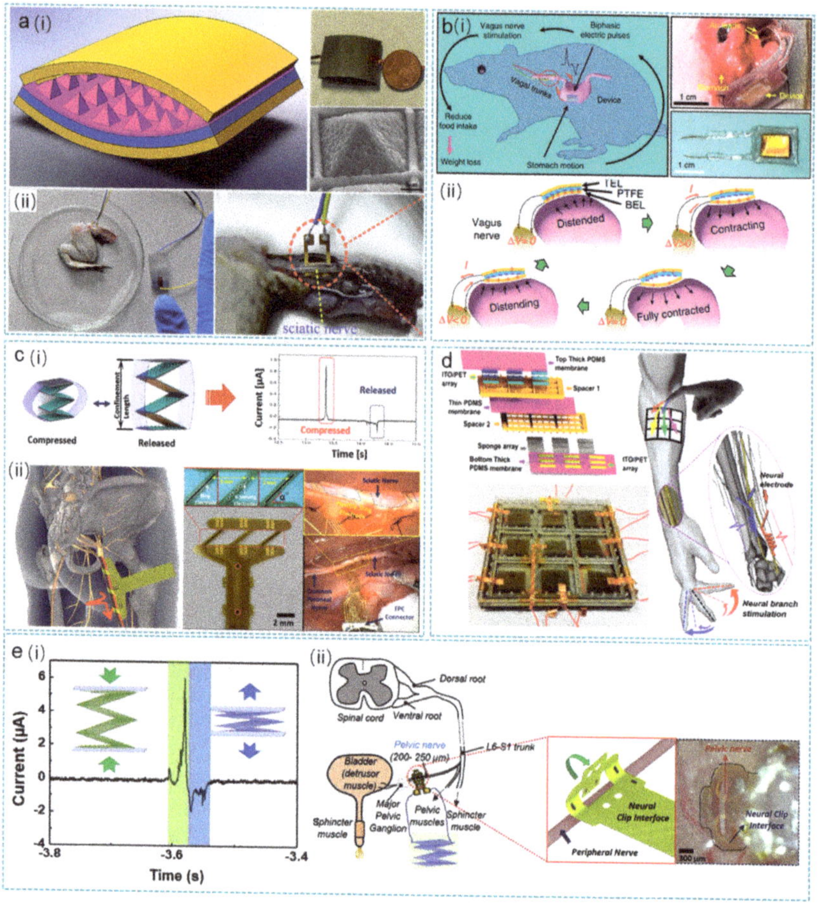

Figure 6. TENGs for nerve stimulation. (**a**) High-performance TENG for frog sciatic nerve stimulation. (i) The structure diagram and image of the high-performance TENG. (ii) The high performance TENG for frog sciatic nerve stimulation. Reprinted with permission from Ref. [169]. Copyright 2014 Elsevier. (**b**) Effective weight control via nerve stimulation (VNS). The schematic diagram of VNS system (i) and the working principle of VNS devices (ii). Reprinted from Ref. [109]. Copyright 2018 Springer Nature. (**c**) Modulated control of tibialis anterior muscle via common peroneal nerve stimulation based on TENGs. (i) The TENG in a compressed state and a released state, outputting voltage pulse. (ii) The TENG for tibialis anterior muscle controlling via common peroneal nerve stimulation. Reprinted with permission from Ref. [107]. Copyright 2017 Elsevier. (**d**) Peripheral nerve direct stimulation based on TENGs. Reprinted with permission from Ref. [170]. Copyright 2018 Elsevier. (**e**) Direct stimulation of autonomic pelvic nerve for restoring bladder functions. Schematically showing the working principle (i) and applications as nerve stimulation power sources (ii) of the TENG. Reprinted with permission from Ref. [171]. Copyright 2019 Elsevier.

5.2. Muscle Stimulation

Muscle function loss exhibits the symptoms of abnormal muscle movement and/or, more seriously, paralysis. In terms of treatment, electrical muscle stimulation to restore muscle functions is feasible as a therapeutic method. Compared to nerves that are normally concentrated in a small nerve bundle, muscles contain numerous excitable motoneurons distributed sparsely within the muscle tissues. Thus, the threshold current for muscle stimulation is much higher than that for nerve stimulation. In this regard, to achieve direct muscle stimulation from TENGs, more effective neural interface designs or higher output-performance TENGs are required, which then have been investigated by researchers [173–176].

To realize a more controllable and effective electric field in muscle stimulation, Wang et al. developed a flexible multi-channel intramuscular electrode powered by a staked-layer TENG (Figure 7a) [177]. The 12 channels (6 channels from each side) of the intramuscular electrode allow the mapping of stimulation efficiency with different channel combinations, from which a more effective stimulation can be determined even with a relatively low short-circuit current of 35 μA from the TENG. Based on the in vivo experiment on rats, it is concluded that the efficiency of electrical muscle stimulation is mainly affected by the applied waveform polarity and the electrode–motoneuron position. Therefore, when implanted in muscle, the multi-channel electrode offers great tunability in the stimulation polarity and position by connecting the outputs from the TENG to different channels, leading to a more effective direct muscle stimulation.

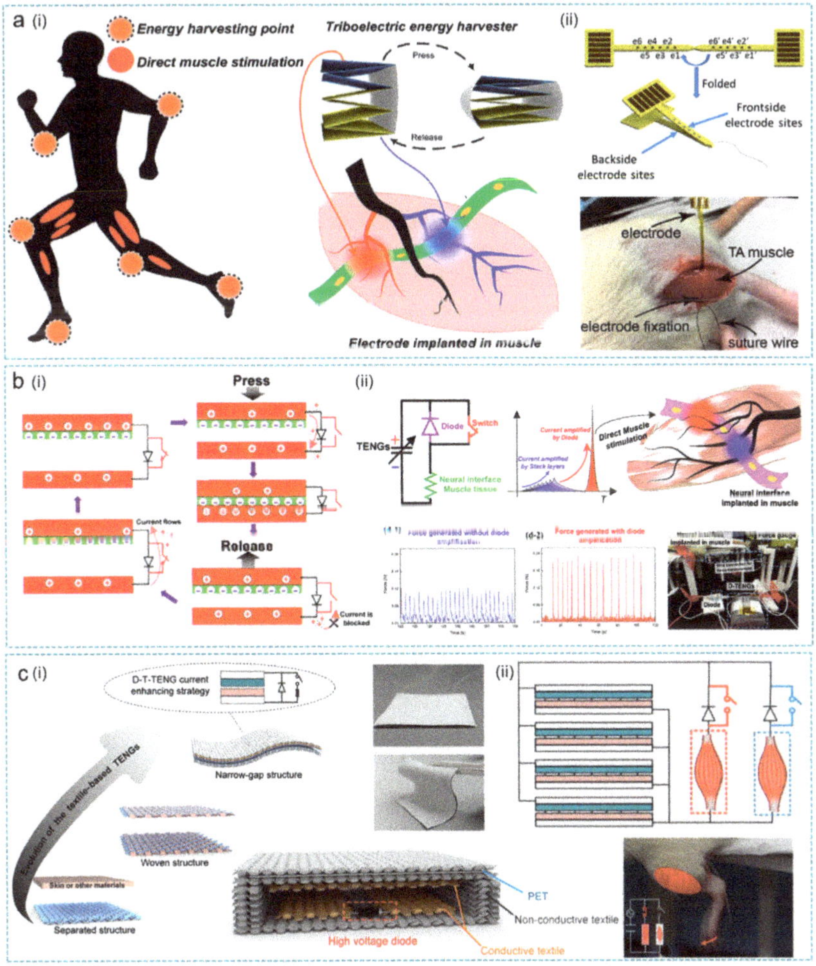

Figure 7. TENGs for muscle stimulation. (**a**) Direct muscle stimulation using a stacked-layer TENG and a multi-channel intramuscular electrode. The working principle (i) and in vivo experiment (ii) of the TENG using directly for muscle stimulation. Reprinted with permission from Ref. [177].

Copyright 2019 American Chemical Society. (**b**) Muscle stimulation method using a switching operation TENG to achieve higher efficiency. (i) Device working principle. (ii) The current amplified principle during muscle stimulation. Reprinted with permission from Ref. [108]. Copyright 2019 Elsevier. (**c**) A soft fabric-based thin-film TENG for selective and controllable muscle stimulation. (i) The fabrication process and (ii) working principle of the thin-film TENG. Reprinted from Ref. [178]. Copyright 2019 John Wiley & Sons.

Along with the neural interface optimization, the performance-enhancing strategy of TENGs has also been proposed to improve muscle stimulation efficiency. Wang et al. reported a new TENG configuration, called diode-amplified TENG (D-TENG), with enhanced TENG outputs for direct muscle stimulation using a similar multi-channel neural interface (Figure 7b) [108]. As indicated, the D-TENG integrates a diode and a switch on the original stacked-TENG; thus, charges can be released instantaneously, generating a higher output current with uni-polarity (~40 μA). In vivo demonstrations show that the more efficient muscle stimulation by the D-TENG can be attributed to three aspects: (1) improvement in the current amplitude; (2) exponential waveform of the current pulse; (3) frequency increment with a shorter pulse that matches the resonance frequency of motoneurons (~500 Hz). Hence, the authors concluded that the D-TENG design is one of the most optimal configurations of TENGs for direct muscle stimulations.

To directly use the harvested energy from body motions for muscle stimulation, wearable TENG devices based on thin and soft textiles are more convenient to be worn on various body parts. In this regard, He et al. developed a narrow-gap diode-enhanced textile-based TENG (D-T-TENG) coupled with textile switches for enhanced output currents (>100 μA), as shown in Figure 7c [178]. The D-T-TENG is fabricated using conductive textile as the electrode and Ecoflex/nitrile as the negative/positive triboelectric materials. The enhanced output currents from the D-T-TENG are then applied for direct stimulation of the anterior tibialis muscle and gastrocnemius muscle through stainless-steel wire electrodes, corresponding to leg movements of forward kicking and backward kicking a rat model. The kicking direction and force can be controlled by selecting the switches and controlling the activation area of the D-T-TENG, indicating its great applications in future prosthesis control.

5.3. Cardiac Pacemaker

Artificial pacemakers provide electrical impulses to regulate the heartbeat, playing a significant role in the treatment of sick sinus syndrome, atrioventricular block, and some bradycardia-type conditions. Due to the limited life span, the battery of the artificial pacemaker needs to be surgically removed and replaced every 3–5 years, which brings great pain and surgical risks to patients.

In 2014, Zheng et al. developed an implantable triboelectric nanogenerator (iTENG) in a living rat to harvest energy from periodic breathing and to be used as the direct power source in powering a prototype pacemaker for the first time. After that, as shown in Figure 8, Ouyang et al. further designed a fully implanted symbiotic pacemaker (SPM), which made progress in terms of optimizing output performance of the energy harvesting unit (iTENG), integrating the power management unit (PMU) and the pacemaker unit all in one system [179]. The iTENG was designed in a core-shell multilayer structure, with hydrophobic PDMS serving as the encapsulation and memory alloy used as the support layer. The iTENG achieved high biomechanical–electrical conversion efficiency with the electrical output of 97.5 V (V_{oc}), 49.1 nC (Q_{sc}), and 10.1 μA (I_{sc}) in vivo, as well as the superior stability after 10^9 times in liquid, moist (humidity of 45–50%), and dry environments. In a big animal model, driven by the heartbeat, the iTENG harvested 0.495 μJ during one cardiac cycle, which is higher than the energy required for one heart pacing (0.262 μJ).

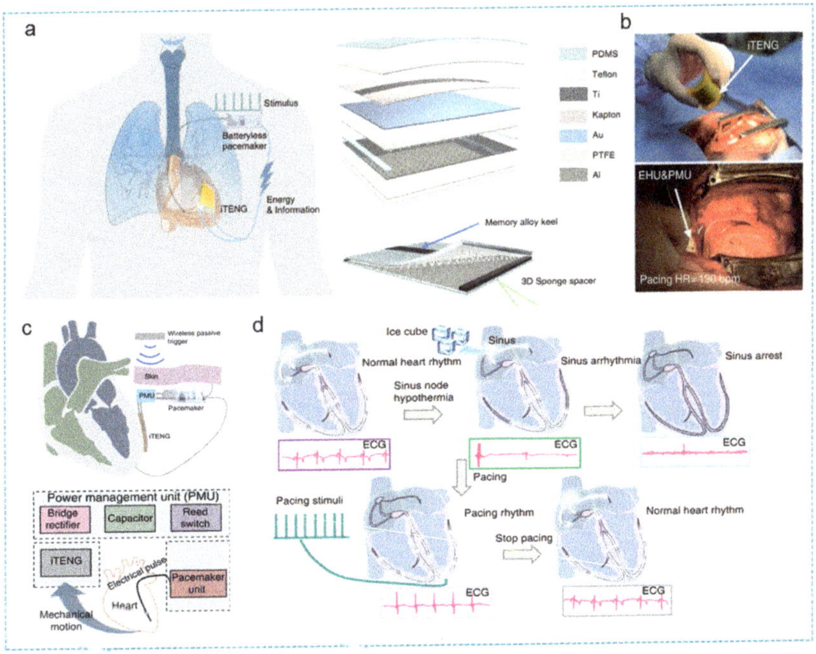

Figure 8. An implantable TENG for a symbiotic pacemaker. (**a**) Illustration of the symbiotic cardiac pacemaker system and structure diagram of the iTENG. (**b**) The implantation pictures of iTENG in a big animal experiment. (**c**) Illustration of a symbiotic cardiac pacemaker system in vivo. (**d**) The symbiotic cardiac pacemaker system is used to correct arrhythmia events. Reprinted from Ref. [179]. Copyright 2019 Springer Nature.

For the cardiac pacing in a large animal model, the PMU managed the charging and peacemaking process through a wireless triggered switch. Electrical energy from iTENG was stored in a 100 μF commercial capacitance, powering the peacemaking unit to produce a 3 V, 0.5 ms electrical pulse in regulating heart rate after a wireless controller outside the body activated the control switch. Furthermore, the PMU validated the diagnosis and correction of sinus arrhythmia events in an adult Yorkshire pig. The authors created a sinus arrhythmia model using ice cubes to create sinus node hypothermia. When the abnormal ECG is detected, the pacing module starts to work, and the sinus arrhythmia is converted into a pacing rhythm. After about one minute, the SMP stopped working, and the pacing rhythm snapped back to a normal rhythm.

The above research progress has demonstrated the TENGs as the direct electrical stimulation source for muscle function recovery, weight control, and cardiac pacing. Further neuromodulation applications, such as deep brain stimulation, vagus nerve stimulation and human functional electrical stimulation, still remain challenges. On the one hand, the frequency of the current pulses (no more than 2 Hz) generated by TENGs could not reach the requirement for neuromodulation applications, which is usually higher than 10 Hz [168]. On the other hand, direct current pulses from the TENGs need to be further amplified to a suitable magnitude for neural stimulation. Thus, the power management switches and amplifiers should be integrated into the neuro-stimulation system.

6. Advanced IMDs Based on Functional TENGs

6.1. Biodegradable and Absorbable TENGs

IMDs play significant roles in patient care and disease management [49,180–183]. However, after their service, almost all implantable electronic devices need to be removed or replaced by invasive, complicated surgery [184–187]. IMDs that are degradable and absorbable in the body provide a feasible solution to avoid secondary invasive procedures [23,188–190]. In 2016, Zheng et al. reported a biodegradable triboelectric nanogenerator (BD-TENG) for in vivo energy harvesting with the fully bio-degradable polymers (BDPs) as the friction and encapsulation layer, the resorbable Mg as electrodes (Figure 9a) [180]. A series of TENGs based on several pairs of BDPs as friction layers compared BDP's relative ability to gain or lose electrodes during the triboelectric process, which achieves a voltage output ranging from 20–40 V and different degradable rates at the same time. In an adult SD rat model, the PLGA-encapsulated BD-TENG output voltage decreases from 4 V to 1 V within 2 weeks, failed in about 4 weeks, and completely lost its structure after 9 weeks, indicating the feasibility of degradable in vivo. After that, in the same group, Jiang et al. further developed fully bioabsorbable natural-materials-based triboelectric (BN-TENG) nanogenerators based on cellulose, chitin, silk fibroin (SF), rice paper (RP), and egg white (EW) (Figure 9b) [189]. The authors verified the triboelectric series of these five materials as EW > SF > Chitin > Cellulose > RP, which provides a reference for future BN-TENGs in materials selection and structure design. With different pairwise combinations, the as-fabricated BN-TENG realized a large range of output performance with V_{oc} ranging from 8 to 55 V and I_{sc} ranging from 0.08 to 0.6 μA. The degradation rate of BN-TENG can be regulated by adjusting the fibroin crystallinity with methanol, achieving device failure time in vivo ranging from 24 h to 11 d, and then complete degradation within 84 days.

In addition, active sensors with biodegradable and absorbable properties also made a breakthrough. As shown in Figure 9c, Ouyang et al. designed a bioresorbable dynamic pressure sensor based on the triboelectric effect (BTS), which can convert biological pressure signals into electrical signals [93]. The BTS has excellent sensing characteristics, achieving sensitivity up to 11 mV/mmHg and good linearity of R^2 = 99.3%. In the large animal model (dog), BTS achieved accurate monitoring of pleural pressure, vascular pressure, and identification of vascular occlusion events. In addition, the poly(lactic acid)-(chitosan 4%), which serves as both friction and adhesion layer, equipped the BTS outstanding antibacterial ability of 99% sterilization, making the implantable BTS free from the risk of tissue infection. The implanted BTS can be completely degraded after its service and completely be absorbed within 21–84 days.

The development of implanted TENGs based on degradable materials has strongly promoted the progress of next-generation customizable IMDs [191,192]. Since degradable materials were first applied to TENGs, more than a dozen degradable materials have been developed [193,194]. They have been applied in energy harvesting, mechanical sensing, and even implantable electrical stimulation [195,196]. In the future, the design of degradable materials and the performance optimization of degradable TENGs are still full of challenges, and more research and exploration are needed.

6.2. *Self-Healing TENGs*

The performance of TENGs relies on the dynamical process of converting bio-mechanical energy into electrical energy [197,198]. At the same time, the long-term repeat mechanical deformation also led to different mechanical and functional damages of TENG: including cracks, mechanical fatigue, degradation of friction surface structure [199,200]. Developing advanced material with self-healing ability helps damaged TENGs recover their mechanical morphology and functions, which contributes to prolonging their service life and usage safety [201–203].

In 2013, Lee et al. demonstrated the first self-healing TENG based on a shape memory polyurethane (SMPU) triboelectric layer, with an original output of 80 V and 3 μA, respectively (Figure 10a) [204]. A series of compression forces were applied on the SMPU-TENG to investigate the degradation of the micropatterned triboelectric layer. With a high compression force of 10 kgf for 15 min, the performance of SMPU-TENG reduced by about 90%. By raising the temperature to T_g = 55 °C to trigger the healing process, the SMPU micro-pyramid pattern recovers its initial shape and performance to its original state.

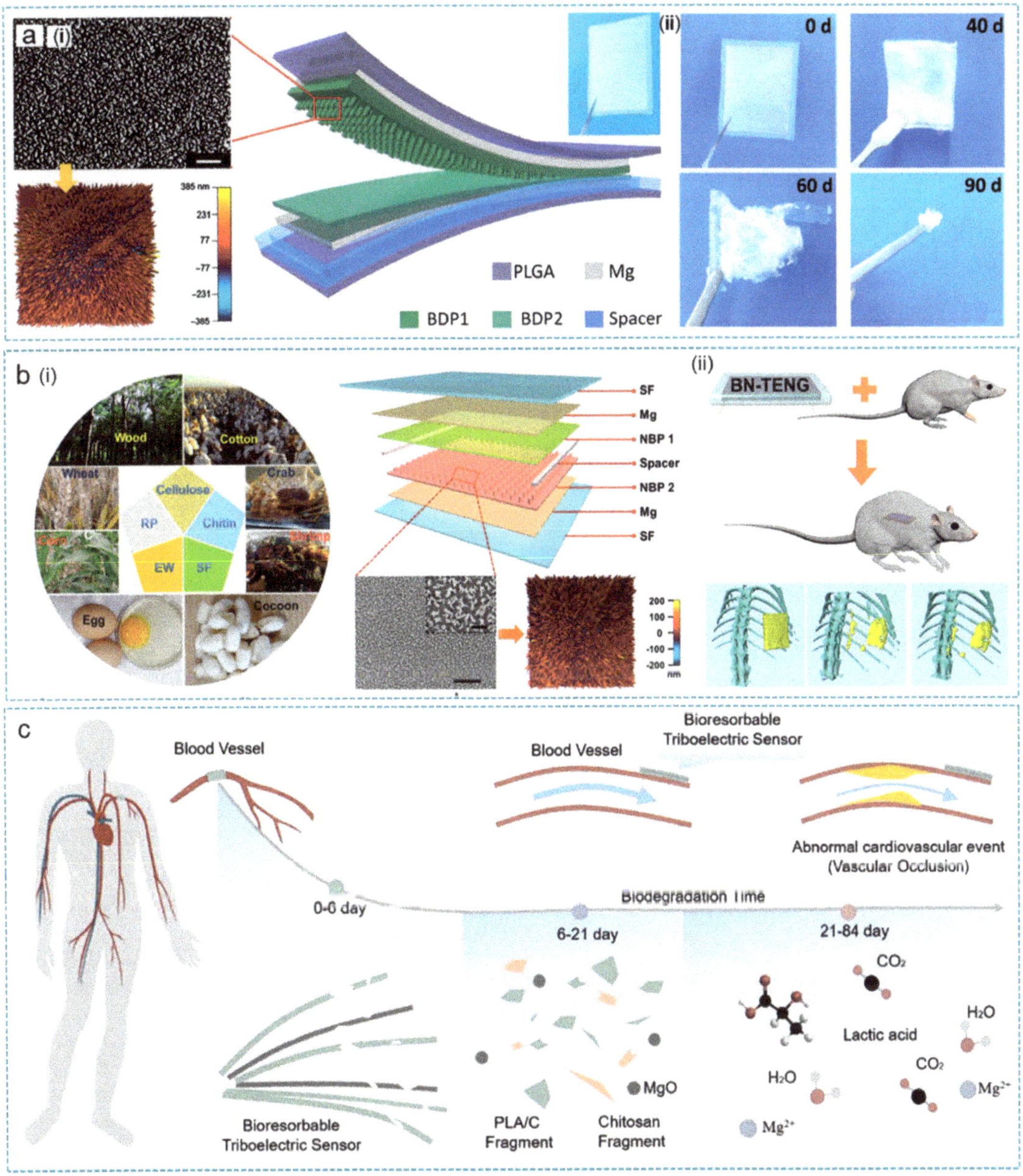

Figure 9. Degradable and absorbable TENGs. (**a**) Biodegradable TENG (BD-TENG) as a lifetime designed implantable power source. The structure diagram (i) and in vitro degradation process (ii) of the BD-TENG. Reprinted from Ref. [180]. Copyright 2016 American Association for the Advancement of Science. (**b**) The structure of fully bioabsorbable natural-materials-based TENGs (i) and the in vivo biodegradation process (ii). Reprinted with permission from Ref. [189]. Copyright 2020 American Chemical Society. (**c**) A bioresorbable dynamic pressure sensor based on TENG. Reprinted with permission from Ref. [93]. Copyright 2021 John Wiley & Sons.

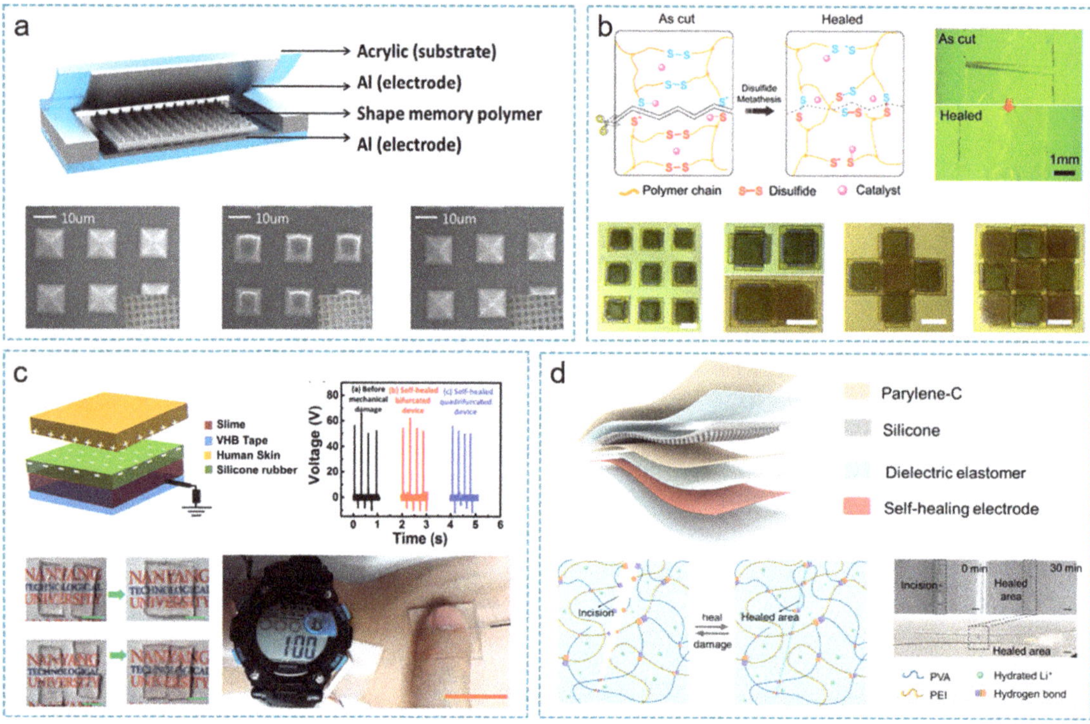

Figure 10. Self-healing TENGs. (**a**) Shape memory polymer-based self-healing TENG. Reprinted with permission from Ref. [204]. Copyright 2013 Royal Society of Chemistry. (**b**) Vitrimer elastomer-based jigsaw puzzle-like healable TENG for self-powered wearable electronics. Reprinted with permission from Ref. [200]. Copyright 2018 John Wiley & Sons. (**c**) Highly transparent, stretchable, and self-healing ionic skin for energy harvesting and touch applications. Reprinted with permission from Ref. [201]. Copyright 2017 John Wiley & Sons. (**d**) Structure of the stretchable, self-healing, and skin-mounted muscle sensor for muscle function assessment. Reprinted with permission from Ref. [205]. Copyright 2021 American Chemical Society.

Deng et al. then developed a self-healing vitrimer elastomer and successfully used it in the single-electrode mode TENG (Figure 10b) [200]. Based on the break and reform of dynamic disulfide bonds, the vitrimer elastomer realized 100% self-healing efficiency after extreme damage when exposed to 90 °C high temperatures within 20 min and intense pulsed light within the 30 s. Embedded with silver wires, the elastomer demonstrated superior conductivity and was utilized as a matrix to construct TENG, which performs outstanding output with I_{sc} reaching 350 nA and V_{oc} reaching 26 V. The self-healing ability of the vitrimer elastomer not only helps TENG recover from accidental injury but also provides a simple method to assemble TENGs into desired 2D and 3D structure free of glue with the mechanism of jigsaw puzzles and building blocks toys. At the same time, the VTENGs in 3D structure increased to about 450% charging efficiency as the unit increased from one to nine and showed satisfactory sensitivity to external motion when used as sensing elements and wearable electronics.

Compared with other components, the electrode of flexible TENGs is more vulnerable to damage, for they are mostly based on a thin layer of metal deposition or conductive composite. In 2017, Parida et al. reported a highly transparent, stretchable, and self-healing TENG (IS-TENG) for energy harvesting and touch application, using a slime-based ionic conductor as the self-healing electrode for the first time (Figure 10c) [201]. The IS-TENG based on the ionic conductor as current collector achieved a charge collecting efficiency 12 times higher than that based on silver, which exhibited 92% transparency and over 700% stretchability at the same time. The IS-TENG behaved with excellent self-healing ability and recovered its performance in energy harvesting and powering small wearable electronic devices after 300 times complete bifurcation and self-healing at room temperature.

Furthermore, Wang et al. designed an ultra-stretchable and fast self-healing ionic hydrogel based on PVA/PEI/LiCl (Figure 10d), which acts as a current collector in the active muscle sensor based on a single electrode mode TENG (TSAS) [205]. The ionic hydrogel exhibited superior self-healing ability at room temperature, achieved complete recovery within 30 min, and further afforded more than 2000% tensile deformation without breaking. At the same time, the conductivity recovered 98.57% within 4 min after bringing damaged samples together and maintained over 96% healing efficiency after 10 times extreme damage. The as-fabricated TSAS performed outstanding sensing properties with a detection limit of 0.2 mN, a response time of 1 ms, and cycle stability of up to 10^5. In the in vitro experiments, the authors set up a quantitative relationship between muscle force and output of TSAS, and they developed a real-time muscle function monitoring system through combining TSAS with wireless transmission and visualization terminal.

Self-healing TENG plays an important role in improving the service life and safety of the IMDs [206–209]. The 2D and 3D modular assembly of the IMDs, with the help of self-healing ability, greatly simplifies the tedious fabrication process [210,211]. However, the current self-healing TENGs mainly focus on wearable electronic devices and remain blank for implantation scenarios [212–215]. The self-healing efficiency in liquid environments, compatibility with other components, and the development of fully self-healing systems are major challenges affecting in vivo applications [216–218]. Therefore, optimizing the self-healing performance in vivo and rationally designing the self-healing devices' structure are feasible solutions for the expanding application scenarios of self-healing TENGs in vivo.

7. Discussion and Challenges

Despite implantable TENGs having made significant progress in self-powered sensors, in vivo energy harvesters, and direct electrical stimulators, aided by improved materials, structural design, performance, and applications, some unsolved challenges still require continuous attention. Here, we discuss the optimization directions of IMDs and their prospects mainly from three aspects of performance, advanced functional materials and the close-looped IMDs system.

7.1. Output Performance

Higher output performance is a key factor for IMDs. We summarized the output of TENG-based implanted energy harvesting devices [102–104,219–222]. As shown in Figure 11, the electrical output is not only related to the performance of the device but also affected by the implanted animal, and implantation site. To increase the output, a variety of methods were proved to be effective, increasing the surface charge density as well as the contact area of two friction materials. Two frictional materials with a large difference in triboelectric sequence generate more triboelectric charges and transfer charges more efficiently [223–225]. The surface charge density could be increased by injecting charges into the friction layers using the corona discharge method [179]. Increasing the relative contact area between two friction layers proved to be another effective method, such as grinding the fractional surface by sandpaper, ICP to form pillar-shaped, pyramid-shaped, and origin-shaped surface structures [226]. The connection of multiple devices is considered a common and effective method to increase output [144]. In addition to the above-mentioned, matching external loads and effective energy management modules should also be considered for better output performance. Furthermore, considering the liquid environments in the human body, the electrostatic shielding effect should be reduced during the design of the IMDs.

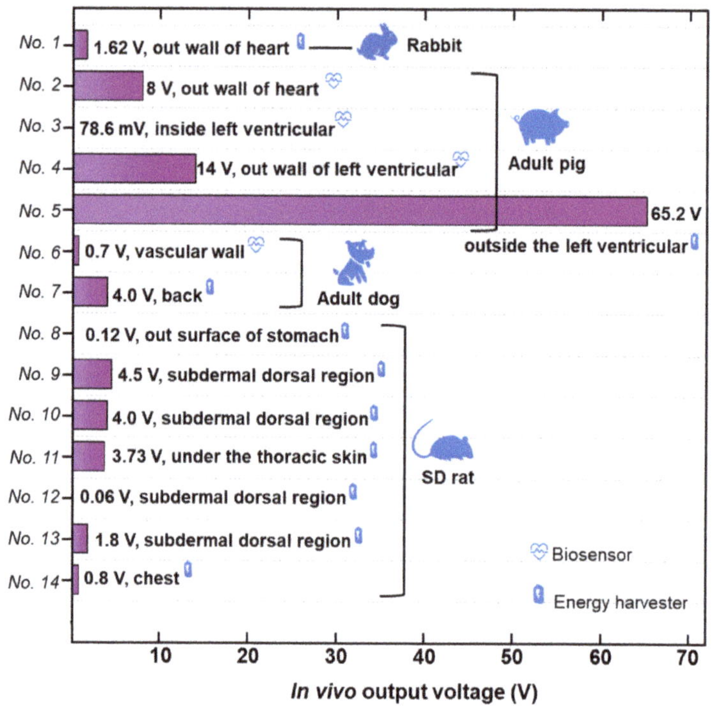

Figure 11. The in vivo output voltage of IMDs. Ref: No. 1-Shi et al., 2019 [92]. No. 2-Ma et al. 2016 [102]. No. 3-Liu et al. 2019 [103]. No. 4-Zheng et al. 2016 [3]. No. 5-Ouyang et al. 2019 [180]. No. 6-Ouyang et al. 2021 [93]. No. 7-Ryu et al. 2021 [5]. No. 8-Yao et al. 2018 [109]. No. 9-Jiang et al. 2018 [190]. No. 10-Zheng et al. 2016 [181]. No. 11-Zheng et al. 2014 [144]. No. 12-He et al. 2019 [58]. No. 13-Li et al. 2018 [191]. No. 14-Li et al. 2018 [105].

Another important point is the implantation site. For cardiac function monitoring, the IMD output varied when implanted on the right ventricular, the auricle of the left atrium, the cardiac base, the lateral wall of the left ventricular, and the inferior wall of the left ventricular, for different motion amplitude and volume of the interspace of these implantation sites. Therefore, a reasonable implantation site is suggested to optimize the output of IMDs and reduce the organism damage as much as possible.

7.2. Advanced Materials with Multifunctions

Another significant issue is the safety, reliability, and high efficiency of IMDs. Significances should be taken into consideration for the large differences among the different application scenarios [48,56,227,228]. In Table 2, we summarized the IMDs based on advanced TENG with various friction materials and made a comparison about their performance and applications. For IMDs working in the stomach, acid-resistant encapsulation is required to resist gastric acid erosion and better tissue-matched modulus because the stomach is constantly peristaltic [109]. Biosensors used for blood pressure monitoring must have anti-hemolytic properties to reduce the impact on the monitoring site [103]. Meanwhile, almost IMDs are generally subject to mechanical losses and accidental damage during operation, resulting in mechanical fatigue and material wear of the device, thereby affecting its service life. Others need surgical removal after service, which increases patient suffering and the risk of infection. Developing functional materials with better mechanical performance, biodegradable and self-healing ability, antibacterial ability, may provide promising solutions.

Table 2. Performance comparison of TENGs with different triboelectric materials.

Property	Friction Materials	Electrodes	Optimal Output	Size	Application	Ref.
Robust	PTFE/caption	Cu/Al	70 V, 0.8 μA, 10.5 nC	6 × 4 × 0.15 cm^3	Energy harvesting	[188]
Implantable	PTFE	Cu/Ti	70 V, 0.55 μA, 25 nC	—	In vivo cancer therapy	[184]
	PFA	Au/Cu	9.71 V, 427 μA	4 × 4 cm^2	Ultrasound driven energy harvesting	[106]
	PTFE	Cr/Cu	2.2 V, 0.1 μW	—	Breath energy harvesting	[105]
	PTFE	Cu/Al	14 V, 5 μW	2.5 × 1 × 0.15 cm^3	Cardiac monitoring	[3]
	caption	Cu/Al	22 V, 0.3 μA	—	Energy harvesting	[51]
	silicone	Au	16.8 mV, 54.4 nW	—	Ultrasound driven energy harvester	[155]
	PTFE	Au/Al	6.2 V	10 × 5 × 1 mm^3	Endocardial pressure sensor	[103]
	caption	Au/Al	12 V, 0.25 μA	1.2 × 1.2 cm^2	Powering prototype pacemaker	[144]
	caption	Au/Al	187 V, 19.5 μA,.80.2 nC	3.9 × 6.1 × 0.099 cm^3	Powering cardiac pacemaker	[179]
	[a] PVA-NH_2/PFA	Au/Cu	136 V, 2 μA/cm^3	15 mm in radius, 2.4 mm in height	Powering cardiac pacemaker	[5]
[b] BA-TENG	[c] cellulose/chitin/SF/RP/EW	Mg	55 V, 0.6 μA, 12 nC	—	Energy harvesting	[189]
	[d] PLA/C	Mg	3.2 V	—	Blood pressure sensor	[93]
	[e] PLGA/PCL/PLA	Au	28 V, 220 nA	1.2 × 1.2 cm^2	Tissue repair	[190]
[f] SH-TENG	SH-PCL/PVDF	Cu/AgNWs	800 V, 28 μA	3 × 3 cm^2	Energy harvesting	[197]
	Parylene-C/silicone	[f] SH-IH	78.44 V, 1.42 μA, 47.48 nC	3 × 5 cm^2	Muscle strength sensor	[205]
	vitrimer elastomer	Al/AgNWs	26 V, 350 nA, 10 nC	10 × 3 cm^2	Energy harvesting, pressure/tactile sensor	[200]
	silicone	slime	50 V, 6.5 μA/cm^2, 17 nC/cm^2	2 × 2 cm^2	Energy harvesting	[201]
	shape memory polymer	Al	100 V, 15 μA	4 × 2.5 cm^2	Energy harvesting	[204]
	SH-PUA/latex	[g] PUA-SF-LM	100 V, 4 μA/cm^2, 12 nC/cm^2	3 × 3 cm^2	Energy harvesting	[209]

[a] PVA-NH2: Amine-functionalized poly (vinyl alcohol); PFA: Perfluoroalkoxy. [b] BA: biodegradable and absorbable. [c] SF: silk fibroin; RP: rice paper; EW: egg white. [d] PLA/C: poly (lactic acid)–(chitosan 4%); [e] PLGA: poly (L-lactide-co-glycolide); PCL: poly (caprolactone). [f] SH: Self-healing; IH: ionic hydrogel. [g] SF: silver flakes; LM: liquid metal

7.3. Close-Looped IMDs System

A complete close-looped IMDs system that starts with accurate real-time physiological information collection and gives corresponding feedback and guidance is growing as the main development demand of modern intelligent electronic medicine [227]. Various health monitoring devices and treatment tools emerge in an endless stream but mainly work independently as active biosensors, power sources, and other functional units [22]. On the other hand, the development of self-powered technology has made IMDs break through the limitation of service life and realize long-term real-time monitoring/treatment of the human body [28]. With the gradually reduced power consumption of electronic devices and the continued development of advanced integration technologies, it is possible to develop a self-driven close-looped bioelectronic system with the human body as the energy source to power the continuous working of the whole system [48,78].

In a close-looped IMDs system, the energy harvesting units collect the various types of bioenergy from the human body and convert them into electrical energy, which is further rectified by the power management module and then stored in the power storage unit or directly supplied to other energy-consuming units. Without additional energy supply, the self-powered sensors perceive the real-time physiological states of the human body and deliver the sensing dates to the processing unit through a wireless transmission module.

Once the dates are diagnosed as anomalous, instructions from the controller will be sent to the relevant downstream functional units so as to perform feedback adjustment on the corresponding tissues, organs, or nerves of the human body until the physiological states are back to normal. With the help of the close-looped IMDs system, the physiological state of the human body can be automatically adjusted to the optimal state, and when a disease occurs, the physiological system can be actively and intelligently repaired.

8. Conclusions

Biomedical systems and advanced nanotechnology are sparking a new revolution in the field of healthcare. The continuous reduced power consumption of microchips and the continuously improved working efficiency of micro–nano energy harvesting devices have made fully self-charged implantable biomedical devices (IMDs) possible. Freeing the IMDs from the bulk batteries is a key step during this process. In this review, we summarized the recent progress of advanced IMDs driven by TENGs and reviewed the research progress of self-powered biosensors, bioenergy harvesting devices, and electrical stimulation therapy devices. For the self-charged IMDs, it is important to improve the output performance of the implantable TENGs and optimize the energy management module. The flexible packaging technology with high stability and high biosafety is required to ensure high performance in liquid environments. Besides, the good adaptability of these IMDs to the shape of human tissues or organs will improve sensor performance and energy harvesting efficiency.

The rapid development of advanced electronic materials and fabrication technology has ushered in a new opportunity to develop a fully self-charged IMD system. The integration of multi-functional components of the whole system is currently the main challenge. Advanced self-charged IMDs with self-healing properties enable a longer lifetime in devices in accidental damage, and also help achieve 3D expansion of multiple functional devices, which breaks the fabrication process limitations of traditional integrated systems [205,229]. At the same time, the proposal and development of the body net provides another distributed integrating method [157,230–233]. Each functional unit communicates through a local wireless network to realize a close-looped system integrating in suit sensing, data analysis, and treatment, which will make the implantable electronic medical system more intelligent.

Author Contributions: Conceptualization, C.L., C.W. and Q.S.; writing of the manuscript, C.L., C.W. and Q.S.; review and editing of the manuscript, C.L., C.W. and Q.S. All authors have read and agreed to the published version of the manuscript.

Funding: This work was supported by the following grants: the Advanced Research and Technology Innovation Centre (ARTIC) with grant No. R-261-518-009-720/A-0005947-20-00; NUS iHealthtech Grant: Smart Sensors and Artificial Intelligence (AI) for Health with grant No. R-263-501-017-133/A-0005180-16-00; the Ministry of Education (MOE) with grant No. R-263-000-F18-112/A-0009520-01-00; the National Key Research and Development Program of China with grant No. 2019YFB2004800 and the Project with grant No. R-2020-S-002.

Institutional Review Board Statement: Not applicable.

Informed Consent Statement: Not applicable.

Data Availability Statement: No new data were created or analyzed in this study. Data sharing is not applicable to this article.

Conflicts of Interest: The authors declare no conflict of interest.

References

1. Dong, L.; Closson, A.B.; Jin, C.; Nie, Y.; Cabe, A.; Escobedo, D.; Huang, S.; Trase, I.; Xu, Z.; Chen, Z.; et al. Multifunctional Pacemaker Lead for Cardiac Energy Harvesting and Pressure Sensing. *Adv. Healthc. Mater.* **2020**, *9*, e2000053. [CrossRef] [PubMed]
2. Han, W.B.; Ko, G.-J.; Jang, T.-M.; Hwang, S.-W. Materials, Devices, and Applications for Wearable and Implantable Electronics. *ACS Appl. Electron. Mater.* **2021**, *3*, 485–503. [CrossRef]

3. Zheng, Q.; Zhang, H.; Shi, B.; Xue, X.; Liu, Z.; Jin, Y.; Ma, Y.; Zou, Y.; Wang, X.; An, Z.; et al. In Vivo Self-Powered Wireless Cardiac Monitoring via Implantable Triboelectric Nanogenerator. *ACS Nano* **2016**, *10*, 6510–6518. [CrossRef] [PubMed]
4. Dong, L.; Closson, A.B.; Oglesby, M.; Escobedo, D.; Han, X.; Nie, Y.; Huang, S.; Feldman, M.D.; Chen, Z.; Zhang, J.X.J. In vivo cardiac power generation enabled by an integrated helical piezoelectric pacemaker lead. *Nano Energy* **2019**, *66*, 104085. [CrossRef]
5. Ryu, H.; Park, H.M.; Kim, M.K.; Kim, B.; Myoung, H.S.; Kim, T.Y.; Yoon, H.J.; Kwak, S.S.; Kim, J.; Hwang, T.H.; et al. Self-rechargeable cardiac pacemaker system with triboelectric nanogenerators. *Nat. Commun.* **2021**, *12*, 4374. [CrossRef]
6. Dagdeviren, C.; Yang, B.D.; Su, Y.; Tran, P.L.; Joe, P.; Anderson, E.; Xia, J.; Doraiswamy, V.; Dehdashti, B.; Feng, X.; et al. Conformal piezoelectric energy harvesting and storage from motions of the heart, lung, and diaphragm. *Proc. Natl. Acad. Sci. USA* **2014**, *111*, 1927–1932. [CrossRef]
7. Helmut, P.; Gillis, A.M.; Holbrook, R.; Hettrick, D.A. Accuracy of Atrial Tachyarrhythmia Detection in Implantable Devices. *Pacing Clin. Electrophysiol.* **2004**, *27*, 983–992. [CrossRef]
8. Zou, Y.; Liao, J.; Ouyang, H.; Jiang, D.; Zhao, C.; Li, Z.; Qu, X.; Liu, Z.; Fan, Y.; Shi, B.; et al. A flexible self-arched biosensor based on combination of piezoelectric and triboelectric effects. *Appl. Mater. Today* **2020**, *20*, 100699. [CrossRef]
9. Kotov, N.A.; Winter, J.O.; Clements, I.P.; Jan, E.; Timko, B.P.; Campidelli, S.; Pathak, S.; Mazzatenta, A.; Lieber, C.M.; Prato, M.; et al. Nanomaterials for Neural Interfaces. *Adv. Mater.* **2009**, *21*, 3970–4004. [CrossRef]
10. Warden, M.R.; Cardin, J.A.; Deisseroth, K. Optical neural interfaces. *Annu. Rev. Biomed. Eng.* **2014**, *16*, 103–129. [CrossRef]
11. Hwang, G.-T.; Kim, Y.; Lee, J.-H.; Oh, S.; Jeong, C.K.; Park, D.Y.; Ryu, J.; Kwon, H.; Lee, S.-G.; Joung, B.; et al. Self-powered deep brain stimulation via a flexible PIMNT energy harvester. *Energy Environ. Sci.* **2015**, *8*, 2677–2684. [CrossRef]
12. Zou, Y.; Tan, P.; Shi, B.; Ouyang, H.; Jiang, D.; Liu, Z.; Li, H.; Yu, M.; Wang, C.; Qu, X.; et al. A bionic stretchable nanogenerator for underwater sensing and energy harvesting. *Nat. Commun.* **2019**, *10*, 2695. [CrossRef] [PubMed]
13. Song, Y.; Min, J.; Gao, W. Wearable and Implantable Electronics: Moving toward Precision Therapy. *ACS Nano* **2019**, *13*, 12280–12286. [CrossRef] [PubMed]
14. Abidian, M.R.; Martin, D.C. Multifunctional Nanobiomaterials for Neural Interfaces. *Adv. Funct. Mater.* **2009**, *19*, 573–585. [CrossRef]
15. Lee, G.H.; Moon, H.; Kim, H.; Lee, G.H.; Kwon, W.; Yoo, S.; Myung, D.; Yun, S.H.; Bao, Z.; Hahn, S.K. Multifunctional materials for implantable and wearable photonic healthcare devices. *Nat. Rev. Mater.* **2020**, *5*, 149–165. [CrossRef]
16. Bock, D.C.; Marschilok, A.C.; Takeuchi, K.J.; Takeuchi, E.S. Batteries used to Power Implantable Biomedical Devices. *Electrochim. Acta.* **2012**, *84*, 155–164. [CrossRef]
17. Joung, Y.H. Development of implantable medical devices: From an engineering perspective. *Int. Neurourol. J.* **2013**, *17*, 98–106. [CrossRef]
18. Yao, G.; Jiang, D.; Li, J.; Kang, L.; Chen, S.; Long, Y.; Wang, Y.; Huang, P.; Lin, Y.; Cai, W.; et al. Self-Activated Electrical Stimulation for Effective Hair Regeneration via a Wearable Omnidirectional Pulse Generator. *ACS Nano* **2019**, *13*, 12345–12356. [CrossRef]
19. Shi, Q.; Zhang, Z.; He, T.; Sun, Z.; Wang, B.; Feng, Y.; Shan, X.; Salam, B.; Lee, C. Deep learning enabled smart mats as a scalable floor monitoring system. *Nat. Commun.* **2020**, *11*, 4609. [CrossRef]
20. Zhang, M.; Tang, Z.; Liu, X.; Van der Spiegel, J. Electronic neural interfaces. *Nat. Electron.* **2020**, *3*, 191–200. [CrossRef]
21. Wen, F.; Wang, H.; He, T.; Shi, Q.; Sun, Z.; Zhu, M.; Zhang, Z.; Cao, Z.; Dai, Y.; Zhang, T.; et al. Battery-free short-range self-powered wireless sensor network (SS-WSN) using TENG based direct sensory transmission (TDST) mechanism. *Nano Energy* **2020**, *67*, 104266. [CrossRef]
22. Ben Amar, A.; Kouki, A.B.; Cao, H. Power Approaches for Implantable Medical Devices. *Sensors* **2015**, *15*, 28889–28914. [CrossRef] [PubMed]
23. Jiang, D.; Shi, B.; Ouyang, H.; Fan, Y.; Wang, Z.L.; Li, Z. Emerging Implantable Energy Harvesters and Self-Powered Implantable Medical Electronics. *ACS Nano* **2020**, *14*, 6436–6448. [CrossRef] [PubMed]
24. Feng, H.; Zhao, C.; Tan, P.; Liu, R.; Chen, X.; Li, Z. Nanogenerator for Biomedical Applications. *Adv. Healthc. Mater.* **2018**, *7*, e1701298. [CrossRef]
25. Luo, R.; Dai, J.; Zhang, J.; Li, Z. Accelerated Skin Wound Healing by Electrical Stimulation. *Adv. Healthc. Mater.* **2021**, *10*, e2100557. [CrossRef]
26. Cha, S.N.; Seo, J.S.; Kim, S.M.; Kim, H.J.; Park, Y.J.; Kim, S.W.; Kim, J.M. Sound-driven piezoelectric nanowire-based nanogenerators. *Adv. Mater.* **2010**, *22*, 4726–4730. [CrossRef]
27. Jeong, C.K.; Park, K.-I.; Son, J.H.; Hwang, G.-T.; Lee, S.H.; Park, D.Y.; Lee, H.E.; Lee, H.K.; Byun, M.; Lee, K.J. Self-powered fully-flexible light-emitting system enabled by flexible energy harvester. *Energy Environ. Sci.* **2014**, *7*, 4035–4043. [CrossRef]
28. Dagdeviren, C.; Joe, P.; Tuzman, O.L.; Park, K.-I.; Lee, K.J.; Shi, Y.; Huang, Y.; Rogers, J.A. Recent progress in flexible and stretchable piezoelectric devices for mechanical energy harvesting, sensing and actuation. *Extreme Mech. Lett.* **2016**, *9*, 269–281. [CrossRef]
29. Zou, Y.; Bo, L.; Li, Z. Recent progress in human body energy harvesting for smart bioelectronic system. *Fundam. Res.* **2021**, *1*, 364–382. [CrossRef]
30. Jiang, L.; Yang, Y.; Chen, R.; Lu, G.; Li, R.; Li, D.; Humayun, M.S.; Shung, K.K.; Zhu, J.; Chen, Y.; et al. Flexible piezoelectric ultrasonic energy harvester array for bio-implantable wireless generator. *Nano Energy* **2019**, *56*, 216–224. [CrossRef]
31. Song, Y.; Wang, N.; Hu, C.; Wang, Z.L.; Yang, Y. Soft triboelectric nanogenerators for mechanical energy scavenging and self-powered sensors. *Nano Energy* **2021**, *84*, 105919. [CrossRef]

32. Park, K.I.; Xu, S.; Liu, Y.; Hwang, G.T.; Kang, S.J.; Wang, Z.L.; Lee, K.J. Piezoelectric BaTiO(3) thin film nanogenerator on plastic substrates. *Nano Lett* **2010**, *10*, 4939–4943. [CrossRef] [PubMed]
33. Rawy, K.; Sharma, R.; Yoon, H.-J.; Khan, U.; Kim, S.-W.; Kim, T.T.-H. A triboelectric nanogenerator energy harvesting system based on load-aware control for input power from 2.4 μW to 15.6 μW. *Nano Energy* **2020**, *74*, 104839. [CrossRef]
34. Liu, Z.; Li, H.; Shi, B.; Fan, Y.; Wang, Z.L.; Li, Z. Wearable and Implantable Triboelectric Nanogenerators. *Adv. Funct. Mater.* **2019**, *29*, 1808820. [CrossRef]
35. Chou, N.; Shin, H.; Kim, K.; Chae, U.; Jang, M.; Jeong, U.J.; Hwang, K.S.; Yi, B.; Lee, S.E.; Woo, J.; et al. A Multimodal Multi-Shank Fluorescence Neural Probe for Cell-Type-Specific Electrophysiology in Multiple Regions across a Neural Circuit. *Adv. Sci.* **2022**, *9*, e2103564. [CrossRef]
36. Tat, T.; Libanori, A.; Au, C.; Yau, A.; Chen, J. Advances in triboelectric nanogenerators for biomedical sensing. *Biosens. Bioelectron.* **2021**, *171*, 112714. [CrossRef]
37. Zhao, L.; Li, H.; Meng, J.; Li, Z. The recent advances in self-powered medical information sensors. *InfoMat* **2019**, *2*, 212–234. [CrossRef]
38. Yoon, H.-J.; Kim, S.-W. Nanogenerators to Power Implantable Medical Systems. *Joule* **2020**, *4*, 1398–1407. [CrossRef]
39. Song, P.; Yang, G.; Lang, T.; Yong, K.-T. Nanogenerators for wearable bioelectronics and biodevices. *J. Phys. D Appl. Phys.* **2019**, *52*, 023002. [CrossRef]
40. Zhao, J.; Ghannam, R.; Htet, K.O.; Liu, Y.; Law, M.K.; Roy, V.A.L.; Michel, B.; Imran, M.A.; Heidari, H. Self-Powered Implantable Medical Devices: Photovoltaic Energy Harvesting Review. *Adv. Healthc. Mater.* **2020**, *9*, e2000779. [CrossRef]
41. Park, K.I.; Son, J.H.; Hwang, G.T.; Jeong, C.K.; Ryu, J.; Koo, M.; Choi, I.; Lee, S.H.; Byun, M.; Wang, Z.L.; et al. Highly-efficient, flexible piezoelectric PZT thin film nanogenerator on plastic substrates. *Adv. Mater.* **2014**, *26*, 2514–2520. [CrossRef] [PubMed]
42. He, T.; Guo, X.; Lee, C. Flourishing energy harvesters for future body sensor network: From single to multiple energy sources. *iScience* **2021**, *24*, 101934. [CrossRef] [PubMed]
43. Zhang, T.; Yang, T.; Zhang, M.; Bowen, C.R.; Yang, Y. Recent Progress in Hybridized Nanogenerators for Energy Scavenging. *iScience* **2020**, *23*, 101689. [CrossRef] [PubMed]
44. Baek, C.; Yun, J.H.; Wang, J.E.; Jeong, C.K.; Lee, K.J.; Park, K.I.; Kim, D.K. A flexible energy harvester based on a lead-free and piezoelectric BCTZ nanoparticle-polymer composite. *Nanoscale* **2016**, *8*, 17632–17638. [CrossRef] [PubMed]
45. Zhao, L.; Li, H.; Meng, J.; Wang, A.C.; Tan, P.; Zou, Y.; Yuan, Z.; Lu, J.; Pan, C.; Fan, Y.; et al. Reversible Conversion between Schottky and Ohmic Contacts for Highly Sensitive, Multifunctional Biosensors. *Adv. Funct. Mater.* **2019**, *30*, 1907999. [CrossRef]
46. Chang, Y.; Xu, S.; Dong, B.; Wei, J.; Le, X.; Ma, Y.; Zhou, G.; Lee, C. Development of triboelectric-enabled tunable Fabry-Pérot photonic-crystal-slab filter towards wearable mid-infrared computational spectrometer. *Nano Energy* **2021**, *89*, 106446. [CrossRef]
47. Lee, D.-M.; Rubab, N.; Hyun, I.; Kang, W.; Kim, Y.-J.; Kang, M.; Choi, B.O.; Kim, S.-W. Ultrasound-mediated triboelectric nanogenerator for powering on-demand transient electronics. *Sci. Adv.* **2022**, *8*, eabl8423. [CrossRef]
48. Hwang, G.T.; Byun, M.; Jeong, C.K.; Lee, K.J. Flexible piezoelectric thin-film energy harvesters and nanosensors for biomedical applications. *Adv. Healthc. Mater.* **2015**, *4*, 646–658. [CrossRef]
49. Li, Z.; Zhu, G.; Yang, R.; Wang, A.C.; Wang, Z.L. Muscle-driven in vivo nanogenerator. *Adv. Mater.* **2010**, *22*, 2534–2537. [CrossRef]
50. Liao, J.; Zou, Y.; Jiang, D.; Liu, Z.; Qu, X.; Li, Z.; Liu, R.; Fan, Y.; Shi, B.; Li, Z.; et al. Nestable arched triboelectric nanogenerator for large deflection biomechanical sensing and energy harvesting. *Nano Energy* **2020**, *69*, 104417. [CrossRef]
51. Li, H.; Zhang, X.; Zhao, L.; Jiang, D.; Xu, L.; Liu, Z.; Wu, Y.; Hu, K.; Zhang, M.R.; Wang, J.; et al. A Hybrid Biofuel and Triboelectric Nanogenerator for Bioenergy Harvesting. *Nanomicro. Lett.* **2020**, *12*, 50. [CrossRef] [PubMed]
52. Zhong, J.; Zhong, Q.; Fan, F.; Zhang, Y.; Wang, S.; Hu, B.; Wang, Z.L.; Zhou, J. Finger typing driven triboelectric nanogenerator and its use for instantaneously lighting up LEDs. *Nano Energy* **2013**, *2*, 491–497. [CrossRef]
53. Tan, P.; Zheng, Q.; Zou, Y.; Shi, B.; Jiang, D.; Qu, X.; Ouyang, H.; Zhao, C.; Cao, Y.; Fan, Y.; et al. A Battery-Like Self-Charge Universal Module for Motional Energy Harvest. *Adv. Energy Mater.* **2019**, *9*, 1901875. [CrossRef]
54. Wang, Z.L. Triboelectric Nanogenerator (TENG)-Sparking an Energy and Sensor Revolution. *Adv. Energy Mater.* **2020**, *10*, 2000137. [CrossRef]
55. Yang, Y.; Xu, L.; Jiang, D.; Chen, B.Z.; Luo, R.; Liu, Z.; Qu, X.; Wang, C.; Shan, Y.; Cui, Y.; et al. Self-Powered Controllable Transdermal Drug Delivery System. *Adv. Funct. Mater.* **2021**, *31*, 2104092. [CrossRef]
56. Tang, W.; Tian, J.; Zheng, Q.; Yan, L.; Wang, J.; Li, Z.; Wang, Z.L. Implantable Self-Powered Low-Level Laser Cure System for Mouse Embryonic Osteoblasts' Proliferation and Differentiation. *ACS Nano* **2015**, *9*, 7867–7873. [CrossRef]
57. Yu, A.; Zhu, Y.; Wang, W.; Zhai, J. Progress in Triboelectric Materials: Toward High Performance and Widespread Applications. *Adv. Funct. Mater.* **2019**, *29*, 1900098. [CrossRef]
58. He, T.; Shi, Q.; Wang, H.; Wen, F.; Chen, T.; Ouyang, J.; Lee, C. Beyond energy harvesting—multi-functional triboelectric nanosensors on a textile. *Nano Energy* **2019**, *57*, 338–352. [CrossRef]
59. Zhu, M.; Sun, Z.; Chen, T.; Lee, C. Low cost exoskeleton manipulator using bidirectional triboelectric sensors enhanced multiple degree of freedom sensory system. *Nat. Commun.* **2021**, *12*, 2692. [CrossRef]
60. Liu, Z.; Nie, J.; Miao, B.; Li, J.; Cui, Y.; Wang, S.; Zhang, X.; Zhao, G.; Deng, Y.; Wu, Y.; et al. Self-Powered Intracellular Drug Delivery by a Biomechanical Energy-Driven Triboelectric Nanogenerator. *Adv. Mater.* **2019**, *31*, e1807795. [CrossRef]
61. Jiang, D.; Shi, B.; Ouyang, H.; Fan, Y.; Wang, Z.L.; Chen, Z.-M.; Li, Z. A 25-year bibliometric study of implantable energy harvesters and self-powered implantable medical electronics researches. *Mater. Today Energy* **2020**, *16*, 100386. [CrossRef]

62. Qu, X.; Liu, Y.; Liu, Z.; Li, Z. Assistive devices for the people with disabilities enabled by triboelectric nanogenerators. *J. Phys. Mater.* **2021**, *4*, 034015. [CrossRef]
63. Li, L.; Xiang, H.; Xiong, Y.; Zhao, H.; Bai, Y.; Wang, S.; Sun, F.; Hao, M.; Liu, L.; Li, T.; et al. Ultrastretchable Fiber Sensor with High Sensitivity in Whole Workable Range for Wearable Electronics and Implantable Medicine. *Adv. Sci.* **2018**, *5*, 1800558. [CrossRef] [PubMed]
64. Zhang, Q.; Jin, T.; Cai, J.; Xu, L.; He, T.; Wang, T.; Tian, Y.; Li, L.; Peng, Y.; Lee, C. Wearable Triboelectric Sensors Enabled Gait Analysis and Waist Motion Capture for IoT-Based Smart Healthcare Applications. *Adv. Sci.* **2021**, *9*, e2103694. [CrossRef]
65. Dhakar, L.; Pitchappa, P.; Tay, F.E.H.; Lee, C. An intelligent skin based self-powered finger motion sensor integrated with triboelectric nanogenerator. *Nano Energy* **2016**, *19*, 532–540. [CrossRef]
66. Dudem, B.; Mule, A.R.; Patnam, H.R.; Yu, J.S. Wearable and durable triboelectric nanogenerators via polyaniline coated cotton textiles as a movement sensor and self-powered system. *Nano Energy* **2019**, *55*, 305–315. [CrossRef]
67. He, T.; Sun, Z.; Shi, Q.; Zhu, M.; Anaya, D.V.; Xu, M.; Chen, T.; Yuce, M.R.; Thean, A.V.-Y.; Lee, C. Self-powered glove-based intuitive interface for diversified control applications in real/cyber space. *Nano Energy* **2019**, *58*, 641–651. [CrossRef]
68. Song, Y.; Wang, N.; Wang, Y.; Zhang, R.; Olin, H.; Yang, Y. Direct Current Triboelectric Nanogenerators. *Adv. Energy Mater.* **2020**, *10*, 2002756. [CrossRef]
69. Wen, F.; Sun, Z.; He, T.; Shi, Q.; Zhu, M.; Zhang, Z.; Li, L.; Zhang, T.; Lee, C. Machine Learning Glove Using Self-Powered Conductive Superhydrophobic Triboelectric Textile for Gesture Recognition in VR/AR Applications. *Adv. Sci.* **2020**, *7*, 2000261. [CrossRef]
70. Arab Hassani, F.; Shi, Q.; Wen, F.; He, T.; Haroun, A.; Yang, Y.; Feng, Y.; Lee, C. Smart materials for smart healthcare– moving from sensors and actuators to self-sustained nanoenergy nanosystems. *Smart Mater. Medic.* **2020**, *1*, 92–124. [CrossRef]
71. Zhu, M.; Yi, Z.; Yang, B.; Lee, C. Making use of nanoenergy from human—Nanogenerator and self-powered sensor enabled sustainable wireless IoT sensory systems. *Nano Today* **2021**, *36*, 101016. [CrossRef]
72. Shi, Q.; Dong, B.; He, T.; Sun, Z.; Zhu, J.; Zhang, Z.; Lee, C. Progress in wearable electronics/photonics—Moving toward the era of artificial intelligence and internet of things. *InfoMat.* **2020**, *2*, 1131–1162. [CrossRef]
73. Liu, Z.; Zheng, Q.; Shi, Y.; Xu, L.; Zou, Y.; Jiang, D.; Shi, B.; Qu, X.; Li, H.; Ouyang, H.; et al. Flexible and stretchable dual mode nanogenerator for rehabilitation monitoring and information interaction. *J. Mater. Chem. B* **2020**, *8*, 3647–3654. [CrossRef]
74. Sun, Z.; Zhu, M.; Lee, C. Progress in the Triboelectric Human–Machine Interfaces (HMIs)-Moving from Smart Gloves to AI/Haptic Enabled HMI in the 5G/IoT Era. *Nanoenergy Adv.* **2021**, *1*, 81–121. [CrossRef]
75. Zhang, Z.; He, T., Zhu, M.; Sun, Z.; Shi, Q.; Zhu, J.; Dong, B.; Yuce, M.R.; Lee, C. Deep learning-enabled triboelectric smart socks for IoT-based gait analysis and VR applications. *Npj Flex. Electron.* **2020**, *4*, 1–12. [CrossRef]
76. Wu, T.; Redoute, J.-M.; Yuce, M.R. A Wireless Implantable Sensor Design With Subcutaneous Energy Harvesting for Long-Term IoT Healthcare Applications. *IEEE Access* **2018**, *6*, 35801–35808. [CrossRef]
77. Jeong, C.K.; Cho, S.B.; Han, J.H.; Park, D.Y.; Yang, S.; Park, K.-I.; Ryu, J.; Sohn, H.; Chung, Y.-C.; Lee, K.J. Flexible highly-effective energy harvester via crystallographic and computational control of nanointerfacial morphotropic piezoelectric thin film. *Nano Res.* **2016**, *10*, 437–455. [CrossRef]
78. Liu, L.; Guo, X.; Liu, W.; Lee, C. Recent Progress in the Energy Harvesting Technology-From Self-Powered Sensors to Self-Sustained IoT, and New Applications. *Nanomaterials* **2021**, *11*, 2975. [CrossRef]
79. Liu, L.; Shi, Q.; Ho, J.S.; Lee, C. Study of thin film blue energy harvester based on triboelectric nanogenerator and seashore IoT applications. *Nano Energy* **2019**, *66*, 104167. [CrossRef]
80. Liu, S.; Wang, H.; He, T.; Dong, S.; Lee, C. Switchable textile-triboelectric nanogenerators (S-TENGs) for continuous profile sensing application without environmental interferences. *Nano Energy* **2020**, *69*, 104462. [CrossRef]
81. Wang, J.; He, T.; Lee, C. Development of neural interfaces and energy harvesters towards self-powered implantable systems for healthcare monitoring and rehabilitation purposes. *Nano Energy* **2019**, *65*, 104039. [CrossRef]
82. Dong, B.; Yang, Y.; Shi, Q.; Xu, S.; Sun, Z.; Zhu, S.; Zhang, Z.; Kwong, D.L.; Zhou, G.; Ang, K.W.; et al. Wearable Triboelectric-Human-Machine Interface (THMI) Using Robust Nanophotonic Readout. *ACS Nano* **2020**, *14*, 8915–8930. [CrossRef] [PubMed]
83. Bai, Z.; He, T.; Zhang, Z.; Xu, Y.; Zhang, Z.; Shi, Q.; Yang, Y.; Zhou, B.; Zhu, M.; Guo, J.; et al. Constructing highly tribopositive elastic yarn through interfacial design and assembly for efficient energy harvesting and human-interactive sensing. *Nano Energy* **2022**, *94*, 106956. [CrossRef]
84. Zhu, M.; He, T.; Lee, C. Technologies toward next generation human machine interfaces: From machine learning enhanced tactile sensing to neuromorphic sensory systems. *Appl. Phys. Rev.* **2020**, *7*, 031305. [CrossRef]
85. Shi, Q.; Lee, C. Self-Powered Bio-Inspired Spider-Net-Coding Interface Using Single-Electrode Triboelectric Nanogenerator. *Adv. Sci.* **2019**, *6*, 1900617. [CrossRef]
86. Dong, K.; Hu, Y.; Yang, J.; Kim, S.-W.; Hu, W.; Wang, Z.L. Smart textile triboelectric nanogenerators: Current status and perspectives. *MRS Bull.* **2021**, *46*, 512–521. [CrossRef]
87. Fang, Y.; Tang, T.; Li, Y.; Hou, C.; Wen, F.; Yang, Z.; Chen, T.; Sun, L.; Liu, H.; Lee, C. A high-performance triboelectric-electromagnetic hybrid wind energy harvester based on rotational tapered rollers aiming at outdoor IoT applications. *iScience* **2021**, *24*, 102300. [CrossRef]
88. Dong, B.; Shi, Q.; Yang, Y.; Wen, F.; Zhang, Z.; Lee, C. Technology evolution from self-powered sensors to AIoT enabled smart homes. *Nano Energy* **2021**, *79*, 105414. [CrossRef]

89. Zhu, M.; Shi, Q.; He, T.; Yi, Z.; Ma, Y.; Yang, B.; Chen, T.; Lee, C. Self-Powered and Self-Functional Cotton Sock Using Piezoelectric and Triboelectric Hybrid Mechanism for Healthcare and Sports Monitoring. *ACS Nano* **2019**, *13*, 1940–1952. [CrossRef]
90. Chen, T.; Shi, Q.; Zhu, M.; He, T.; Sun, L.; Yang, L.; Lee, C. Triboelectric Self-Powered Wearable Flexible Patch as 3D Motion Control Interface for Robotic Manipulator. *ACS Nano* **2018**, *12*, 11561–11571. [CrossRef]
91. Liu, Z.; Xu, L.; Zheng, Q.; Kang, Y.; Shi, B.; Jiang, D.; Li, H.; Qu, X.; Fan, Y.; Wang, Z.L.; et al. Human Motion Driven Self-Powered Photodynamic System for Long-Term Autonomous Cancer Therapy. *ACS Nano* **2020**, *14*, 8074–8083. [CrossRef] [PubMed]
92. Shi, B.; Liu, Z.; Zheng, Q.; Meng, J.; Ouyang, H.; Zou, Y.; Jiang, D.; Qu, X.; Yu, M.; Zhao, L.; et al. Body-Integrated Self-Powered System for Wearable and Implantable Applications. *ACS Nano* **2019**, *13*, 6017–6024. [CrossRef] [PubMed]
93. Ouyang, H.; Li, Z.; Gu, M.; Hu, Y.; Xu, L.; Jiang, D.; Cheng, S.; Zou, Y.; Deng, Y.; Shi, B.; et al. A Bioresorbable Dynamic Pressure Sensor for Cardiovascular Postoperative Care. *Adv. Mater.* **2021**, *33*, e2102302. [CrossRef] [PubMed]
94. Zou, Y.; Gai, Y.; Tan, P.; Jiang, D.; Qu, X.; Xue, J.; Ouyang, H.; Shi, B.; Li, L.; Luo, D.; et al. Stretchable graded multichannel self-powered respiratory sensor inspired by shark gill. *Fundam. Res.* **2022**, accepted. [CrossRef]
95. Shi, Q.; Wang, H.; Wang, T.; Lee, C. Self-powered liquid triboelectric microfluidic sensor for pressure sensing and finger motion monitoring applications. *Nano Energy* **2016**, *30*, 450–459. [CrossRef]
96. Sun, C.; Shi, Q.; Hasan, D.; Yazici, M.S.; Zhu, M.; Ma, Y.; Dong, B.; Liu, Y.; Lee, C. Self-powered multifunctional monitoring system using hybrid integrated triboelectric nanogenerators and piezoelectric microsensors. *Nano Energy* **2019**, *58*, 612–623. [CrossRef]
97. Vera Anaya, D.; He, T.; Lee, C.; Yuce, M.R. Self-powered eye motion sensor based on triboelectric interaction and near-field electrostatic induction for wearable assistive technologies. *Nano Energy* **2020**, *72*, 104675. [CrossRef]
98. Haroun, A.; Le, X.; Gao, S.; Dong, B.; He, T.; Zhang, Z.; Wen, F.; Xu, S.; Lee, C. Progress in micro/nano sensors and nanoenergy for future AIoT-based smart home applications. *Nano Express* **2021**, *2*, 022005. [CrossRef]
99. Wang, S.; He, M.; Weng, B.; Gan, L.; Zhao, Y.; Li, N.; Xie, Y. Stretchable and Wearable Triboelectric Nanogenerator Based on Kinesio Tape for Self-Powered Human Motion Sensing. *Nanomaterials* **2018**, *8*, 657. [CrossRef]
100. Yang, J.; Chen, J.; Su, Y.; Jing, Q.; Li, Z.; Yi, F.; Wen, X.; Wang, Z.; Wang, Z.L. Eardrum-inspired active sensors for self-powered cardiovascular system characterization and throat-attached anti-interference voice recognition. *Adv. Mater.* **2015**, *27*, 1316–1326. [CrossRef]
101. Wang, Z.L. Energy Harvesting Using Piezoelectric Nanowires-A Correspondence on "Energy Harvesting Using Nanowires?" by Alexe et al. *Adv. Mater.* **2009**, *21*, 1311–1315. [CrossRef]
102. Ma, Y.; Zheng, Q.; Liu, Y.; Shi, B.; Xue, X.; Ji, W.; Liu, Z.; Jin, Y.; Zou, Y.; An, Z.; et al. Self-Powered, One-Stop, and Multifunctional Implantable Triboelectric Active Sensor for Real-Time Biomedical Monitoring. *Nano Lett.* **2016**, *16*, 6042–6051. [CrossRef] [PubMed]
103. Liu, Z.; Ma, Y.; Ouyang, H.; Shi, B.; Li, N.; Jiang, D.; Xie, F.; Qu, D.; Zou, Y.; Huang, Y.; et al. Transcatheter Self-Powered Ultrasensitive Endocardial Pressure Sensor. *Adv. Func. Mater.* **2019**, *29*, 1807560. [CrossRef]
104. Arab Hassani, F.; Mogan, R.P.; Gammad, G.G.L.; Wang, H.; Yen, S.C.; Thakor, N.V.; Lee, C. Toward Self-Control Systems for Neurogenic Underactive Bladder: A Triboelectric Nanogenerator Sensor Integrated with a Bistable Micro-Actuator. *ACS Nano* **2018**, *12*, 3487–3501. [CrossRef] [PubMed]
105. Li, J.; Kang, L.; Long, Y.; Wei, H.; Yu, Y.; Wang, Y.; Ferreira, C.A.; Yao, G.; Zhang, Z.; Carlos, C.; et al. Implanted Battery-Free Direct-Current Micro-Power Supply from *In Vivo* Breath Energy Harvesting. *ACS Appl. Mater. Interfaces* **2018**, *10*, 42030–42038. [CrossRef]
106. Hinchet, R.; Yoon, H.-J.; Ryu, H.; Kim, M.-K.; Choi, E.-K.; Kim, S.-W. Transcutaneous ultrasound energy harvesting using capacitive triboelectric technology. *Science* **2019**, *365*, 491–494. [CrossRef]
107. Lee, S.; Wang, H.; Shi, Q.; Dhakar, L.; Wang, J.; Thakor, N.V.; Yen, S.-C.; Lee, C. Development of battery-free neural interface and modulated control of tibialis anterior muscle via common peroneal nerve based on triboelectric nanogenerators (TENGs). *Nano Energy* **2017**, *33*, 1–11. [CrossRef]
108. Wang, H.; Wang, J.; He, T.; Li, Z.; Lee, C. Direct muscle stimulation using diode-amplified triboelectric nanogenerators (TENGs). *Nano Energy* **2019**, *63*, 103844. [CrossRef]
109. Yao, G.; Kang, L.; Li, J.; Long, Y.; Wei, H.; Ferreira, C.A.; Jeffery, J.J.; Lin, Y.; Cai, W.; Wang, X. Effective weight control via an implanted self-powered vagus nerve stimulation device. *Nat. Commun.* **2018**, *9*, 5349. [CrossRef]
110. Zhang, W.; Liu, Q.; Chao, S.; Liu, R.; Cui, X.; Sun, Y.; Ouyang, H.; Li, Z. Ultrathin Stretchable Triboelectric Nanogenerators Improved by Postcharging Electrode Material. *ACS Appl. Mater. Interfaces* **2021**, *13*, 42966–42976. [CrossRef]
111. Bai, Z.; Xu, Y.; Lee, C.; Guo, J. Autonomously Adhesive, Stretchable, and Transparent Solid-State Polyionic Triboelectric Patch for Wearable Power Source and Tactile Sensor. *Adv. Funct. Mater.* **2021**, *31*, 2104365. [CrossRef]
112. Wen, Z.; Yang, Y.; Sun, N.; Li, G.; Liu, Y.; Chen, C.; Shi, J.; Xie, L.; Jiang, H.; Bao, D.; et al. A Wrinkled PEDOT:PSS Film Based Stretchable and Transparent Triboelectric Nanogenerator for Wearable Energy Harvesters and Active Motion Sensors. *Adv. Funct. Mater.* **2018**, *28*, 1803684. [CrossRef]
113. Zhou, H.; Zhang, Y.; Qiu, Y.; Wu, H.; Qin, W.; Liao, Y.; Yu, Q.; Cheng, H. Stretchable piezoelectric energy harvesters and self-powered sensors for wearable and implantable devices. *Biosens. Bioelectron.* **2020**, *168*, 112569. [CrossRef] [PubMed]
114. Jiang, D.; Ouyang, H.; Shi, B.; Zou, Y.; Tan, P.; Qu, X.; Chao, S.; Xi, Y.; Zhao, C.; Fan, Y.; et al. A wearable noncontact free-rotating hybrid nanogenerator for self-powered electronics. *InfoMat* **2020**, *2*, 1191–1200. [CrossRef]

115. Wang, L.; He, T.; Zhang, Z.; Zhao, L.; Lee, C.; Luo, G.; Mao, Q.; Yang, P.; Lin, Q.; Li, X.; et al. Self-sustained autonomous wireless sensing based on a hybridized TENG and PEG vibration mechanism. *Nano Energy* **2021**, *80*, 105555. [CrossRef]
116. Wang, H.; Xiang, Z.; Giorgia, P.; Mu, X.; Yang, Y.; Wang, Z.L.; Lee, C. Triboelectric liquid volume sensor for self-powered lab-on-chip applications. *Nano Energy* **2016**, *23*, 80–88. [CrossRef]
117. Li, D.; Zhou, H.; Hui, X.; He, X.; Huang, H.; Zhang, J.; Mu, X.; Lee, C.; Yang, Y. Multifunctional Chemical Sensing Platform Based on Dual-Resonant Infrared Plasmonic Perfect Absorber for On-Chip Detection of Poly(ethyl cyanoacrylate). *Adv. Sci.* **2021**, *8*, e2101879. [CrossRef]
118. Xu, C.; Zi, Y.; Wang, A.C.; Zou, H.; Dai, Y.; He, X.; Wang, P.; Wang, Y.C.; Feng, P.; Li, D.; et al. On the Electron-Transfer Mechanism in the Contact-Electrification Effect. *Adv. Mater.* **2018**, *30*, e1706790. [CrossRef]
119. Niu, S.; Wang, S.; Lin, L.; Liu, Y.; Zhou, Y.S.; Hu, Y.; Wang, Z.L. Theoretical study of contact-mode triboelectric nanogenerators as an effective power source. *Energy Environm. Sci.* **2013**, *6*, 3576. [CrossRef]
120. Wu, P.; Xiao, A.; Zhao, Y.; Chen, F.; Ke, M.; Zhang, Q.; Zhang, J.; Shi, X.; He, X.; Chen, Y. An implantable and versatile piezoresistive sensor for the monitoring of human-machine interface interactions and the dynamical process of nerve repair. *Nanoscale* **2019**, *11*, 21103–21118. [CrossRef]
121. Wen, F.; Zhang, Z.; He, T.; Lee, C. AI enabled sign language recognition and VR space bidirectional communication using triboelectric smart glove. *Nat. Commun.* **2021**, *12*, 5378. [CrossRef] [PubMed]
122. Shi, Q.; Zhang, Z.; Chen, T.; Lee, C. Minimalist and multi-functional human machine interface (HMI) using a flexible wearable triboelectric patch. *Nano Energy* **2019**, *62*, 355–366. [CrossRef]
123. Wen, F.; He, T.; Liu, H.; Chen, H.-Y.; Zhang, T.; Lee, C. Advances in chemical sensing technology for enabling the next-generation self-sustainable integrated wearable system in the IoT era. *Nano Energy* **2020**, *78*, 105155. [CrossRef]
124. Wang, Z.L. On the first principle theory of nanogenerators from Maxwell's equations. *Nano Energy* **2020**, *68*, 104272. [CrossRef]
125. Chen, D.; Wang, D.; Yang, Y.; Huang, Q.; Zhu, S.; Zheng, Z. Self-Healing Materials for Next-Generation Energy Harvesting and Storage Devices. *Adv. Energy Mater.* **2017**, *7*, 1700890. [CrossRef]
126. Ershad, F.; Sim, K.; Thukral, A.; Zhang, Y.S.; Yu, C. Invited Article: Emerging soft bioelectronics for cardiac health diagnosis and treatment. *APL Mater.* **2019**, *7*, 031301. [CrossRef]
127. Cho, Y.; Park, J.; Lee, C.; Lee, S. Recent progress on peripheral neural interface technology towards bioelectronic medicine. *Bioelectron. Med.* **2020**, *6*, 23. [CrossRef]
128. Ganesh, R.S.; Yoon, H.J.; Kim, S.W. Recent trends of biocompatible triboelectric nanogenerators toward self-powered e-skin. *EcoMat* **2020**, *2*, e12065. [CrossRef]
129. Zhang, S.; Yen, S.-C.; Xiang, Z.; Liao, L.-D.; Kwong, D.-L.; Lee, C. Development of Silicon Probe With Acute Study on In Vivo Neural Recording and Implantation Behavior Monitored by Integrated Si-Nanowire Strain Sensors. *J. Microelectrom. Syst.* **2015**, *24*, 1303–1313. [CrossRef]
130. Slabov, V.; Kopyl, S.; Soares Dos Santos, M.P.; Kholkin, A.L. Natural and Eco-Friendly Materials for Triboelectric Energy Harvesting. *Nanomicro Lett.* **2020**, *12*, 42. [CrossRef]
131. Li, Z.; Zheng, Q.; Wang, Z.L.; Li, Z. Nanogenerator-Based Self-Powered Sensors for Wearable and Implantable Electronics. *Res. Wash D C* **2020**, *2020*, 8710686. [CrossRef] [PubMed]
132. Shi, Q.; Wang, T.; Lee, C. MEMS Based Broadband Piezoelectric Ultrasonic Energy Harvester (PUEH) for Enabling Self-Powered Implantable Biomedical Devices. *Sci. Rep.* **2016**, *6*, 24946. [CrossRef] [PubMed]
133. Lee, S.; Sheshadri, S.; Xiang, Z.; Delgado-Martinez, I.; Xue, N.; Sun, T.; Thakor, N.V.; Yen, S.-C.; Lee, C. Selective stimulation and neural recording on peripheral nerves using flexible split ring electrodes. *Sens. Actuators B Chem.* **2017**, *242*, 1165–1170. [CrossRef]
134. Niu, S.; Liu, Y.; Wang, S.; Lin, L.; Zhou, Y.S.; Hu, Y.; Wang, Z.L. Theory of Sliding-Mode Triboelectric Nanogenerators. *Adv. Mater.* **2013**, *25*, 6184–6193. [CrossRef]
135. Bai, Y.; Chen, S.; Wang, H.; Wang, E.; Kong, X.; Gai, Y.; Qu, X.; Li, Q.; Xue, S.; Guo, P.; et al. Chemical warfare agents decontamination via air mircoplasma excited by a triboelectric nanogenerator. *Nano Energy* **2022**, *95*, 106992. [CrossRef]
136. Feng, H.; Bai, Y.; Qiao, L.; Li, Z.; Wang, E.; Chao, S.; Qu, X.; Cao, Y.; Liu, Z.; Han, X.; et al. An Ultra-Simple Charge Supplementary Strategy for High Performance Rotary Triboelectric Nanogenerators. *Small* **2021**, *17*, e2101430. [CrossRef]
137. Niu, S.; Wang, S.; Liu, Y.; Zhou, Y.S.; Lin, L.; Hu, Y.; Pradel, K.C.; Wang, Z.L. A theoretical study of grating structured triboelectric nanogenerators. *Energy Environ. Sci.* **2014**, *7*, 2339–2349. [CrossRef]
138. Lin, Z.-H.; Cheng, G.; Lee, S.; Pradel, K.C.; Wang, Z.L. Harvesting Water Drop Energy by a Sequential Contact-Electrification and Electrostatic-Induction Process. *Adv. Mater.* **2014**, *26*, 4690–4696. [CrossRef]
139. Lin, L.; Xie, Y.; Niu, S.; Wang, S.; Yang, P.-K.; Wang, Z.L. Robust Triboelectric Nanogenerator Based on Rolling Electrification and Electrostatic Induction at an Instantaneous Energy Conversion Efficiency of ~55%. *ACS Nano* **2015**, *9*, 922–930. [CrossRef]
140. Jung, Y.H.; Hong, S.K.; Wang, H.S.; Han, J.H.; Pham, T.X.; Park, H.; Kim, J.; Kang, S.; Yoo, C.D.; Lee, K.J. Flexible Piezoelectric Acoustic Sensors and Machine Learning for Speech Processing. *Adv. Mater.* **2020**, *32*, e1904020. [CrossRef]
141. Nazempour, R.; Liu, C.; Chen, Y.; Ma, C.; Sheng, X. Performance evaluation of an implantable sensor for deep brain imaging: An analytical investigation. *Optical. Mater. Express* **2019**, *9*, 3729. [CrossRef]
142. Tang, G.; Shi, Q.; Zhang, Z.; He, T.; Sun, Z.; Lee, C. Hybridized wearable patch as a multi-parameter and multi-functional human-machine interface. *Nano Energy* **2021**, *81*, 105582. [CrossRef]

143. Gao, S.; He, T.; Zhang, Z.; Ao, H.; Jiang, H.; Lee, C. A Motion Capturing and Energy Harvesting Hybridized Lower-Limb System for Rehabilitation and Sports Applications. *Adv. Sci.* **2021**, *8*, e2101834. [CrossRef] [PubMed]
144. Zheng, Q.; Shi, B.; Fan, F.; Wang, X.; Yan, L.; Yuan, W.; Wang, S.; Liu, H.; Li, Z.; Wang, Z.L. In vivo powering of pacemaker by breathing-driven implanted triboelectric nanogenerator. *Adv. Mater.* **2014**, *26*, 5851–5856. [CrossRef]
145. Hossain, N.A.; Yamomo, G.G.; Willing, R.; Towfighian, S. Characterization of a Packaged Triboelectric Harvester under Simulated Gait Loading for Total Knee Replacement. *IEEE ASME Trans. Mechatron.* **2021**, *26*, 2967–2976. [CrossRef]
146. Chen, M.; Zhou, Y.; Lang, J.; Li, L.; Zhang, Y. Triboelectric nanogenerator and artificial intelligence to promote precision medicine for cancer. *Nano Energy* **2022**, *92*, 106783. [CrossRef]
147. Kim, D.H.; Shin, H.J.; Lee, H.; Jeong, C.K.; Park, H.; Hwang, G.-T.; Lee, H.-Y.; Joe, D.J.; Han, J.H.; Lee, S.H.; et al. In Vivo Self-Powered Wireless Transmission Using Biocompatible Flexible Energy Harvesters. *Adv. Funct. Mater.* **2017**, *27*, 1700341. [CrossRef]
148. Ibrahim, A.; Jain, M.; Salman, E.; Willing, R.; Towfighian, S. A smart knee implant using triboelectric energy harvesters. *Smart Mater. Struct.* **2019**, *28*, 025040. [CrossRef]
149. Sheng, H.; Zhang, X.; Liang, J.; Shao, M.; Xie, E.; Yu, C.; Lan, W. Recent Advances of Energy Solutions for Implantable Bioelectronics. *Adv. Healthc. Mater.* **2021**, *10*, e2100199. [CrossRef]
150. Lee, J.; Jang, J.; Song, Y.-K. A review on wireless powering schemes for implantable microsystems in neural engineering applications. *Biomed. Engin. Lett.* **2016**, *6*, 205–215. [CrossRef]
151. Maleki, T.; Cao, N.; Song, S.H.; Kao, C.; Ko, S.; Ziaie, B. An Ultrasonically Powered Implantable Micro-Oxygen Generator (IMOG). *IEEE Trans. Biomed. Eng.* **2011**, *58*, 3104–3111. [CrossRef] [PubMed]
152. Shi, Q.; Wang, T.; Kobayashi, T.; Lee, C. Investigation of geometric design in piezoelectric microelectromechanical systems diaphragms for ultrasonic energy harvesting. *Appl. Phys. Lett.* **2016**, *108*, 193902. [CrossRef]
153. Seo, D.; Neely, R.M.; Shen, K.; Singhal, U.; Alon, E.; Rabaey, J.M.; Carmena, J.M.; Maharbiz, M.M. Wireless Recording in the Peripheral Nervous System with Ultrasonic Neural Dust. *Neuron* **2016**, *91*, 529–539. [CrossRef] [PubMed]
154. Deng, W.; Libanori, A.; Xiao, X.; Fang, J.; Zhao, X.; Zhou, Y.; Chen, G.; Li, S.; Chen, J. Computational investigation of ultrasound induced electricity generation via a triboelectric nanogenerator. *Nano Energy* **2022**, *91*, 106656. [CrossRef]
155. Chen, C.; Wen, Z.; Shi, J.; Jian, X.; Li, P.; Yeow, J.T.W.; Sun, X. Micro triboelectric ultrasonic device for acoustic energy transfer and signal communication. *Nat. Commun.* **2020**, *11*, 4143. [CrossRef]
156. Choi, S.; Lee, H.; Ghaffari, R.; Hyeon, T.; Kim, D.-H. Recent Advances in Flexible and Stretchable Bio-Electronic Devices Integrated with Nanomaterials. *Adv. Mater.* **2016**, *28*, 4203–4218. [CrossRef]
157. Shin, H.; Kang, M.; Lee, S. Mechanism of peripheral nerve modulation and recent applications. *Int. J. Optomechatron.* **2021**, *15*, 182–198. [CrossRef]
158. Grill, W.M.; Norman, S.E.; Bellamkonda, R.V. Implanted neural interfaces: Biochallenges and engineered solutions. *Annu Rev. Biomed. Eng.* **2009**, *11*, 1–24. [CrossRef]
159. Song, K.I.; Seo, H.; Seong, D.; Kim, S.; Yu, K.J.; Kim, Y.C.; Kim, J.; Kwon, S.J.; Han, H.S.; Youn, I.; et al. Adaptive self-healing electronic epineurium for chronic bidirectional neural interfaces. *Nat. Commun.* **2020**, *11*, 4195. [CrossRef]
160. Johnson, M.I. *Transcutaneous Electrical Nerve Stimulation (TENS)*; eLS: Leeds, UK, 2012. [CrossRef]
161. Shi, Y.; Liu, R.; He, L.; Feng, H.; Li, Y.; Li, Z. Recent development of implantable and flexible nerve electrodes. *Smart Mater. Med.* **2020**, *1*, 131–147. [CrossRef]
162. Shan, Y.; Feng, H.; Li, Z. Electrical Stimulation for Nervous System Injury: Research Progress and Prospects. *Acta Phys.-Chim. Sin.* **2020**, *36*, 2005038. [CrossRef]
163. Lee, S.; Peh, W.Y.X.; Wang, J.; Yang, F.; Ho, J.S.; Thakor, N.V.; Yen, S.C.; Lee, C. Toward Bioelectronic Medicine-Neuromodulation of Small Peripheral Nerves Using Flexible Neural Clip. *Adv. Sci.* **2017**, *4*, 1700149. [CrossRef] [PubMed]
164. Wang, H.; Wang, J.; Thow, X.Y.; Lee, S.; Peh, W.Y.X.; Ng, K.A.; He, T.; Thakor, N.V.; Lee, C. Unveiling Stimulation Secrets of Electrical Excitation of Neural Tissue Using a Circuit Probability Theory. *Front. Comput. Neurosci.* **2020**, *14*, 50. [CrossRef] [PubMed]
165. Wang, J.; Thow, X.Y.; Wang, H.; Lee, S.; Voges, K.; Thakor, N.V.; Yen, S.C.; Lee, C. A Highly Selective 3D Spiked Ultraflexible Neural (SUN) Interface for Decoding Peripheral Nerve Sensory Information. *Adv. Healthc. Mater.* **2018**, *7*, 1700987. [CrossRef] [PubMed]
166. Wang, J.; Wang, H.; Lee, C. Mechanism and Applications of Electrical Stimulation Disturbance on Motoneuron Excitability Studied Using Flexible Intramuscular Electrode. *Adv. Biosyst.* **2019**, *3*, e1800281. [CrossRef]
167. Lee, S.; Shi, Q.; Lee, C. From flexible electronics technology in the era of IoT and artificial intelligence toward future implanted body sensor networks. *APL Mater.* **2019**, *7*, 031302. [CrossRef]
168. Wang, H.; Wu, T.; Zeng, Q.; Lee, C. A Review and Perspective for the Development of Triboelectric Nanogenerator (TENG)-Based Self-Powered Neuroprosthetics. *Micromachines* **2020**, *11*, 865. [CrossRef]
169. Zhang, X.-S.; Han, M.-D.; Wang, R.-X.; Meng, B.; Zhu, F.-Y.; Sun, X.-M.; Hu, W.; Wang, W.; Li, Z.-H.; Zhang, H.-X. High-performance triboelectric nanogenerator with enhanced energy density based on single-step fluorocarbon plasma treatment. *Nano Energy* **2014**, *4*, 123–131. [CrossRef]
170. Lee, S.; Wang, H.; Wang, J.; Shi, Q.; Yen, S.-C.; Thakor, N.V.; Lee, C. Battery-free neuromodulator for peripheral nerve direct stimulation. *Nano Energy* **2018**, *50*, 148–158. [CrossRef]

171. Lee, S.; Wang, H.; Xian Peh, W.Y.; He, T.; Yen, S.-C.; Thakor, N.V.; Lee, C. Mechano-neuromodulation of autonomic pelvic nerve for underactive bladder: A triboelectric neurostimulator integrated with flexible neural clip interface. *Nano Energy* **2019**, *60*, 449–456. [CrossRef]
172. Lee, S.; Lee, C. Toward advanced neural interfaces for the peripheral nervous system (PNS) and their future applications. *Curr. Opin. Biomed. Eng.* **2018**, *6*, 130–137. [CrossRef]
173. Dunkelberger, N.; Schearer, E.M.; O'Malley, M.K. A review of methods for achieving upper limb movement following spinal cord injury through hybrid muscle stimulation and robotic assistance. *Exp. Neurol.* **2020**, *328*, 113274. [CrossRef] [PubMed]
174. Pfeiffer, M.; Kröger, T.; Seifert, J.; Somaskantharajan, S.; Jahnich, L.; Steinblum, T.; Speckamp, J.; Navas Medrano, S. WONDER—Enhancing VR Training with Electrical Muscle Stimulation. In Proceedings of the 2019 CHI Conference on Human Factors in Computing Systems, (CHI 2019), Glasgow, UK, 4–9 May 2019; pp. 1–6. [CrossRef]
175. Phokaewvarangkul, O.; Vateekul, P.; Wichakam, I.; Anan, C.; Bhidayasiri, R. Using Machine Learning for Predicting the Best Outcomes With Electrical Muscle Stimulation for Tremors in Parkinson's Disease. *Front. Aging. Neurosci.* **2021**, *13*, 727654. [CrossRef] [PubMed]
176. Wang, J.; Wang, H.; He, T.; He, B.; Thakor, N.V.; Lee, C. Investigation of Low-Current Direct Stimulation for Rehabilitation Treatment Related to Muscle Function Loss Using Self-Powered TENG System. *Adv. Sci.* **2019**, *6*, 1900149. [CrossRef]
177. Wang, J.; Wang, H.; Thakor, N.V.; Lee, C. Self-Powered Direct Muscle Stimulation Using a Triboelectric Nanogenerator (TENG) Integrated with a Flexible Multiple-Channel Intramuscular Electrode. *ACS Nano* **2019**, *13*, 3589–3599. [CrossRef]
178. He, T.; Wang, H.; Wang, J.; Tian, X.; Wen, F.; Shi, Q.; Ho, J.S.; Lee, C. Self-Sustainable Wearable Textile Nano-Energy Nano-System (NENS) for Next-Generation Healthcare Applications. *Adv. Sci.* **2019**, *6*, 1901437. [CrossRef]
179. Ouyang, H.; Liu, Z.; Li, N.; Shi, B.; Zou, Y.; Xie, F.; Ma, Y.; Li, Z.; Li, H.; Zheng, Q.; et al. Symbiotic cardiac pacemaker. *Nat. Commun.* **2019**, *10*, 1821. [CrossRef]
180. Zheng, Q.; Zou, Y.; Zhang, Y.; Liu, Z.; Shi, B.; Wang, X.; Jin YOuyang HLi, Z.; Wang, Z.L. Biodegradable triboelectric nanogenerator as a life-time designed implantable power source. *Sci. Adv.* **2016**, *2*, e1501478. [CrossRef]
181. Liu, Z.; Liu, Y.; Li, H.; Wang, Z.L.; Li, Z. Biodegradable energy devices and applications. *Sci. Tech. Rev.* **2020**, *38*, 102–110. [CrossRef]
182. Shi, B.; Li, Z.; Fan, Y. Implantable Energy-Harvesting Devices. *Adv. Mater.* **2018**, *30*, e1801511. [CrossRef]
183. Ashammakhi, N.; Hernandez, A.L.; Unluturk, B.D.; Quintero, S.A.; Barros, N.R.; Hoque Apu, E.; Bin Shams, A.; Ostrovidov, S.; Li, J.; Contag, C.; et al. Biodegradable Implantable Sensors: Materials Design, Fabrication, and Applications. *Adv. Func. Mater.* **2021**, *31*, 2104149. [CrossRef]
184. Zhao, C.; Feng, H.; Zhang, L.; Li, Z.; Zou, Y.; Tan, P.; Ouyang, H.; Jiang, D.; Yu, M.; Wang, C.; et al. Highly Efficient *In Vivo* Cancer Therapy by an Implantable Magnet Triboelectric Nanogenerator. *Adv. Funct. Mater.* **2019**, *29*, 1808640. [CrossRef]
185. Chao, S.; Ouyang, H.; Jiang, D.; Fan, Y.; Li, Z. Triboelectric nanogenerator based on degradable materials. *EcoMat* **2020**, *3*, e12072. [CrossRef]
186. Cao, Y.; Uhrich, K.E. Biodegradable and biocompatible polymers for electronic applications: A review. *J. Bioact. Compat. Polym.* **2018**, *34*, 3–15. [CrossRef]
187. Sun, L.; Zhou, Z.; Zhong, J.; Shi, Z.; Mao, Y.; Li, H.; Cao, J.; Tao, T.H. Implantable, Degradable, Therapeutic Terahertz Metamaterial Devices. *Small* **2020**, *16*, e2000294. [CrossRef]
188. Zheng, Q.; Jin, Y.; Liu, Z.; Ouyang, H.; Li, H.; Shi, B.; Jiang, W.; Zhang, H.; Li, Z.; Wang, Z.L. Robust Multilayered Encapsulation for High-Performance Triboelectric Nanogenerator in Harsh Environment. *ACS Appl. Mater. Interfaces* **2016**, *8*, 26697–26703. [CrossRef]
189. Jiang, W.; Li, H.; Liu, Z.; Li, Z.; Tian, J.; Shi, B.; Zou, Y.; Ouyang, H.; Zhao, C.; Zhao, L.; et al. Fully Bioabsorbable Natural-Materials-Based Triboelectric Nanogenerators. *Adv. Mater.* **2018**, *30*, e1801895. [CrossRef]
190. Li, Z.; Feng, H.; Zheng, Q.; Li, H.; Zhao, C.; Ouyang, H.; Noreen, S.; Yu, M.; Su, F.; Liu, R.; et al. Photothermally tunable biodegradation of implantable triboelectric nanogenerators for tissue repairing. *Nano Energy* **2018**, *54*, 390–399. [CrossRef]
191. Li, C.; Guo, C.; Fitzpatrick, V.; Ibrahim, A.; Zwierstra, M.J.; Hanna, P.; Lechtig, A.; Nazarian, A.; Lin, S.J.; Kaplan, D.L. Design of biodegradable, implantable devices towards clinical translation. *Nat. Rev. Mater.* **2019**, *5*, 61–81. [CrossRef]
192. Curry, E.J.; Ke, K.; Chorsi, M.T.; Wrobel, K.S.; Miller, A.N., 3rd; Patel, A.; Kim, I.; Feng, J.; Yue, L.; Wu, Q.; et al. Biodegradable Piezoelectric Force Sensor. *Proc. Natl. Acad. Sci. USA* **2018**, *115*, 909–914. [CrossRef]
193. De Santis, M.; Cacciotti, I. Wireless implantable and biodegradable sensors for postsurgery monitoring: Current status and future perspectives. *Nanotechnology* **2020**, *31*, 252001. [CrossRef] [PubMed]
194. Kang, S.K.; Murphy, R.K.; Hwang, S.W.; Lee, S.M.; Harburg, D.V.; Krueger, N.A.; Shin, J.; Gamble, P.; Cheng, H.; Yu, S.; et al. Bioresorbable silicon electronic sensors for the brain. *Nature* **2016**, *530*, 71–76. [CrossRef] [PubMed]
195. Tian, J.; Shi, R.; Liu, Z.; Ouyang, H.; Yu, M.; Zhao, C.; Zou, Y.; Jiang, D.; Zhang, J.; Li, Z. Self-powered implantable electrical stimulator for osteoblasts' proliferation and differentiation. *Nano Energy* **2019**, *59*, 705–714. [CrossRef]
196. Curry, E.J.; Le, T.T.; Das, R.; Ke, K.; Santorella, E.M.; Paul, D.; Chorsi, M.T.; Tran, K.T.M.; Baroody, J.; Borges, E.R.; et al. Biodegradable nanofiber-based piezoelectric transducer. *Proc. Natl. Acad. Sci. USA* **2020**, *117*, 214–220. [CrossRef]
197. Luo, N.; Feng, Y.; Wang, D.; Zheng, Y.; Ye, Q.; Zhou, F.; Liu, W. New Self-Healing Triboelectric Nanogenerator Based on Simultaneous Repair Friction Layer and Conductive Layer. *ACS Appl. Mater. Interfaces* **2020**, *12*, 30390–30398. [CrossRef]

198. Sun, J.; Pu, X.; Liu, M.; Yu, A.; Du, C.; Zhai, J.; Hu, W.; Wang, Z.L. Self-Healable, Stretchable, Transparent Triboelectric Nanogenerators as Soft Power Sources. *ACS Nano* **2018**, *12*, 6147–6155. [CrossRef]
199. Yang, D.; Ni, Y.; Kong, X.; Li, S.; Chen, X.; Zhang, L.; Wang, Z.L. Self-Healing and Elastic Triboelectric Nanogenerators for Muscle Motion Monitoring and Photothermal Treatment. *ACS Nano* **2021**, *15*, 14653–14661. [CrossRef]
200. Deng, J.; Kuang, X.; Liu, R.; Ding, W.; Wang, A.C.; Lai, Y.C.; Dong, K.; Wen, Z.; Wang, Y.; Wang, L.; et al. Vitrimer Elastomer-Based Jigsaw Puzzle-Like Healable Triboelectric Nanogenerator for Self-Powered Wearable Electronics. *Adv. Mater.* **2018**, *30*, e1705918. [CrossRef]
201. Parida, K.; Kumar, V.; Jiangxin, W.; Bhavanasi, V.; Bendi, R.; Lee, P.S. Highly Transparent, Stretchable, and Self-Healing Ionic-Skin Triboelectric Nanogenerators for Energy Harvesting and Touch Applications. *Adv. Mater.* **2017**, *29*, 1702181. [CrossRef]
202. Xu, C.; Ma, B.; Yuan, S.; Zhao, C.; Liu, H. High-Resolution Patterning of Liquid Metal on Hydrogel for Flexible, Stretchable, and Self-Healing Electronics. *Adv. Electron. Mater.* **2019**, *6*, 1900721. [CrossRef]
203. Lai, Y.C.; Wu, H.M.; Lin, H.C.; Chang, C.L.; Chou, H.H.; Hsiao, Y.C.; Wu, Y.C. Entirely, Intrinsically, and Autonomously Self-Healable, Highly Transparent, and Superstretchable Triboelectric Nanogenerator for Personal Power Sources and Self-Powered Electronic Skins. *Adv. Funct. Mater.* **2019**, *29*, 1904626. [CrossRef]
204. Lee, J.H.; Hinchet, R.; Kim, S.K.; Kim, S.; Kim, S.-W. Shape memory polymer-based self-healing triboelectric nanogenerator. *Energy Environ. Sci.* **2015**, *8*, 3605–3613. [CrossRef]
205. Wang, C.; Qu, X.; Zheng, Q.; Liu, Y.; Tan, P.; Shi, B.; Ouyang, H.; Chao, S.; Zou, Y.; Zhao, C.; et al. Stretchable, Self-Healing, and Skin-Mounted Active Sensor for Multipoint Muscle Function Assessment. *ACS Nano* **2021**, *15*, 10130–10140. [CrossRef] [PubMed]
206. Yang, P.-K.; Zhu, X.-L.; Huang, L.; Zhou, Y.; Song, G.-Q.; Qin, C.-C.; Wang, D.; Wu, Y.-P. Skin-Inspired Electret Nanogenerator with Self-Healing Abilities. *Cell Rep. Phys. Sci.* **2020**, *1*, 100185. [CrossRef]
207. Mai, W.; Yu, Q.; Han, C.; Kang, F.; Li, B. Self-Healing Materials for Energy-Storage Devices. *Adv. Funct. Mater.* **2020**, *30*, 1909912. [CrossRef]
208. Mou, L.; Qi, J.; Tang, L.; Dong, R.; Xia, Y.; Gao, Y.; Jiang, X. Highly Stretchable and Biocompatible Liquid Metal-Elastomer Conductors for Self-Healing Electronics. *Small* **2020**, *16*, e2005336. [CrossRef]
209. Wang, C.; Hu, K.; Zhao, C.; Zou, Y.; Liu, Y.; Qu, X.; Jiang, D.; Li, Z.; Zhang, M.R.; Li, Z. Customization of Conductive Elastomer Based on PVA/PEI for Stretchable Sensors. *Small* **2020**, *16*, e1904758. [CrossRef]
210. Han, X.; Jiang, D.; Qu, X.; Bai, Y.; Cao, Y.; Luo, R.; Li, Z. A Stretchable, Self-Healable Triboelectric Nanogenerator as Electronic Skin for Energy Harvesting and Tactile Sensing. *Materials* **2021**, *14*, 1689. [CrossRef]
211. Guan, Q.; Dai, Y.; Yang, Y.; Bi, X.; Wen, Z.; Pan, Y. Near-infrared irradiation induced remote and efficient self-healable triboelectric nanogenerator for potential implantable electronics. *Nano Energy* **2018**, *51*, 333–339. [CrossRef]
212. Huang, L.-B.; Dai, X.; Sun, Z.; Wong, M.-C.; Pang, S.-Y.; Han, J.; Zheng, Q.; Zhao, C.-H.; Kong, J.; Hao, J. Environment-resisted flexible high performance triboelectric nanogenerators based on ultrafast self-healing non-drying conductive organohydrogel. *Nano Energy* **2021**, *82*, 105724. [CrossRef]
213. Kim, J.-N.; Lee, J.; Lee, H.; Oh, I.-K. Stretchable and self-healable catechol-chitosan-diatom hydrogel for triboelectric generator and self-powered tremor sensor targeting at Parkinson disease. *Nano Energy* **2021**, *82*, 105705. [CrossRef]
214. Wu, Z.; Chen, J.; Boukhvalov, D.W.; Luo, Z.; Zhu, L.; Shi, Y. A new triboelectric nanogenerator with excellent electric breakdown self-healing performance. *Nano Energy* **2021**, *85*, 105990. [CrossRef]
215. Liao, W.; Liu, X.; Li, Y.; Xu, X.; Jiang, J.; Lu, S.; Bao, D.; Wen, Z.; Sun, X. Transparent, stretchable, temperature-stable and self-healing ionogel-based triboelectric nanogenerator for biomechanical energy collection. *Nano Res.* **2021**, *15*, 2060–2068. [CrossRef]
216. Parida, K.; Thangavel, G.; Cai, G.; Zhou, X.; Park, S.; Xiong, J.; Lee, P.S. Extremely stretchable and self-healing conductor based on thermoplastic elastomer for all-three-dimensional printed triboelectric nanogenerator. *Nat. Commun.* **2019**, *10*, 2158. [CrossRef]
217. Dzhardimalieva, G.I.; Yadav, B.C.; Kudaibergenov, S.E.; Uflyand, I.E. Basic Approaches to the Design of Intrinsic Self-Healing Polymers for Triboelectric Nanogenerators. *Polymers* **2020**, *12*, 2594. [CrossRef]
218. Gai, Y.; Li, H.; Li, Z. Self-Healing Functional Electronic Devices. *Small* **2021**, *17*, e2101383. [CrossRef]
219. Chen, X.; Xie, X.; Liu, Y.; Zhao, C.; Wen, M.; Wen, Z. Advances in Healthcare Electronics Enabled by Triboelectric Nanogenerators. *Adv. Funct. Mater.* **2020**, *30*, 2004673. [CrossRef]
220. Sun, J.; Yang, A.; Zhao, C.; Liu, F.; Li, Z. Recent progress of nanogenerators acting as biomedical sensors in vivo. *Sci. Bull.* **2019**, *64*, 1336–1347. [CrossRef]
221. Zheng, Q.; Tang, Q.; Wang, Z.L.; Li, Z. Self-powered cardiovascular electronic devices and systems. *Nat. Rev. Cardiol.* **2021**, *18*, 7–21. [CrossRef]
222. Liu, Z.; Shi, B.; Zheng, Q.; Zou, Y.; Ouyang, H.; Li, Z.; Tian, J. Self-powered implantable electronic medical devices research based on triboelectric nanogenerator. *Sci. Sin. Technol.* **2017**, *47*, 1075–1080. [CrossRef]
223. Zhao, Z.; Zhou, L.; Li, S.; Liu, D.; Li, Y.; Gao, Y.; Liu, Y.; Dai, Y.; Wang, J.; Wang, Z.L. Selection rules of triboelectric materials for direct-current triboelectric nanogenerator. *Nat. Commun.* **2021**, *12*, 4686. [CrossRef] [PubMed]
224. Choi, Y.S.; Kim, S.W.; Kar-Narayan, S. Materials-Related Strategies for Highly Efficient Triboelectric Energy Generators. *Adv. Energy Mater.* **2021**, *11*, 2003802. [CrossRef]
225. Chen, J.; Wang, Z.L. Reviving Vibration Energy Harvesting and Self-Powered Sensing by a Triboelectric Nanogenerator. *Joule* **2017**, *1*, 480–521. [CrossRef]

226. Zhao, L.; Zheng, Q.; Ouyang, H.; Li, H.; Yan, L.; Shi, B.; Li, Z. A size-unlimited surface microstructure modification method for achieving high performance triboelectric nanogenerator. *Nano Energy* **2016**, *28*, 172–178. [CrossRef]
227. Hannan, M.A.; Mutashar, S.; Salina, S.A.; Hussain, A. Energy harvesting for the implantable biomedical devices: Issues and challenges. *Biomed. Eng.* **2014**, *13*, 79. [CrossRef]
228. Wang, J.; Li, S.; Yi, F.; Zi, Y.; Lin, J.; Wang, X.; Xu, Y.; Wang, Z.L. Sustainably powering wearable electronics solely by biomechanical energy. *Nat. Commun.* **2016**, *7*, 12744. [CrossRef]
229. Zhang, Q.; Liu, L.; Pan, C.; Li, D. Review of recent achievements in self-healing conductive materials and their applications. *J. Mater. Sci.* **2017**, *53*, 27–46. [CrossRef]
230. Yakovlev, A.; Kim, S.; Poon, A. Implantable biomedical devices: Wireless powering and communication. *IEEE Commun. Mag.* **2012**, *50*, 152–159. [CrossRef]
231. He, T.; Lee, C. Evolving Flexible Sensors, Wearable and Implantable Technologies Towards BodyNET for Advanced Healthcare and Reinforced Life Quality. *IEEE Open J. Circ. Syst.* **2021**, *2*, 702–720. [CrossRef]
232. Zhang, Z.; Wen, F.; Sun, Z.; Guo, X.; He, T.; Lee, C. Artificial Intelligence-Enabled Sensing Technologies in the 5G/Internet of Things Era: From Virtual Reality/Augmented Reality to the Digital Twin. *Adv. Intell. Syst.* **2022**, *2100228*, 1–23. [CrossRef]
233. Shi, Q.; Yang, Y.; Sun, Z.; Lee, C. Progress of Advanced Devices and Internet of Things Systems as Enabling Technologies for Smart Homes and Health Care. *ACS Mater. Au* **2022**, accepted. [CrossRef]

nanomaterials

MDPI

Article

A Non-Resonant Piezoelectric–Electromagnetic–Triboelectric Hybrid Energy Harvester for Low-Frequency Human Motions

Gang Tang [1], Zhen Wang [1], Xin Hu [1], Shaojie Wu [1], Bin Xu [1], Zhibiao Li [1], Xiaoxiao Yan [1], Fang Xu [1], Dandan Yuan [1], Peisheng Li [1], Qiongfeng Shi [2,3,*] and Chengkuo Lee [2,*]

[1] Jiangxi Province Key Laboratory of Precision Drive & Control, Nanchang Institute of Technology, Nanchang 330099, China; tanggang@nit.edu.cn (G.T.); 15797625058@163.com (Z.W.); xinh9602@163.com (X.H.); wu2020313111@163.com (S.W.); 2010994268@nit.edu.cn (B.X.); lizhibiao@nit.edu.cn (Z.L.); qwwxiaoy@163.com (X.Y.); 2021994767@nit.edu.cn (F.X.); yuandandan@nit.edu.cn (D.Y.); lipeisheng@ncu.edu.cn (P.L.)

[2] Department of Electrical & Computer Engineering, National University of Singapore, 4 Engineering Drive 3, Singapore 117583, Singapore

[3] School of Electronic Science and Engineering, Southeast University, Nanjing 210096, China

* Correspondence: eleshiq@nus.edu.sg (Q.S.); elelc@nus.edu.sg (C.L.); Tel.: +65-65-16-58-65 (C.L.)

Abstract: With the rapid development of wireless communication and micro-power technologies, smart wearable devices with various functionalities appear more and more in our daily lives. Nevertheless, they normally possess short battery life and need to be recharged with external power sources with a long charging time, which seriously affects the user experience. To help extend the battery life or even replace it, a non-resonant piezoelectric–electromagnetic–triboelectric hybrid energy harvester is presented to effectively harvest energy from low-frequency human motions. In the designed structure, a moving magnet is used to simultaneously excite the three integrated energy collection units (i.e., piezoelectric, electromagnetic, and triboelectric) with a synergistic effect, such that the overall output power and energy-harvesting efficiency of the hybrid device can be greatly improved under various excitations. The experimental results show that with a vibration frequency of 4 Hz and a displacement of 200 mm, the hybrid energy harvester obtains a maximum output power of 26.17 mW at 70 kΩ for one piezoelectric generator (PEG) unit, 87.1 mW at 500 Ω for one electromagnetic generator (EMG) unit, and 63 μW at 140 MΩ for one triboelectric nanogenerator (TENG) unit, respectively. Then, the generated outputs are adopted for capacitor charging, which reveals that the performance of the three-unit integration is remarkably stronger than that of individual units. Finally, the practical energy-harvesting experiments conducted on various body parts such as wrist, calf, hand, and waist indicate that the proposed hybrid energy harvester has promising application potential in constructing a self-powered wearable system as the sustainable power source.

Keywords: triboelectric; piezoelectric; electromagnetic; self-powered; hybrid

Citation: Tang, G.; Wang, Z.; Hu, X.; Wu, S.; Xu, B.; Li, Z.; Yan, X.; Xu, F.; Yuan, D.; Li, P.; et al. A Non-Resonant Piezoelectric–Electromagnetic–Triboelectric Hybrid Energy Harvester for Low-Frequency Human Motions. *Nanomaterials* **2022**, *12*, 1168. https://doi.org/10.3390/nano12071168

Academic Editor: Federico Cesano

Received: 19 February 2022
Accepted: 29 March 2022
Published: 31 March 2022

Publisher's Note: MDPI stays neutral with regard to jurisdictional claims in published maps and institutional affiliations.

1. Introduction

In recent years, with the advancement of sensing and fabrication technologies, numerous micro-power Internet of Things (IoT) sensors and electronic devices have been developed rapidly [1,2], but how to achieve sustainable energy supply to such a large number of devices remains a grand challenge. The bulky size, limited service life, and environmental pollution issue of traditional chemical batteries have gradually become the bottlenecks restricting the further development of micro-power devices. In this regard, scavenging ambient energies such as light, heat, wind, and mechanical vibration using energy harvesters appears to be a promising alternative [3,4]. At present, researchers across the world have performed detailed investigations of using energy harvesters to power wearable devices. In 2011, Yun et al., proposed a self-powered wireless sensor node based on solar energy harvesting, and the output power can reach 20 mW under outdoor conditions [5]. Although the output of solar energy harvesting is relatively large, it is also easily affected by the environmental parameters, leading to no electrical output in the absence of light. In addition, another form of energy—thermal energy—has also been studied by researchers for energy harvesting. For example, Wang et al., connected 4700 thermocouples in series and attached them to the human body, which can produce 0.15 V open-circuit voltage and 0.3 nW output power at room temperature [6]. Although the collection of thermal energy could be weather independent, the output power is too low due to the limited temperature difference between the human body and the environment, which is normally insufficient to power wearable devices. The above-discussed two mechanisms exhibit high dependence on the ambient conditions and could be greatly influenced by external factors; thus, they are not ideal as a steady power supply for wearable devices. In this regard, a lot of researchers have turned to use mechanical vibration energy, which is more ubiquitous in the environment and human activities for electrical energy generation. This could be an ideal approach, since it is not easily interfered with by other external factors such as weather and temperature. Thus, it exhibits a broad application prospect in supplying energy for wearable and/or microscale electronics.

To effectively harvest the mechanical energy existing in the base excitations, biological motions, and flow waves [7,8], a large variety of vibrational energy harvesters have been proposed based on different mechanisms, e.g., piezoelectric [9–14], electromagnetic [15–19], electrostatic [20,21], triboelectric [22–26], and magnetostriction energy harvesters [27]. Moreover, two or more mechanisms can also be synergistically combined, which is called a hybrid mechanism [28–30]. With optimized design, the integrated mechanisms can work synergistically and complementarily, leading to an improvement in the output performance, transducing efficiency, and device's adaptability [31,32]. To realize a hybrid mechanism, the triboelectric mechanism is usually adopted to be integrated with other mechanisms due to its great advantages of simple structure, easy fabrication, broad materials choice, diverse operation modes, low cost, and high scalability [33–35]. In 2012, the first triboelectric nanogenerator (TENG) was invented by Wang's team based on the coupling of contact electrification and electrostatic induction between two different materials [36]. After that, TENG has received tremendous research efforts and has been proven as a promising technology for energy harvesting and self-powered sensing [37–41], for the application scenarios involved with body motions [42,43], wind [44,45], water waves [46,47], vibrations [48,49], rotations [50,51], and even acoustic waves [52,53].

In 2015, Li et al., designed a triboelectric–piezoelectric composite energy harvester on clothes containing several carbon fibers for human energy harvesting, with zinc oxide nanorods as the piezoelectric material and polydimethylsiloxane (PDMS) as the triboelectric material [54]. In 2020, Rahman et al., reported a miniaturized freestanding hybrid generator based on TENG and an electromagnetic generator for harvesting the human vibrational energy during motions [55]. Then, Li et al., proposed a flexible motion monitoring device based on a magnetic microneedle array [56]. The magnetic microneedles in this structure not only act as the curved magnetic poles of the electromagnetic generator but also act as a triboelectric friction layer. The closed and curved characteristics of the microneedles are used to complete the energy-harvesting action. These examples have proven that the hybrid strategy is feasible and effective for output performance and efficiency enhancement. Yet, most of them only utilize two mechanisms, and a more efficient way to harvest the low-frequency human motions is still highly desirable.

In order to harvest the mechanical energy from human motions more efficiently and improve the output performance, the synergistic effect between the adopted mechanisms in the hybrid energy harvester should be carefully designed. Generally speaking, piezoelectric and triboelectric energy harvesters can generate large output voltage with large internal impedance, and electromagnetic energy harvesters can produce large output current with low internal impedance. Therefore, these three energy-harvesting mechanisms can complement each other and improve the overall output performance of a hybrid device. Herein, we propose a novel non-resonant hybrid energy harvester with the piezoelectric–electromagnetic–triboelectric mechanism for effectively harvesting the low-frequency energy from various human activities. The structure of the device is configured into a cylindrical shape, which is convenient to be held in the hand or fixed to a certain part of the body. When the human body moves, the Al-coated magnet in the cylindrical structure starts to reciprocate inside the cavity, inducing power generation on the polyvinylidene fluoride (PVDF) piezoelectric elements at the two ends, the triboelectric element on the cavity sidewall, as well as the electromagnetic element winding around the middle part simultaneously. Under the conditions of an excitation frequency of 4 Hz and a movement displacement of 200 mm, each mechanism achieves the following output voltage and current: 9.4 V and 21.4 mA for the electromagnetic generator (EMG); 95 V and 106 μA for the piezoelectric generator (PEG); and 150 V and 1.4 μA for the TENG. Then, the generated outputs are regulated through an energy-harvesting circuit and finally stored in a capacitor for powering IoT sensors. The integrated outputs from the hybrid energy harvester can charge a 2.2 μF capacitor up to 13 V in only 1 s. Therefore, this hybrid structure proposed in this work shows great potential as a sustainable power source for various IoT sensor nodes and wearable devices in the IoT era.

2. Results and Discussion

2.1. Device Structure

Figure 1a shows the 3D schematic view of the hybrid energy harvester, i.e., a hybrid generator, which is mainly a symmetrical cylinder structure. Two lightweight and hollow half-cylinders with a dimension of 117 mm × 36 mm (length × diameter) are prepared by 3D printing and used as the supporting frame. Two caps with whorls are also fabricated for covering both ends of the tunnel with the same 3D printing method. Overall, the hybrid generator consists of one EMG at the middle, two PEGs at the two ends, and one TENG along the inner surface. A cylindrical NdFeB permanent magnet (with a diameter of 10 mm and height of 30 mm) is adopted as the proof mass to respond to environmental perturbations and trigger all three energy-harvesting units with synergistic output generation. To construct the EMG unit, the middle part of the hollow cylinder is wound with copper (Cu) wires (with a diameter of 70 μm) for 1000 turns, forming the coil structure to capture the magnetic flux variation. In terms of the PEG unit, two commercially available PVDF thin films with high piezoelectric coefficient and high durability are used. In order to make use of the large mechanical impact from the magnet more efficiently, PVDF films are attached to the top and bottom of the cylinder frame with both ends fixed, forming the suspended membrane configuration that is more sensitive to impact force. The PVDF film is coated with silver on both sides as upper and lower electrodes, which is further encapsulated with polymer films for insulation and protection. When subjected to external vibration, the cylindrical magnet oscillates and impacts the PVDF films, causing structural deformation, thereby generating piezoelectric outputs accordingly. To construct the TENG unit, two aluminum (Al) electrodes are first attached to the upper part and the lower part of the hollow cylinder, which is followed by the attachment of a poly tetra fluoroethylene (PTFE) thin film as the negative triboelectric material. To improve the triboelectric output performance by the magnet friction, a more positive Al foil is attached to the surface of the magnet, acting as the positive triboelectric material. Figure 1 shows a digital photograph of the assembled device, while its diameter (≈36 mm) and length (≈117 mm) are indicated in Figure 1e,f, respectively.

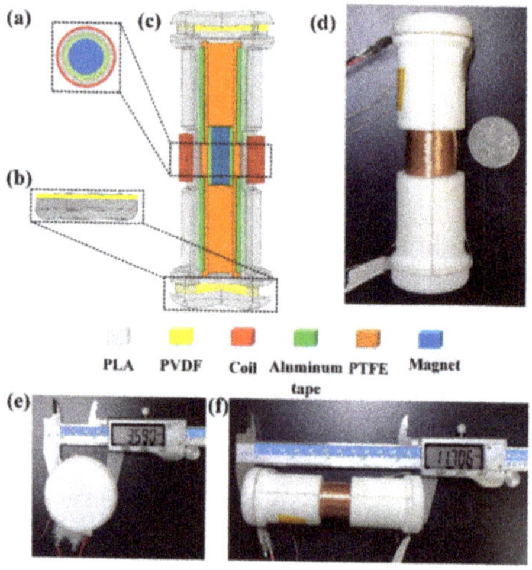

Figure 1. Structure of the piezoelectric–electromagnetic–triboelectric hybrid energy harvester. (**a**) TENG-EMG unit. (**b**) PEG unit simply supported beam structure. (**c**) Three-dimensional (3D) schematic diagram of the proposed hybrid energy harvester. (**d**) Photograph of the hybrid energy harvester. (**e**,**f**) The diameter and length of the hybrid energy harvester.

2.2. Working Principle

According to the structural design of the hybrid generator, synergistic outputs could be generated by the three integrated units: EMG, PEG, and TENG. The EMG unit uses the movement of the magnet to change the magnetic flux of the Cu coil and generates current based on Faraday's law of electromagnetic induction. When the permanent magnet moves to the ends and impacts the PVDF films, a piezoelectric current is generated from the PEG unit based on the piezoelectric effect. Meanwhile, the triboelectric charges on the magnet surface induce potential difference on the two triboelectric electrodes during this reciprocating movement, leading to current generation on the TENG unit. All these three units are triggered by the magnet movement simultaneously, generating synchronized outputs that can be further regulated to increase the overall output power of the device.

As shown in Figure 2, the detailed working principle of the hybrid generator is illustrated. When the overall structure of the device is not subject to external vibration, it is in an equilibrium state and has no output. As shown in Figure 2a, when the device is shaken to the right, the permanent magnet hits the PVDF film on the left end due to inertia, which deforms the PVDF film and generates a corresponding piezoelectric current. Meanwhile, due to the electrostatic equilibrium of triboelectric charges, there is no current generated from the two Al electrodes. Similarly, the magnetic flux through the coil does not change, and there is no corresponding electromotive force and induction current in the Cu coil. When the external force changes direction, the magnet separates from the PVDF film at the left end and moves toward the right end, as shown in Figure 2b. Upon separation, a small piezoelectric current is generated when the PVDF film restore to its original state. Then, during this movement of the magnet, the magnetic flux of the Cu coil changes significantly. According to Faraday's law of electromagnetic induction, an electromotive force is generated, giving rise to an induction current in the Cu coil. In the TENG unit, the movement of the magnet breaks the previous electrostatic balance and induces potential difference on the two Al electrodes. Thus, the current flows from the high potential end (right) to the low potential end (left), generating a corresponding triboelectric output. When the magnet continues to move to the right, it finally hits the other PVDF film, as shown in Figure 2c. At this time, a corresponding piezoelectric current is generated, and the outputs of the EMG unit and the TENG unit come to rest due to electromagnetic and electrostatic balance. Then, with the applied force changing direction again, the magnet will move away from the right PVDF film toward the left end, as shown in Figure 2d. Similar to the process in Figure 2b, a small piezoelectric output current is first generated, which is followed by the generation of EMG and TENG outputs in a reverse direction, until a new equilibrium is achieved. After that, another new cycle of output generation starts. With the reciprocating movement of external force, this output generation process from these three units will repeat and continue.

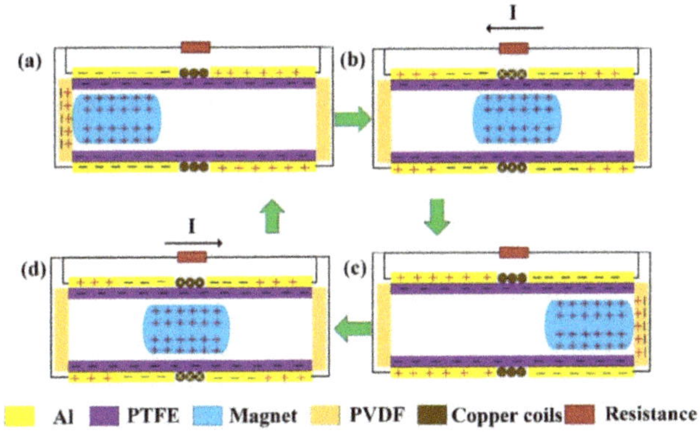

Figure 2. Schematic diagrams showing the operation mechanism and the corresponding current flow from the three generator units, i.e., EMG, PEG, and TENG. (**a**) The magnet on the left end and impacting the left PVDF film. (**b**) The magnet moving from left to right. (**c**) The magnet on the right end and impacting the right PVDF film. (**d**) The magnet moving from right to left.

2.3. Output Characterization

In this section, the output performance of the developed piezoelectric–electromagnetic–triboelectric hybrid generator is analyzed quantitatively. When the magnet moves, all the three generator units can generate synchronized outputs, which will be investigated one by one. As shown in Figure 3, with a vibration frequency of 4 Hz and a vibration displacement of 200 mm, the output performance of the TENG unit is measured when using magnets with different lengths.

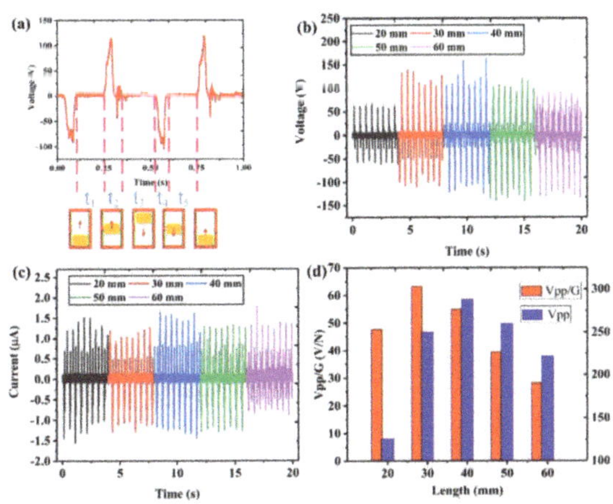

Figure 3. Output characteristics of the TENG unit. (**a**) Triboelectric output voltage corresponding to different stages within one cycle. (**b**) Output voltages and (**c**) output currents of the TENG unit under different magnet lengths. (**d**) Performance comparison of the output voltage and voltage per unit weight for different magnet lengths.

Under this excitation condition, it can be seen from Figure 3a that the triboelectric output voltage can reach 125 V. In order to further investigate the output characteristics of the TENG unit during this process, the output voltage in one cycle is analyzed. It can be divided into the following five stages. (1) The t_1 stage is the initial stage. At this stage, the movement direction of the magnet is vertically upward, while the upper end of the magnet is still within the lower triboelectric electrode range. Thus, there is no charge transfer between the two triboelectric electrodes due to electrostatic balance, leading to zero output voltage. (2) In the t_2 stage, the magnet continues to move upward, beginning to cross the electrode boundary and enter the upper electrode range. Hence, charge transfer occurs due to the arising potential difference, and a positive voltage signal is then produced. (3) At the t_3 stage, the magnet fully enters the upper electrode range, reaches the top, and moves downward. Similar to the t_1 stage, there is no charge transfer and output voltage before the magnet reaches the lower electrode range. (4) At the t_4 stage, the magnet continues to move downward and begins to enter the lower electrode range. At this moment, charge transfer occurs in the opposite direction, producing a negative voltage signal. (5) During the t_5 stage, the magnet fully enters the lower electrode range and returns to the original position as in the t_1 stage, ready for another new cycle to start. The charge transfer and output generation process will continue if the excitation is repeated.

According to the output characteristics of TENG, the length of the magnet could be an important factor affecting the output performance; thus, it is investigated here. As shown in Figure 3b, it can be found that as the length of the magnet increases, the positive voltage peak first increases and then decreases, with a maximum value of 163 V achieved at 40 mm length. For the negative voltage peak, it gradually increases. The corresponding current outputs can be seen in Figure 3c. Based on the overall outputs of the voltage and current, the appropriate magnet size should be 40 mm. In order to further determine the appropriate magnet length, we introduce a new comparison parameter, i.e., Vpp/G (unit: V/N), which is the ratio of the peak-to-peak output voltage to the device weight in gravity. As shown in Figure 3d, the output voltage per unit weight of the TENG shows a trend of first increasing and then decreasing. Although the peak-to-peak output voltage is 288 V when the magnet length is 40 mm, the output voltage per unit weight of 30 mm is greater, 63.168 V/N. Since the device aims at wearable applications, a higher output voltage per unit weight will be more efficient in energy harvesting and more convenient for users, and thus, the most suitable length of the magnets for the TENG unit is 30 mm.

As shown in Figure 4, the output performance of the EMG unit is measured with the same vibration frequency of 4 Hz and displacement of 200 mm. Here, the influence of the length of the magnet on the output performance of the EMG unit of the device is also discussed. Figure 4a shows the output voltage of the EMG unit from 0.1 to 1.1 s, where the positive voltage is 9.4 V, and the negative voltage is −11 V. This asymmetric output is because in the adopted manual excitation, the magnet is accelerating from one end to the other, meaning that the approaching speed toward the Cu coil is less than the leaving speed. Thereby, the negative peak due to the magnet leaving is relatively larger than the positive peak due to the magnet approaching.

To further study the output characteristics of the EMG unit, the output waveforms are investigated in detail. In the magnet movement process from one end to the other, the output signal of the EMG unit can be divided into the following four stages. (1) In the t_1 stage, the magnet has just started to move and is far from the position of the coil, and thus, there is no signal output due to the barely changing magnetic field. (2) During the t_2 stage, the magnet is gradually approaching the coil, and the induced magnetic flux in the coil is increased, leading to the generation of the positive peak. (3) In the t_3 stage, the magnetic in the coil reaches the largest, and the output current thus gradually comes to zero. After that, the magnet starts to leave the coil, resulting in a decrement in the magnetic flux. Then, an opposite output, i.e., the negative output peak, is generated. (4) At the t_4 stage, the magnet is far from the coil again; thus, there is no signal output as well. When the magnet moves from right to left, a similar positive–negative output will be generated due to the same increment–decrement trend of the magnetic flux.

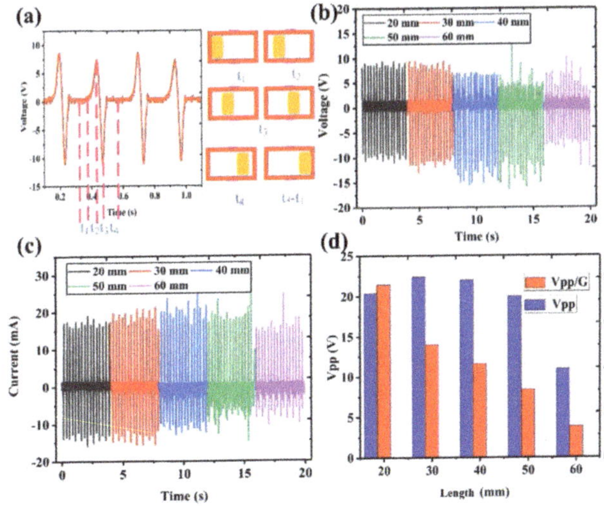

Figure 4. The output characteristics of the EMG unit. (a) The output voltage corresponding to different stages within one cycle. (b) Output voltages and (c) output currents of the EMG unit under different magnet lengths. (d) Performance comparison of the output voltage and voltage per unit weight for different magnet lengths.

Under the premise of the same conditions, the change of the magnetic flux will be affected by the length of the magnet. Thus, the influence of different magnet lengths on the induced outputs of the EMG unit is studied. The output signals using different magnet lengths are measured at the same vibration frequency of 4 Hz and displacement of 200 mm, as shown in Figure 4b,c. It can be seen that for the positive voltages, it first increases slightly, reaching a maximum of 9.8 V at 30 mm, and then decreases. For the negative voltages, it reaches a maximum of 15 V at 40 mm. When the length is 20 mm or 30 mm, the body of the magnet at the ends is not overlapping with the coil position in the initial position (t_1 stage). The main influencing factor here is the magnetic flux difference when the magnet is at the middle position and the initial position. Since the magnet is not overlapping with the coil in the initial position, the magnetic flux can be considered very small for both the 20 mm and 30 mm cases. Then, at the middle position, the 30 mm magnet produces a stronger magnetic flux, which leads to a higher output. When the length of the magnet is 40 mm, 50 mm, or 60 mm, the magnet has begun to overlap with the coil. The initial magnetic flux increases with the magnet length, and thus, the magnetic flux difference will be decreased. Moreover, the acceleration time of the longer magnet is also reduced, resulting in a smaller moving speed, so that the output will gradually decrease.

Similarly, the output voltage per unit weight is also used to evaluate the EMG's working efficiency of magnets with different lengths. As indicated in Figure 4d, when the magnet length is 30 mm, the output voltage is the largest (22.4 V), followed by 40 mm. However, as far as Vpp/G is concerned, when the length of the magnet increases, the output voltage per unit weight shows a gradually decreasing trend. The maximum output voltage per unit weight is 7.718 V/N for the 20 mm magnet, which is followed by 5.611 V/N for the 30 mm magnet. Therefore, in terms of the EMG output, the optimal length is 20 mm, followed by 30 mm.

As shown in Figure 5, the output performance of the PEG unit is analyzed. As discussed above, to ensure that the TENG and EMG units have larger outputs, the magnet length is determined to be 30 mm. The voltage signal of the PEG unit measured under 4 Hz and 200 mm displacement is shown in Figure 5a, while Figure 5b shows the movement of the magnet in one signal generation period of the PVDF film. It can be divided into four stages. (1) During the t_1–t_2 stage, the magnet is not in contact with the PVDF film, and thus, no signal is generated. (2) In the t_2–t_3 stage, the magnet collides with the PVDF film and deforms it in a very short time, inducing a sharp positive output. (3) At the t_3–t_4 stage, the magnet is separating from the PVDF film, which gradually restores to its original state by its own elastic force, resulting in the generation of a small negative output. (4) During the t_4–t_1 stage, the magnet moves downward and starts to prepare to collide with the other PVDF film.

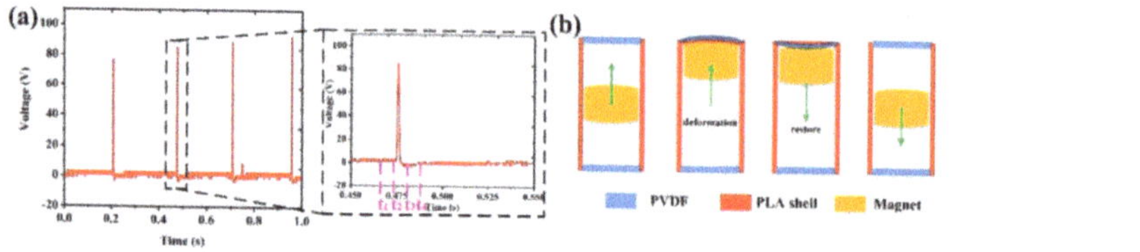

Figure 5. The output characteristics of the PEG unit. (**a**) The output voltage corresponding to different stages within one cycle. (**b**) Magnet movement and PVDF status in different stages.

Through the above investigations, it can be concluded that the most suitable magnet length for the hybrid generator is 30 mm. Since the device is mainly used in light activities such as human walking, it is necessary to analyze the device's output performance with varying frequencies and displacements. From Figure 6a,d,g, it can be observed that the outputs of the TENG, PEG, and EMG units all increase with the vibration frequency. This is because when the frequency increases, the period for charge transfer and output generation is shortened, leading to the increment of output currents and voltages. Then, as shown in Figure 6b,e,h, the influence of movement displacement on the output performance is investigated. A similar increment trend can be observed, which is because larger displacement induces a higher moving speed of the magnet. As a result, the output performance of each generator unit is increased.

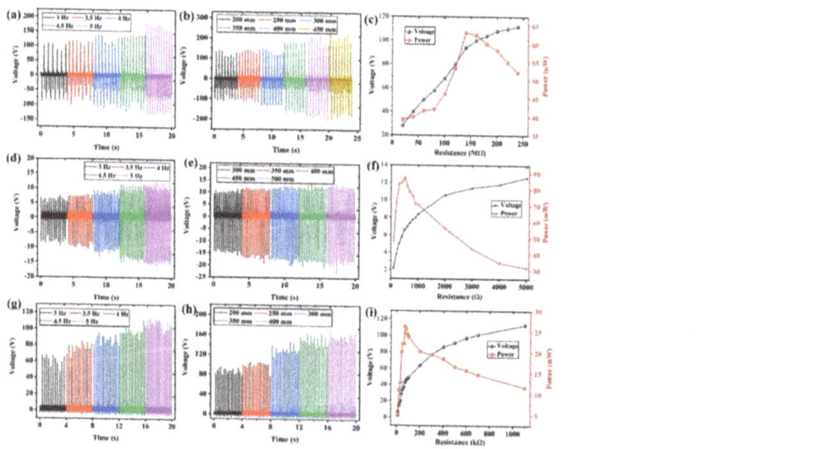

Figure 6. The output voltages of the TENG unit at (**a**) different frequencies and (**b**) different displacements, and (**c**) the output voltage and power curve under different external loads. The output voltages of the EMG unit at (**d**) different frequencies and (**e**) different displacements, and (**f**) the output voltage and power curve under different external loads. The output voltages of the PEG unit at (**g**) different frequencies and (**h**) different displacements, and (**i**) the output voltage and power curve under different external loads.

Figure 6c,f,i show the output voltage and power curve for each generator unit when connecting to different external loads, with a magnet length of 30 mm, vibration frequency of 4 Hz, and displacement of 200 mm. The output power is calculated from the output voltage, using the following equation:

$$P = \frac{U_P^2}{R} \quad (1)$$

Among them, R refers to the resistance of the external load, U_P refers to the output voltage on that load, and P is the corresponding output power. It can be seen that all the output voltage curves first increase continuously with the external resistance and finally stabilize. This phenomenon leads to a maximum output power when the internal resistance of the TENG, EMG, or PEG unit matches the external resistance. By comparison, the internal resistance of TENG is the largest, which is followed by the PEG, and finally the EMG. It is calculated that under this excitation condition, the hybrid generator obtains the maximum output power of 63 μW at 140 MΩ for one TENG unit, 87.1 mW at 500 Ω for one EMG unit, and 26.17 mW at 70 kΩ for one PEG unit.

2.4. Device Application

The hybrid generator is designed to harvest kinetic energy from the human body, so it is necessary to investigate its performance under various human activities. As shown in Figure 7, four positions at the wrist, hand, calf, and waist are selected to wear the hybrid generator for testing, following the exercise conditions: wrists and hands in situ swinging; calf in situ stepping; and waist in situ twisting. Under these conditions, the peak-to-peak output voltages at each position are summarized in Table 1. According to the results, the hybrid generator can effectively capture the kinetic energy from the body's routine movements.

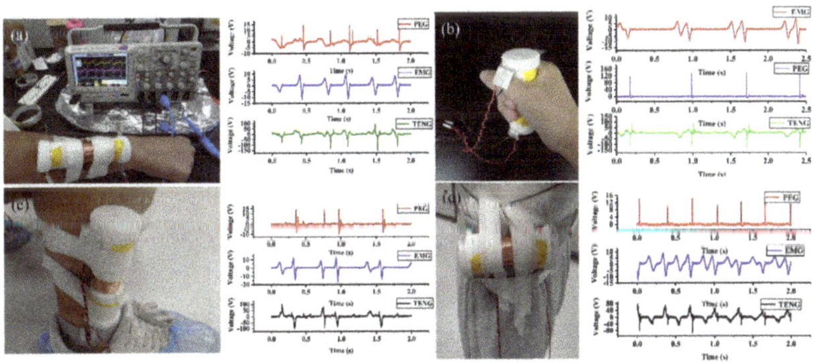

Figure 7. Practical implementation of the hybrid generator to harvest the human activity energy on various body parts: (**a**) wrist; (**b**) hand; (**c**) calf; (**d**) waist.

Table 1. Output voltage generated from each wearing position (Vpp is the peak-to-peak voltage in volts).

Position	PEG (Vpp)	EMG (Vpp)	TENG (Vpp)
Wrist	21	22	180
Calf	20	24	200
Hand	120	15	220
Waist	14	17.5	140

In practical applications, constant current and voltage from a power source are more usable for driving electrical sensors and other components. Therefore, the pulse-like outputs generated from the hybrid generator need to be regulated and stored in an energy storage unit first, such as a capacitor or chemical battery [57–59].

A customized circuit is designed for the hybrid generator. The outputs from each generator unit are first rectified, which are then connected in parallel to the inputs of a commercial power management chip (LTC3588). The outputs from the chip are further connected to a capacitor for energy storage, as shown in Figure 8a. The capacitor charging capability of each generator unit and their combination is compared using a 2.2 μF capacitor. As illustrated in Figure 8b, the charging speed of the three-unit combination is remarkably faster than that of a single unit. It indicates that the hybridization of multiple mechanisms is indeed a good strategy to improve the energy-harvesting efficiency. The voltage of the capacitor at 100 s when charging with the TENG unit is only 10.5 V, while 11–13 V can be achieved when other units or combination are used. This can be attributed to the fact that TENG normally has a large inner impedance and low output current, hindering its charging capability. In this regard, the EMG and PEG units with relatively large output current can be good complements to the TENG unit, enhancing the charging capability when configured in a hybrid generator.

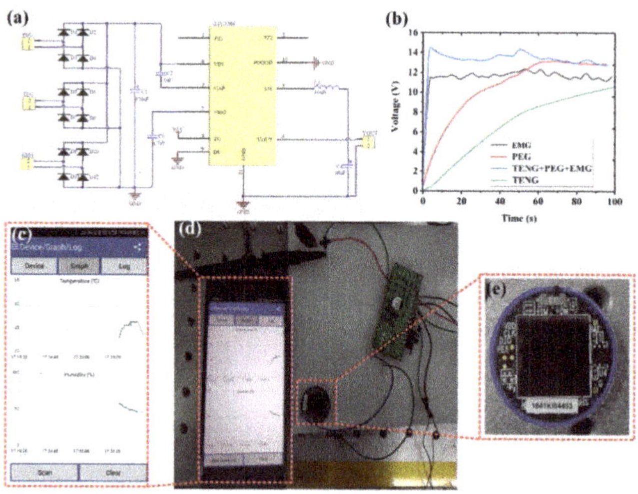

Figure 8. (**a**) Power management circuit for the hybrid generator. (**b**) Charging a capacitor of 2.2 μF. (**c**–**e**) Powering a wireless IoT module by the hybrid generator.

After the power management circuit, the stored energy in the capacitor can be used to drive IoT sensors. Here, a wireless temperature/humidity sensor is used as an example to show the capability of the hybrid generator. As depicted in Figure 8c–e, the sensor is successfully powered by the outputs from the hybrid generator when it is subjected to a vibration frequency of 4 Hz. The sensor starts to work in about 10 s after the generator device is actuated, and it feeds back the temperature and humidity signal to the mobile phone interface through Bluetooth in a real-time and intermittent manner. The entire process can last for 18 s.

3. Conclusions

In this work, a piezoelectric–electromagnetic–triboelectric hybrid generator is developed to harvest energy from various human activities. Benefited by its facile structure, non-resonant nature, and broad responsiveness, the hybrid generator is able to perform effective energy harvesting even with low-frequency human motions. The hybrid generator has a small dimension of ∅36 mm × 117 mm and a light weight, which is suitable to power intelligent mobile devices and wearable electronics. The design integrates three energy-harvesting mechanisms in a synergistic manner to achieve greater output performance. In addition, it also shows good adaptability for wearable applications; for example, it can be worn on a person's wrist, calf, waist, etc., to effectively harvest the kinetic energy from human activities. The dynamic response and performance of each mechanism, as well as the optimal conditions such as magnet length, have been analyzed in detail. Furthermore, the output performance of capacitor charging has been measured for every single unit and three-unit combination. The results reveal that the charging performance of the unit combination is remarkably higher than that of an individual unit, showing the feasibility and advantages of using the hybrid design. Benefited by the superior output performance, the hybrid generator can drive a wireless IoT module in real time, which can collect both the temperature and humidity information and send it to a mobile phone via Bluetooth. The designed hybrid generator in this work demonstrates a promising strategy to integrate multiple working mechanisms into a hybridized structure, for more effective energy harvesting and self-powered monitoring in wearable and IoT applications.

Author Contributions: Conceptualization, G.T., Q.S. and C.L.; methodology, G.T., Z.W. and X.H.; software, S.W.; validation, X.H. and Z.W.; formal analysis, Z.L., F.X. and B.X.; investigation, G.T. and Z.W.; data curation, Z.W., X.H., S.W. and D.Y.; writing—original draft preparation, G.T. and Z.W.; writing—review and editing, G.T., C.L., Q.S. and Z.W.; project administration, G.T., B.X., Z.L., P.L., X.Y. and D.Y.; funding acquisition, G.T., B.X. and X.Y. All authors have read and agreed to the published version of the manuscript.

Funding: This research was funded by the National Natural Science Foundation of China, grant number 62064008, 51565038 and the Science and Technology Project of Jiangxi, grant number 20202BABL204041, GJJ170987, GJJ180936, GJJ180938 and GJJ211939.

Institutional Review Board Statement: Not applicable.

Informed Consent Statement: Informed consent was obtained from all subjects involved in the study.

Data Availability Statement: The data of our study are available upon request.

Conflicts of Interest: The authors declare no conflict of interest.

References

1. Wu, Y.; Li, Y.; Zou, Y.; Rao, W.; Gai, Y.; Xue, J., Wu, L., Qu, X.; Liu, Y., Xu, G.; et al. A multi-mode triboelectric nanogenerator for energy harvesting and biomedical monitoring. *Nano Energy* **2022**, *92*, 106715. [CrossRef]
2. He, T.; Lee, C. Evolving Flexible Sensors, Wearable and Implantable Technologies Towards BodyNET for Advanced Healthcare and Reinforced Life Quality. *IEEE Open J. Circuits Syst.* **2021**, *2*, 702–720. [CrossRef]
3. He, X.; Wen, Q.; Sun, Y.; Wen, Z. A low-frequency piezoelectric-electromagnetic-triboelectric hybrid broadband vibration energy harvester. *Nano Energy* **2017**, *40*, 300–307. [CrossRef]
4. Guo, X.; Liu, L.; Zhang, Z.; Gao, S.; He, T.; Shi, Q.; Lee, C. Technology evolution from micro-scale energy harvesters to nanogenerators. *J. Micromech. Microeng.* **2021**, *31*, 093002. [CrossRef]
5. Danesh, M.; Long, J.R. An Autonomous Wireless Sensor Node Incorporating a Solar Cell Antenna for Energy Harvesting. *IEEE Trans. Microw. Theory Technol.* **2011**, *59*, 3546–3555. [CrossRef]
6. Wang, Z.; Fiorini, P.; Leonov, V.; Van Hoof, C. Characterization and optimization of polycrystalline $Si_{70\%}Ge_{30\%}$ for surface micromachined thermopiles in human body applications. *J. Micromech. Microeng.* **2009**, *19*, 094011. [CrossRef]
7. Yun, Y.; Jang, S.; Cho, S.; Lee, S.H.; Hwang, H.J.; Choi, D. Exo-shoe triboelectric nanogenerator: Toward high-performance wearable biomechanical energy harvester. *Nano Energy* **2021**, *80*, 105525. [CrossRef]
8. Yang, Z.; Zhou, S.; Zu, J.; Inman, D. High-Performance Piezoelectric Energy Harvesters and Their Applications. *Joule* **2018**, *2*, 642–697. [CrossRef]
9. Liu, H.; Zhong, J.; Lee, C.; Lee, S.-W.; Lin, L. A comprehensive review on piezoelectric energy harvesting technology: Materials, mechanisms, and applications. *Appl. Phys. Rev.* **2018**, *5*, 041306. [CrossRef]
10. Liu, H.; Lee, C.; Kobayashi, T.; Tay, C.J.; Quan, C. Investigation of a MEMS piezoelectric energy harvester system with a frequency-widened-bandwidth mechanism introduced by mechanical stoppers. *Smart Mater. Struct.* **2012**, *21*, 035005. [CrossRef]
11. Wu, M.; Ou, Y.; Mao, H.; Li, Z.; Liu, R.; Ming, A.; Ou, W. Multi-resonant wideband energy harvester based on a folded asymmetric M-shaped cantilever. *AIP Adv.* **2015**, *5*, 077149. [CrossRef]
12. Hwang, G.-T.; Park, H.; Lee, J.-H.; Oh, S.; Park, K.-I.; Byun, M.; Park, H.; Ahn, G.; Jeong, C.K.; No, K.; et al. Self-Powered Cardiac Pacemaker Enabled by Flexible Single Crystalline PMN-PT Piezoelectric Energy Harvester. *Adv. Mater.* **2014**, *26*, 4880–4887. [CrossRef]

13. Shi, Q.; Wang, T.; Kobayashi, T.; Lee, C. Investigation of geometric design in piezoelectric microelectromechanical systems diaphragms for ultrasonic energy harvesting. *Appl. Phys. Lett.* **2016**, *108*, 193902. [CrossRef]
14. Shi, Q.; Wang, T.; Lee, C. MEMS Based Broadband Piezoelectric Ultrasonic Energy Harvester (PUEH) for Enabling Self-Powered Implantable Biomedical Devices. *Sci. Rep.* **2016**, *6*, 24946. [CrossRef]
15. Takahashi, T.; Suzuki, M.; Nishida, T.; Yoshikawa, Y.; Aoyagi, S. Vertical capacitive energy harvester positively using contact between proof mass and electret plate-Stiffness matching by spring support of plate and stiction prevention by stopper mechanism. In Proceedings of the 2015 28th IEEE International Conference on Micro Electro Mechanical Systems (MEMS), Estoril, Portugal, 18–22 January 2015; IEEE: Piscataway, NJ, USA, 2015; Volume 2015, pp. 1145–1148.
16. Le, C.P.; Halvorsen, E.; Søråsen, O.; Yeatman, E.M. Microscale electrostatic energy harvester using internal impacts. *J. Intell. Mater. Syst. Struct.* **2012**, *23*, 1409–1421. [CrossRef]
17. Liu, H.; Zhang, J.; Shi, Q.; He, T.; Chen, T.; Sun, L.; Dziuban, J.A.; Lee, C. Development of a thermoelectric and electromagnetic hybrid energy harvester from water flow in an irrigation system. *Micromachines* **2018**, *9*, 395. [CrossRef]
18. Liu, H.; Ji, Z.; Chen, T.; Sun, L.; Menon, S.C.; Lee, C. An Intermittent Self-Powered Energy Harvesting System from Low-Frequency Hand Shaking. *IEEE Sens. J.* **2015**, *15*, 4782–4790. [CrossRef]
19. Liu, H.; Gudla, S.; Hassani, F.A.; Heng, C.H.; Lian, Y.; Lee, C. Investigation of the nonlinear electromagnetic energy harvesters from hand shaking. *IEEE Sens. J.* **2015**, *15*, 2356–2364. [CrossRef]
20. Qiu, J.; Liu, X.; Chen, H.; Xu, X.; Wen, Y.; Li, P. A Low-Frequency Resonant Electromagnetic Vibration Energy Harvester Employing the Halbach Arrays for Intelligent Wireless Sensor Networks. *IEEE Trans. Magn.* **2015**, *51*, 8600604. [CrossRef]
21. Qiu, J.; Chen, H.; Wen, Y.; Li, P. Magnetoelectric and electromagnetic composite vibration energy harvester for wireless sensor networks. *J. Appl. Phys.* **2015**, *117*, 17A331. [CrossRef]
22. Huang, L.-B.; Bai, G.; Wong, M.-C.; Yang, Z.; Xu, W.; Hao, J. Magnetic-Assisted Noncontact Triboelectric Nanogenerator Converting Mechanical Energy into Electricity and Light Emissions. *Adv. Mater.* **2016**, *28*, 2744–2751. [CrossRef]
23. Tang, G.; Cheng, F.; Hu, X.; Huang, B.; Xu, B.; Li, Z.; Yan, X.; Yuan, D.; Wu, W.; Shi, Q. A Two-Degree-of-Freedom Cantilever-Based Vibration Triboelectric Nanogenerator for Low-Frequency and Broadband Operation. *Electronics* **2019**, *8*, 1526. [CrossRef]
24. He, T.; Wang, H.; Wang, J.; Tian, X.; Wen, F.; Shi, Q.; Ho, J.S.; Lee, C. Self-Sustainable Wearable Textile Nano-Energy Nano-System (NENS) for Next-Generation Healthcare Applications. *Adv. Sci.* **2019**, *6*, 1901437. [CrossRef]
25. Liu, L.; Shi, Q.; Ho, J.S.; Lee, C. Study of thin film blue energy harvester based on triboelectric nanogenerator and seashore IoT applications. *Nano Energy* **2019**, *66*, 104167. [CrossRef]
26. Shi, Q.; Wang, H.; Wu, H.; Lee, C. Self-powered triboelectric nanogenerator buoy ball for applications ranging from environment monitoring to water wave energy farm. *Nano Energy* **2017**, *40*, 203–213. [CrossRef]
27. Mizukawa, Y.; Ahmed, U.; Zucca, M.; Blažević, D.; Rasilo, P. Small-signal modeling and optimal operating condition of magnetostrictive energy harvester. *J. Magn. Magn. Mater.* **2022**, *547*, 168819. [CrossRef]
28. Hou, C.; Chen, T.; Li, Y.; Huang, M.; Shi, Q.; Liu, H.; Sun, L.; Lee, C. A rotational pendulum based electromagnetic/triboelectric hybrid-generator for ultra-low-frequency vibrations aiming at human motion and blue energy applications. *Nano Energy* **2019**, *63*, 103871. [CrossRef]
29. Gupta, R.K.; Shi, Q.; Dhakar, L.; Wang, T.; Heng, C.H.; Lee, C. Broadband Energy Harvester Using Non-linear Polymer Spring and Electromagnetic/Triboelectric Hybrid Mechanism. *Sci. Rep.* **2017**, *7*, 41396. [CrossRef]
30. Gao, S.; He, T.; Zhang, Z.; Ao, H.; Jiang, H.; Lee, C. A Motion Capturing and Energy Harvesting Hybridized Lower-Limb System for Rehabilitation and Sports Applications. *Adv. Sci.* **2021**, *8*, 2101834. [CrossRef]
31. Liu, L.; Shi, Q.; Lee, C. A hybridized electromagnetic-triboelectric nanogenerator designed for scavenging biomechanical energy in human balance control. *Nano Res.* **2021**, *14*, 4227–4235. [CrossRef]
32. Liu, L.; Shi, Q.; Lee, C. A novel hybridized blue energy harvester aiming at all-weather IoT applications. *Nano Energy* **2020**, *76*, 105052. [CrossRef]
33. Shi, Q.; Sun, Z.; Zhang, Z.; Lee, C. Triboelectric Nanogenerators and Hybridized Systems for Enabling Next-Generation IoT Applications. *Research* **2021**, *2021*, 6849171. [CrossRef] [PubMed]
34. He, T.; Guo, X.; Lee, C. Flourishing energy harvesters for future body sensor network: From single to multiple energy sources. *IScience* **2021**, *24*, 101934. [CrossRef] [PubMed]
35. Zi, Y.; Guo, H.; Wen, Z.; Yeh, M.-H.; Hu, C.; Wang, Z.L. Harvesting Low-Frequency (<5 Hz) Irregular Mechanical Energy: A Possible Killer Application of Triboelectric Nanogenerator. *ACS Nano* **2016**, *10*, 4797–4805. [CrossRef] [PubMed]
36. Fan, F.R.; Tian, Z.Q.; Lin Wang, Z. Flexible triboelectric generator. *Nano Energy* **2012**, *1*, 328–334. [CrossRef]
37. Wu, C.; Wang, A.C.; Ding, W.; Guo, H.; Wang, Z.L. Triboelectric Nanogenerator: A Foundation of the Energy for the New Era. *Adv. Energy Mater.* **2019**, *9*, 1802906. [CrossRef]
38. Zhu, J.; Zhu, M.; Shi, Q.; Wen, F.; Liu, L.; Dong, B.; Haroun, A.; Yang, Y.; Vachon, P.; Guo, X.; et al. Progress in TENG technology—A journey from energy harvesting to nanoenergy and nanosystem. *EcoMat* **2020**, *2*, 1–45. [CrossRef]
39. Shi, Q.; He, T.; Lee, C. More than energy harvesting–Combining triboelectric nanogenerator and flexible electronics technology for enabling novel micro-/nano-systems. *Nano Energy* **2019**, *57*, 851–871. [CrossRef]
40. Chen, L.; Shi, Q.; Sun, Y.; Nguyen, T.; Lee, C.; Soh, S. Controlling Surface Charge Generated by Contact Electrification: Strategies and Applications. *Adv. Mater.* **2018**, *30*, 1802405. [CrossRef]

41. Wang, J.; He, T.; Lee, C. Development of neural interfaces and energy harvesters towards self-powered implantable systems for healthcare monitoring and rehabilitation purposes. *Nano Energy* **2019**, *65*, 104039. [CrossRef]
42. Kim, K.N.; Chun, J.; Kim, J.W.; Lee, K.Y.; Park, J.-U.; Kim, S.-W.; Wang, Z.L.; Baik, J.M. Highly Stretchable 2D Fabrics for Wearable Triboelectric Nanogenerator under Harsh Environments. *ACS Nano* **2015**, *9*, 6394–6400. [CrossRef] [PubMed]
43. Shi, Q.; Wu, H.; Wang, H.; Wu, H.; Lee, C. Self-Powered Gyroscope Ball Using a Triboelectric Mechanism. *Adv. Energy Mater.* **2017**, *7*, 1701300. [CrossRef]
44. Chen, B.; Yang, Y.; Wang, Z.L. Scavenging Wind Energy by Triboelectric Nanogenerators. *Adv. Energy Mater.* **2018**, *8*, 1702649. [CrossRef]
45. Wang, S.; Mu, X.; Yang, Y.; Sun, C.; Gu, A.Y.; Wang, Z.L. Flow-Driven Triboelectric Generator for Directly Powering a Wireless Sensor Node. *Adv. Mater.* **2015**, *27*, 240–248. [CrossRef]
46. Su, Y.; Wen, X.; Zhu, G.; Yang, J.; Chen, J.; Bai, P.; Wu, Z.; Jiang, Y.; Lin Wang, Z. Hybrid triboelectric nanogenerator for harvesting water wave energy and as a self-powered distress signal emitter. *Nano Energy* **2014**, *9*, 186–195. [CrossRef]
47. Chen, J.; Yang, J.; Li, Z.; Fan, X.; Zi, Y.; Jing, Q.; Guo, H.; Wen, Z.; Pradel, K.C.; Niu, S.; et al. Networks of Triboelectric Nanogenerators for Harvesting Water Wave Energy: A Potential Approach toward Blue Energy. *ACS Nano* **2015**, *9*, 3324–3331. [CrossRef]
48. Wen, X.; Yang, W.; Jing, Q.; Wang, Z.L. Harvesting broadband kinetic impact energy from mechanical triggering/vibration and water waves. *ACS Nano* **2014**, *8*, 7405–7412. [CrossRef]
49. Yang, W.; Chen, J.; Jing, Q.; Yang, J.; Wen, X.; Su, Y.; Zhu, G.; Bai, P.; Wang, Z.L. 3D stack integrated triboelectric nanogenerator for harvesting vibration energy. *Adv. Funct. Mater.* **2014**, *24*, 4090–4096. [CrossRef]
50. Zhu, J.; Wang, H.; Zhang, Z.; Ren, Z.; Shi, Q.; Liu, W.; Lee, C. Continuous direct current by charge transportation for next-generation IoT and real-time virtual reality applications. *Nano Energy* **2020**, *73*, 104760. [CrossRef]
51. Xie, Y.; Wang, S.; Niu, S.; Lin, L.; Jing, Q.; Su, Y.; Wu, Z.; Wang, Z.L. Multi-layered disk triboelectric nanogenerator for harvesting hydropower. *Nano Energy* **2014**, *6*, 129–136. [CrossRef]
52. Hinchet, R.; Yoon, H.J.; Ryu, H.; Kim, M.K.; Choi, E.K.; Kim, D.S.; Kim, S.W. Transcutaneous ultrasound energy harvesting using capacitive triboelectric technology. *Science* **2019**, *365*, 491–494. [CrossRef] [PubMed]
53. Yang, J.; Chen, J.; Liu, Y.; Yang, W.; Su, Y.; Wang, Z.L. Triboelectrification-based organic film nanogenerator for acoustic energy harvesting and self-powered active acoustic sensing. *ACS Nano* **2014**, *8*, 2649–2657. [CrossRef] [PubMed]
54. Li, X.; Lin, Z.H.; Cheng, G.; Wen, X.; Liu, Y.; Niu, S.; Wang, Z.L. 3D fiber-based hybrid nanogenerator for energy harvesting and as a self-powered pressure sensor. *ACS Nano* **2014**, *8*, 10674–10681. [CrossRef] [PubMed]
55. Toyabur Rahman, M.; Sohel Rana, S.; Salauddin, M.; Maharjan, P.; Bhatta, T.; Kim, H.; Cho, H.; Park, J.Y. A highly miniaturized freestanding kinetic-impact-based non-resonant hybridized electromagnetic-triboelectric nanogenerator for human induced vibrations harvesting. *Appl. Energy* **2020**, *279*, 115799. [CrossRef]
56. Li, Y.; Chen, Z.; Zheng, G.; Zhong, W.; Jiang, L.; Yang, Y.; Jiang, L.; Chen, Y.; Wong, C.P. A magnetized microneedle-array based flexible triboelectric-electromagnetic hybrid generator for human motion monitoring. *Nano Energy* **2020**, *69*, 104415. [CrossRef]
57. Yang, Y.; Zhang, H.; Liu, Y.; Lin, Z.-H.; Lee, S.; Lin, Z.; Wong, C.P.; Wang, Z.L. Silicon-Based Hybrid Energy Cell for Self-Powered Electrodegradation and Personal Electronics. *ACS Nano* **2013**, *7*, 2808–2813. [CrossRef]
58. Xing, L.; Nie, Y.; Xue, X.; Zhang, Y. PVDF mesoporous nanostructures as the piezo-separator for a self-charging power cell. *Nano Energy* **2014**, *10*, 44–52. [CrossRef]
59. Kim, W.J.; Vivekananthan, V.; Khandelwal, G.; Chandrasekhar, A.; Kim, S.J. Encapsulated triboelectric–electromagnetic hybrid generator for a sustainable blue energy harvesting and self-powered oil spill detection. *Acs Appl. Electron. Mater.* **2020**, *2*, 3100–3108. [CrossRef]

nanomaterials

MDPI

Article

Physical Operations of a Self-Powered IZTO/β-Ga_2O_3 Schottky Barrier Diode Photodetector

Madani Labed [1], Hojoong Kim [2,3], Joon Hui Park [2], Mohamed Labed [4], Afak Meftah [1], Nouredine Sengouga [1] and You Seung Rim [2,*]

[1] Laboratory of Semiconducting and Metallic Materials (LMSM), University of Biskra, Biskra 07000, Algeria; madani.labed@univ-biskra.dz (M.L.); af.meftah@univ-biskra.dz (A.M.); n.sengouga@univ-biskra.dz (N.S.)

[2] Department of Intelligent Mechatronics Engineering, and Convergence Engineering for Intelligent Drone, Sejong University, Seoul 05006, Korea; hkim3023@gatech.edu (H.K.); julia980406@gmail.com (J.H.P.)

[3] George W. Woodruff School of Mechanical Engineering, Institute for Electronics and Nanotechnology, Georgia Institute of Technology, Atlanta, GA 30332, USA

[4] High Collage of Food Sciences and Food Industries, Algiers 16200, Algeria; labed@essaia.dz

* Correspondence: youseung@sejong.ac.kr

Abstract: In this work, a self-powered, solar-blind photodetector, based on InZnSnO (IZTO) as a Schottky contact, was deposited on the top of Si-doped β-Ga_2O_3 by the sputtering of two-faced targets with InSnO (ITO) as an ohmic contact. A detailed numerical simulation was performed by using the measured J–V characteristics of IZTO/β-Ga_2O_3 Schottky barrier diodes (SBDs) in the dark. Good agreement between the simulation and the measurement was achieved by studying the effect of the IZTO workfunction, β-Ga_2O_3 interfacial layer (IL) electron affinity, and the concentrations of interfacial traps. The IZTO/β-Ga_2O_3 (SBDs) was tested at a wavelength of 255 nm with the photo power density of 1 mW/cm^2. A high photo-to-dark current ratio of 3.70×10^5 and a photoresponsivity of 0.64 mA/W were obtained at 0 V as self-powered operation. Finally, with increasing power density the photocurrent increased, and a 17.80 mA/W responsivity under 10 mW/cm^2 was obtained.

Keywords: IZTO/β-Ga_2O_3 Schottky diode; solar-blind; self-powered; photodetector; modeling

Citation: Labed, M.; Kim, H.; Park, J.H.; Labed, M.; Meftah, A.; Sengouga, N.; Rim, Y.S. Physical Operations of a Self-Powered IZTO/β-Ga_2O_3 Schottky Barrier Diode Photodetector. *Nanomaterials* **2022**, *12*, 1061. https://doi.org/10.3390/nano12071061

Academic Editors: Qiongfeng Shi, Jianxiong Zhu and Uroš Cvelbar

Received: 29 December 2021
Accepted: 21 March 2022
Published: 24 March 2022

1. Introduction

Gallium oxide (Ga_2O_3) is an oxide semiconductor material with a long, rich history [1–3]. It has an ultra wide bandgap (UWBG) of ~4.8 eV, a high breakdown electric field of ~8 MV/cm, and a high saturation velocity of 1×10^7 cm/s, and these properties have brought Ga_2O_3 to the fore once again [1,2,4]. Ga_2O_3 has six polymorphs, i.e., α, β, γ, δ, ε, and k, with β-Ga_2O_3 being the most stable [1]. Unipolar devices based on β-Ga_2O_3, such as the metal–oxide–semiconductor field-effect transistor (MOSFET) [5], thin film transistor (TFT) [6], field emission (FE) [7], and Schottky barrier diode (SBD) [1–4,8,9], have been studied extensively. It is also used for deep ultraviolet (DUV) photodetectors (PDs) for solar-blind applications [10,11]. DUV PDs work in the solar-blind spectrum with wavelengths shorter than 280 nm, which means that they can be applied to optical communication, chemical analysis, missile tracking, and harsh environmental monitoring [12]. Different types of solar-blind PD structures based on β-Ga_2O_3, such as metal–semiconductor–metal [13], heterojunction [14], and Schottky [15], have been reported. The Schottky barrier diode solar-blind has some advantages, including low dark current and low cost in comparison with heterojunctions [11,16]. Self-powered solar-blind PDs are of special interest, because they can work in the absence of an external power supply. A strong built-in electric field ensures that this high-performance, self-powered, solar-blind Schottky barrier diode photodetector can operate at zero bias voltage [14]. Different metals have been used as the Schottky contacts of β-Ga_2O_3, e.g., Au, Ti, Ni, Pt, Cu, and Pb [17]. For example, Chen et al. [18] reported a self-powered photodetector based on a Au/β-Ga_2O_3 nanowire array film Schottky, in which the responsivity reached 0.01 (mA/W) during 254 nm light illumination with 2 mW/cm^2 at a bias of 0 V. Zhi et al. [19] studied the Au/β-Ga_2O_3 Schottky solar-blind photodetector, and they achieved a responsivity of 0.4 (mA/W) for 0 V bias and 254 nm illumination. Peng et al. [10] reported a Pt/β-Ga_2O_3 Schottky barrier diode solar-blind photodetector with nearly a 10^4 of light to dark current ratio at 0 V bias and a wavelength of 254 nm. Liu et al. [20] studied a Ni/β-Ga_2O_3 Schottky barrier diode solar-blind photodetector tested under 254 nm light, and they obtained responsivities of about 806.02 (A/W) and 1372.92 (A/W) under −5 V and 5 V, respectively. In the publications mentioned above, various metals were used, such as Pt, Ni, and Au, for Schottky contact formation with β-Ga_2O_3. In addition, transparent materials and oxides are used for the formation of the Schottky contact with β-Ga_2O_3. However, these types of materials formed a low Schottky barrier height with β-Ga_2O_3. For example, Zhuo et al. [14] reported a MoS_2/β-Ga_2O_3 heterojunction self-powered photodetector with a 670 ratio of light current to dark current at 0 V, and this result was achieved because of the very low value of the Schottky barrier. Chen et al. [17] studied a self-powered MXenes/β-Ga_2O_3 photodetector under 254 nm wavelength with a light illumination of 115.1 μW.cm^{-2} and a 1.6×10^4 ratio of light current to dark current at 0 V and an extracted Schottky barrier height of about 0.9 eV. Then, Cui et al. [21] published a flexible solar-blind amorphous β-Ga_2O_3 photodetector with an indium tin oxide (ITO) transparent conducting electrode, with a photocurrent that is less effected at 0 V and under 254 nm when it is exposed to an oxygen flux of 0.14 SCCM; this result was related to the low Schottky barrier height of about 0.97 eV. In addition, Kim et al. [11] used InZnSnO (IZTO) for Schottky contact formation with β-Ga_2O_3, and a Schottky barrier height greater than 1.06 eV was obtained; these results indicate that this photodetector can work in the absence of an external power supply.

Here, we constructed simulation–experiment combination of ITZO/β-Ga_2O_3 SBD-based UV photodiodes under the illumination to reveal the physics behind the behavior of the J–V characteristics in terms of workfunction, IL electron affinity and interfacial traps. The different conduction mechanisms we performed were taken into consideration either collectively or individually. Their parameters were scanned over a physically acceptable range so that an acceptable comparison of measurement to simulation was achieved. Good agreement between simulation and experiment was observed clearly with the consideration of the effect of different IZTO workfunctions, IZTO/β-Ga_2O_3 interfacial layer electron affinity, and the effect of interfacial traps. Additionally, the IZTO/β-Ga_2O_3 self-powered solar-blind PDs were tested and compared with measurements at the photo power density of 1 mW/cm^2 and the wavelength of 255 nm.

2. Experiment

The Solar-blind Schottky photodetector was fabricated on a 650 μm, Sn-doped, bulk β-Ga_2O_3 single-crystal wafer ((N_d−N_a) = 1×10^{18} cm^{-3}, Novel Crystal Technology, Inc., Saitama, Japan) with (001) surface orientation. The epitaxial layer of Si-doped β-Ga_2O_3 (10 μm thick, 1×10^{18} cm^{-3}) was grown by halide vapor phase epitaxy (HVPE). Si-doped β-Ga_2O_3 was used as the active layer in this solar-blind Schottky photodetector, as it provides a high purity and a low resistance [2]. The ITO electrode was deposited by sputtering on the bottom of Sn-doped β-Ga_2O_3 as an ohmic contact. IZTO was deposited on the top of the Si-doped β-Ga_2O_3 as a Schottky contact by two-faced target co-sputtering with ITO (In_2O_3:SnO = 9:1) and IZO (In_2O_3:ZnO = 9:1) at room temperature. Figure 1 shows a schematic representation of this SBD structure. After the deposition of the layers, the device was annealed at 600 °C in Ar for 1 min using rapid thermal annealing. The electrical J–V in dark and light was measured using a semiconductor analyzer and a source meter (SCS-4200A and 2410 Source meter, Keithley, Beaverton, OR, USA). Further details can be found in our previous publication [11].

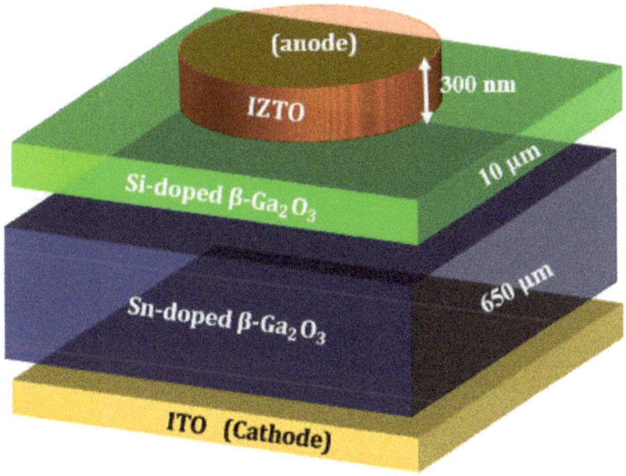

Figure 1. A schematic representation of the β-Ga_2O_3 Schottky barrier diode (SBD) structure.

3. Simulation Methodology

For the simulation, we considered thermionic emission, Shockley–Read–Hall, Auger recombination, and image force lowering models. The physical parameters of different layers and related traps are presented in Tables 1 and 2, respectively.

SILVACO TCAD (Version 5.24.1.R, Silvaco Inc.: Santa Clara, CA, USA) was used to model the above structure. It solves the basic drift–diffusion semiconductor Poisson and continuity equations, which are [2,3,22]:

Table 1. Properties of each layer of the studied SBD [2,3,23].

Parameters	Sn: β-Ga_2O_3	Si: β-Ga_2O_3
Bandgap (eV)	4.8	4.8
Affinity (eV)	4	4
Hole mobility ($cm^2V^{-1}s^{-1}$)	10	10
Electron mobility ($cm^2V^{-1}s^{-1}$)	172	300
m_e^*/m_0	0.28	0.28
m_h^*/m_0	0.35	0.35
Relative permittivity	12.6	11
N_c (cm^{-3})	3.7×10^{18}	3.7×10^{18}
N_v (cm^{-3})	5×10^{18}	5×10^{18}
N_d (cm^{-3})	1×10^{18}	3×10^{16}
Minority carrier diffusion length (nm)	450	450
Saturation velocity (cm s^{-1})	10^7	10^7

Table 2. Characteristics of the Sn-doped and Si-doped β-Ga_2O_3 traps considered in this work [1–3,24].

Traps	Trap Level (E_c−E) (eV)	Concentration (cm^{-3})	Capture Cross Section σ_n (cm^2)	σ_n/σ_p
Sn-doped β-Ga_2O_3 Bulk layer	0.55	3×10^{13}	2×10^{-14}	100
	0.74	2×10^{16}	2×10^{-14}	100
	1.04	4×10^{16}	2×10^{-14}	10
Si-doped β-Ga_2O_3 thin layer	0.60	3.6×10^{13}	2×10^{-14}	100
	0.75	4.6×10^{13}	2×10^{-14}	100
	0.72	4.6×10^{13}	2×10^{-14}	100
	1.05	1.1×10^{14}	2×10^{-14}	10

Poisson equation is given by [2,3,22]:

$$div(\varepsilon\nabla\psi) = -q(p - n + N_d \pm N_t^{\pm}) \quad (1)$$

where ψ is the electrostatic potential, ε is the permittivity, p and n are the concentrations of the free holes and electrons, respectively, and $N_t^{\pm}$ is the density of the ionized traps (cm^{-3}).

The continuity equations for electrons and holes as defined in steady states are given by [2,3,22]:

$$0 = \frac{1}{q} div\vec{J_n} + G_n - R_n \quad (2)$$

$$0 = -\frac{1}{q} div\vec{J_p} + G_p - R_p \quad (3)$$

where G_n and G_p are the generation rates for electrons and holes, respectively, and R_n and R_p are the recombination rates for electrons and holes, respectively. $\vec{J}_n$ and $\vec{J}_p$ are the electron density and the hole current density, respectively, which are given in terms of the free electron and hole density (n and p), electric field (E) and mobility (μ_n and μ_p) [25]:

$$\vec{J}_n = q\mu_n nE + \mu_n K_B T\nabla n \quad (4)$$

$$\vec{J}_p = q\mu_p pE - \mu_p K_B T\nabla p \quad (5)$$

Traps are represented by their ionized density, $N_t^{\pm}$. The sign $\pm$ depends on whether the trap is an acceptor or a donor, so that $N_t^{+} = fN_t$ and $N_t^{-} = (1-f)N_t$, f is the occupancy function given by $f = \frac{\sigma_n n + \sigma_p p}{\sigma_n(n+n_t)+\sigma_p(p+p_t)}$, and $\sigma_{n(p)}$ is the trap capture cross-section for electrons (holes). The recombination rate is related to traps through the well-known SRH formula, i.e., $R_{n,p} = \frac{pn - n_i^2}{\tau_{0n}(p+p_t)+\tau_{0p}(n+n_t)}$, where $n_t = n_i exp(-(E_i - E_t)/kT)$ and $p_t = n_i exp(-(E_t - E_i)/kT)$, and τ_{0n} and τ_{0p} are the minority carrier lifetimes which are also related to the traps through $\tau_{0n(p)} = \frac{1}{v_{thn(p)}\sigma_{n(p)}N_t}$, where $v_{thn(p)}$ is the thermal velocity of electrons (holes).

According to several publications [11,26,27], ITO forms an Ohmic contact with β-Ga_2O_3. In this simulation, an ideal ohmic contact was considered for ITO/β-Ga_2O_3 interface. A low ITO/β-Ga_2O_3 barrier was achieved when highly doped β-Ga_2O_3 substrate was used.

4. Results and Discussion

4.1. Optical and Electrical Properties of IZTO Thin Film

The obtained resistivity, carrier concentration, workfunction, and mobility of IZTO were 4.86×10^{-4} Ω cm, 2.80×10^{20} cm^{-3}, 4.79 eV, and 10.83 cm^2/V s, respectively. Figure 2a shows the optical transmission ($T(\lambda)$) of IZTO thin film as a function of the wavelength in the 250–1200 nm range. The average transmittance of the IZTO films in the visible wavelength range was over 87%.

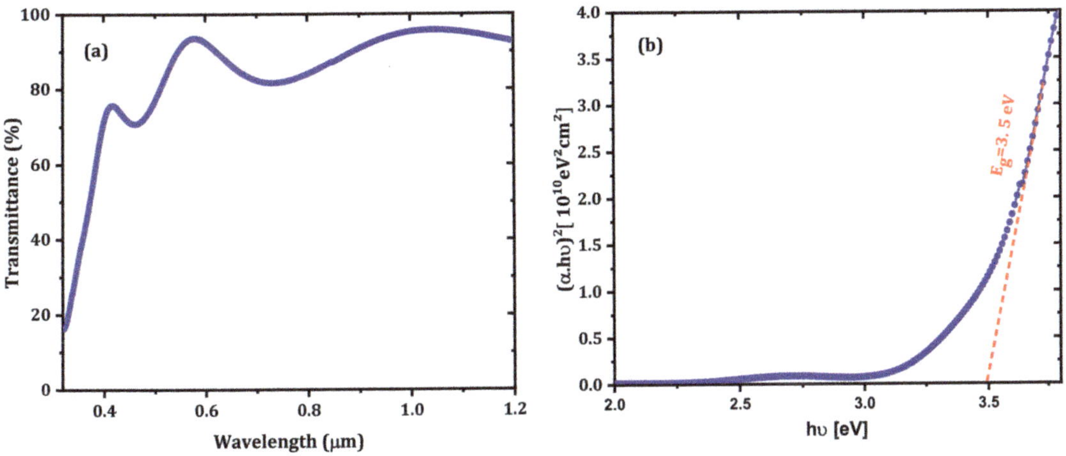

Figure 2. (**a**) The transmittance of IZTO thin film; (**b**) $(\alpha\,(\lambda).h\nu)^2$ of the IZTO thin film.

For direct bandgap semiconductors, the absorption can be obtained from the following Equation [15]:

$$(\alpha h\nu)^2 = C\cdot(h\nu - E_g) \tag{6}$$

where α is the absorption coefficient, C is a constant, h is Planck's constant, and ν is the frequency of the incident light. By plotting $(\alpha\,(\lambda)\cdot h\nu)^2$ versus $h\nu$, the optical bandgap of the IZTO thin film was determined to be 3.5 eV, as shown in Figure 2b. This value agrees with the published value [28]. The refractive index has a significant importance in the design of optical devices. It reflects the crystallinity and optical quality of thin films. The extinction coefficient (k) and the refractive index (n) of IZTO thin film are calculated by [29]:

$$k(\lambda) = \frac{\alpha.\lambda}{4\pi} \tag{7}$$

$$n(\lambda) = \frac{(1 - R(\lambda))}{(1 + R(\lambda))} + \sqrt{\frac{4R(\lambda)}{1 - R(\lambda)^2} - k(\lambda)^2} \tag{8}$$

where $R(\lambda)$ is the reflectance of the thin film, which can be calculated by the following Equation [30]:

$$R(\lambda) = 1 - \sqrt{T(\lambda)e^{\frac{\alpha.t}{2}}} \tag{9}$$

where t is the IZTO thin film thickness, which is evaluated by ellipsometry at $\approx$300 nm. The extracted refractive index (n) and the extinction coefficient (k) are presented in Figure 3.

As mentioned above, the photodetector consisted of 300 nm IZTO deposited on the top of Si-doped β-Ga_2O_3, and the ITO layer was considered as an ohmic contact on the bottom of the Sn-doped β-Ga_2O_3 (Figure 1). Before simulating the proposed photodetector, the measurement of the dark output current density of SBD was reproduced, and details are provided in the next section.

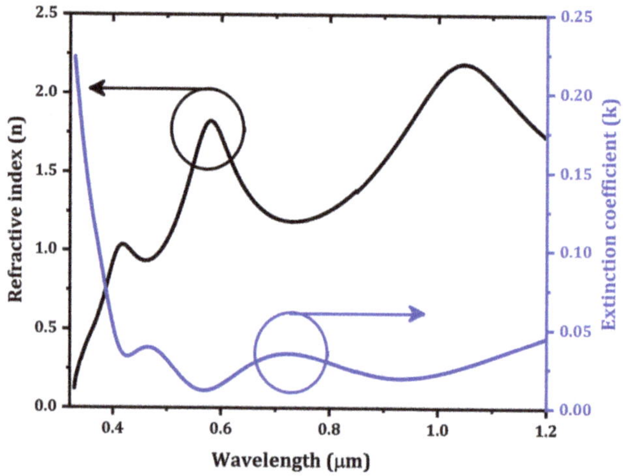

Figure 3. The extracted refractive index (n) and the extinction coefficient (k) of IZTO thin film.

4.2. Modeling the Dark Current of IZTO/β-Ga_2O_3 SBD

As presented in Figure 4, when the tunneling transport mechanism was not considered, the shape of the IZTO/β-Ga_2O_3 SBD current is parallel to the measurement current; this indicates that the thermionic transport mechanism dominates in the forward bias. However, when the properties presented in Table 1 and the traps presented in Table 2 were considered, a huge disagreement was obtained between the simulation and the measurement. This disagreement is related to the IZTO workfunction, i.e., the IZTO/β-Ga_2O_3 IL electron affinity (conduction band minimum), in addition to the effect of the surface traps between IZTO and the β-Ga_2O_3 drift layer.

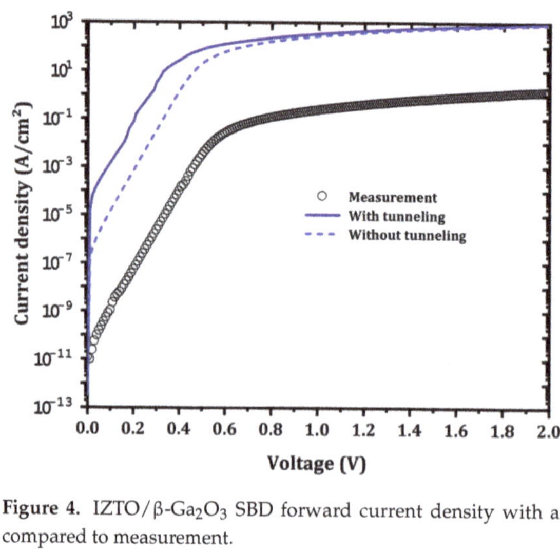

Figure 4. IZTO/β-Ga_2O_3 SBD forward current density with and without the tunneling model compared to measurement.

4.3. Effect of IL Electron Affinity

Next, an IL (10 nm thickness) between Si-doped β-Ga_2O_3 and IZTO was considered for modeling the effect IZTO/β-Ga_2O_3 conduction band offset on the SBD performance. The effects of the IL electron affinity on the SBD J–V characteristics were studied, and these effects are shown in Figure 5. Experimentally, the IL electron affinity is related to the chemical composition of the surface [2] and surface polarization [31] as well as external effects, such as argon (Ar) bombardment, plasma, etc. The current density decreases with decreasing IL electron affinity of Si-doped β-Ga_2O_3, i.e., from 4 to 3.5 e V [32], but the IL electron affinity has a more pronounced effect in the high voltage domain. This is due to the increase in the height of the Schottky barrier (ϕ_B) with decreasing IL electron affinity according to the Schottky–Mott rule and the increase in the series resistance [2]. As presented in Figure 6, with decreasing IL electron affinity, the barrier between IZTO and β-Ga_2O_3 increases. An agreement between measurement and simulation occurs for voltages higher than 1 V for a 3.556 eV IL electron affinity. However, a disagreement between simulation and measurement was noticed in the low-voltage domain, and this is related to the effect of the IZTO workfunction and the concentrations of the interfacial traps, which are addressed in the next two subsections.

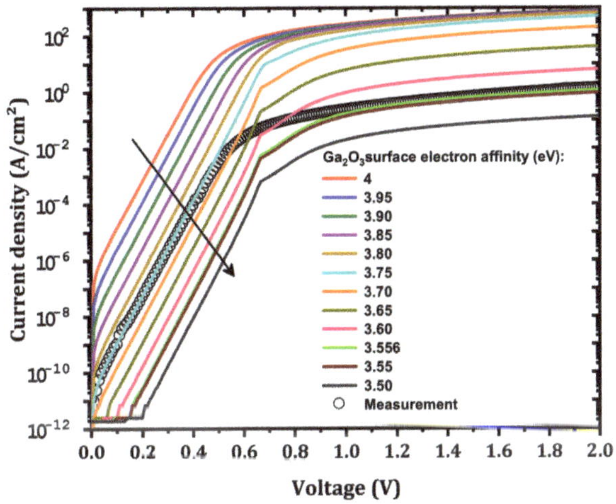

Figure 5. Effect of the IL electron affinity on the simulated J–V characteristics compared to measurement.

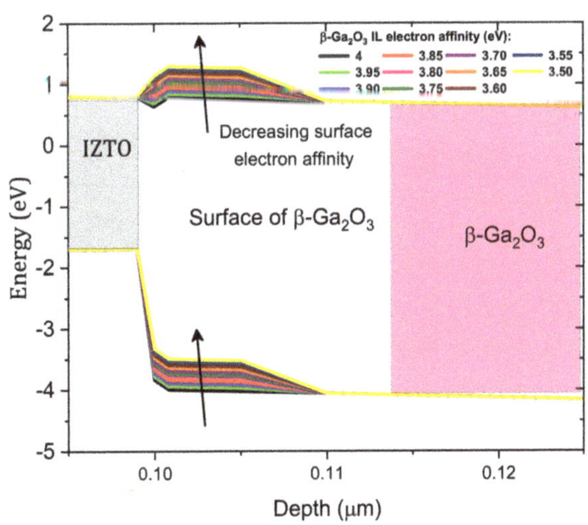

Figure 6. Equilibrium band diagram variation with the IL electron affinity.

4.4. *Effect of the IZTO Workfunction*

In addition to the IL electron affinity, the IZTO workfunction will have an effect. As shown in Figure 7, when the IZTO workfunction decreases from 5 to 4.5 eV [33,34] the current density increases. This increase in the current density is related to the decrease in ϕ_B as presented in Figure 8; the formed barrier between IZTO and Si-doped β-Ga_2O_3 increased. The best agreement between simulation and measurement was achieved for $\phi_{IZTO} = 4.6$ eV. The small deviation from measurement, i.e., in the range of 0.4–0.8 V, was due to the parameters of the traps that are considered (Table 2), which may not be accurate. It also may be related to the effect of the different compositions of the materials (In, Zn, and Sn) [35].

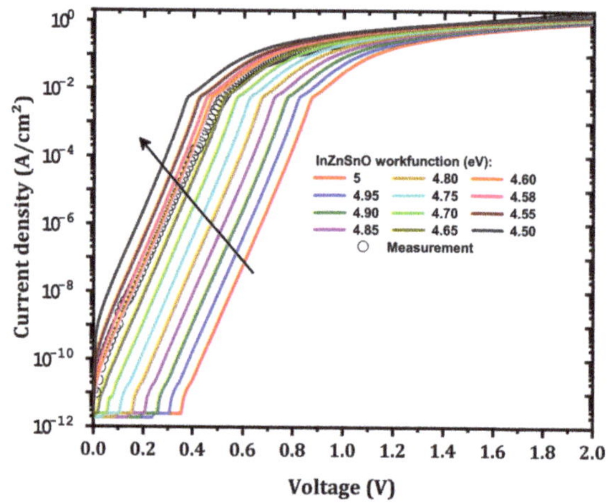

Figure 7. Simulated J–V characteristics for different IZTO workfunctions compared to measurements.

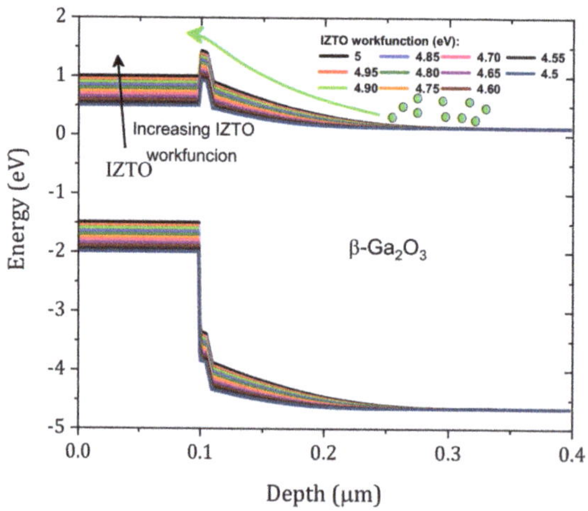

Figure 8. Equilibrium band diagram variation with the IZTO workfunction.

4.5. *Effect of the Concentration of Traps at the IL*

We studied the effect of the concentration of the ($E_c-1.05$) traps at IL on the characteristics of the SBD J–V. The traps that we considered were the most affected, especially given that the surface of the β-Ga_2O_3 is exposed to plasma and Ar bombardment [2]. For the four traps, these effects are shown in Figure 9a–d, respectively. First, all of the defects that were considered have a significant effect on the current density of the SBD. Among the defect concentrations that were considered, those that gave the best comparison with the measurements were 3.6×10^{16}, 4.6×10^{16}, 4.6×10^{16} and 1.1×10^{15} cm^{-3}, respectively. The dark current was affected at high trap densities, and this result was related to electrons being captured by the traps. Figure 10 shows the effect of traps on the equilibrium band diagram. When traps were considered with the concentration mentioned above, the difference of conduction band and Fermi level ($E_c - E_f$) increased, and this meant a decrease in the free electron density. A good comparison between simulation and measurement is obtained as presented in Figure 11 and the extracted SBD parameters are presented in Table 3. The Schottky barrier height (ϕ_B) and the R_s were extracted using the Sato and Yasumura method [3,36]. A high ϕ_B was obtained with a low ideality factor close to unity in addition to the low densities of the interfacial traps, as expected in our previous publication [11]. In addition, a very low saturation current was obtained.

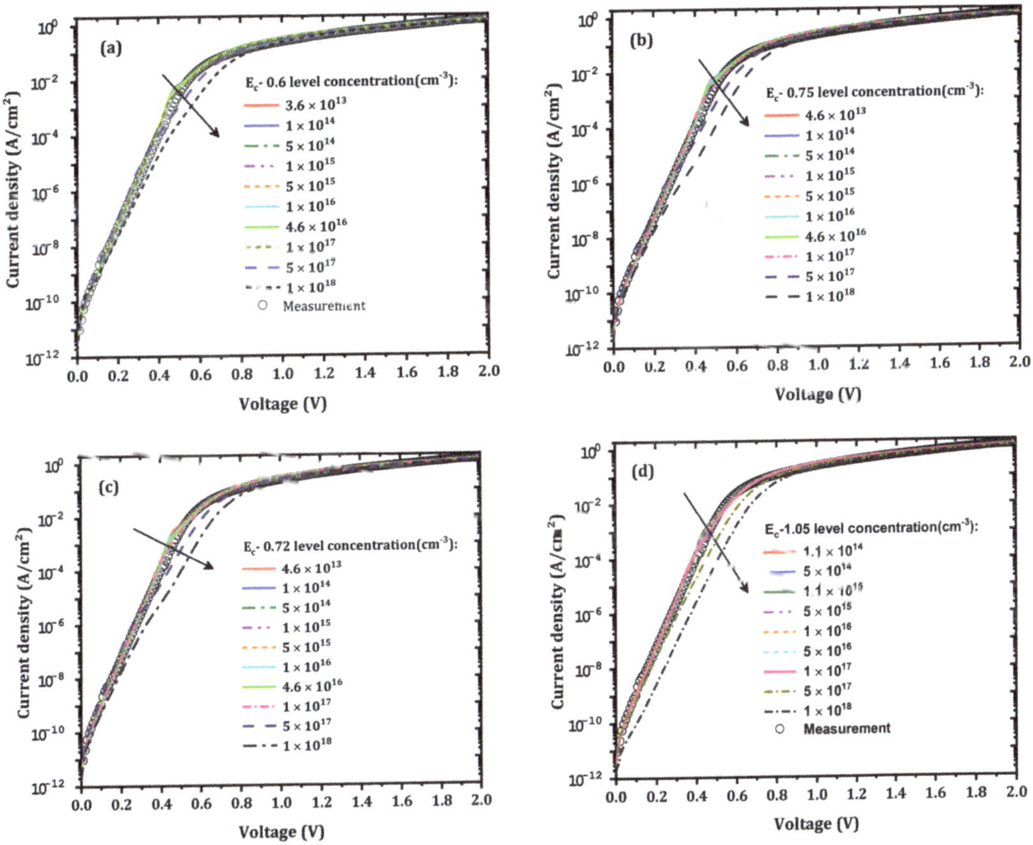

Figure 9. Effects of the density of the traps on SBD, i.e., the (**a**) $E_c - 0.6$ eV, (**b**) $E_c - 0.75$ eV, (**c**) $E_c - 0.72$ eV, and (**d**) $E_c - 1.05$ eV traps.

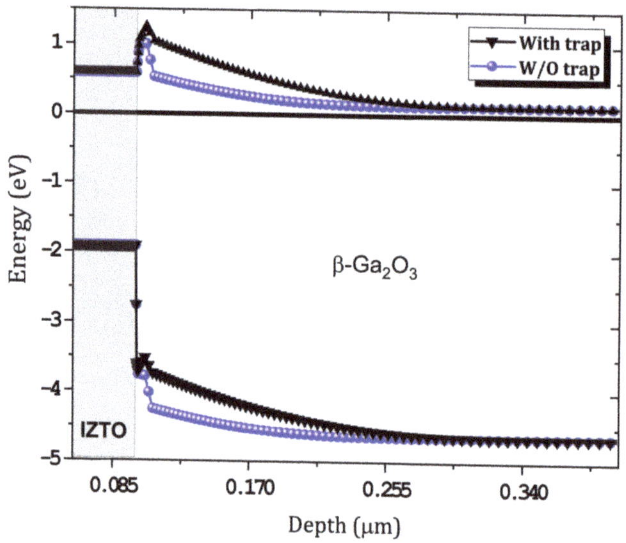

Figure 10. Equilibrium band diagram variation with and without traps.

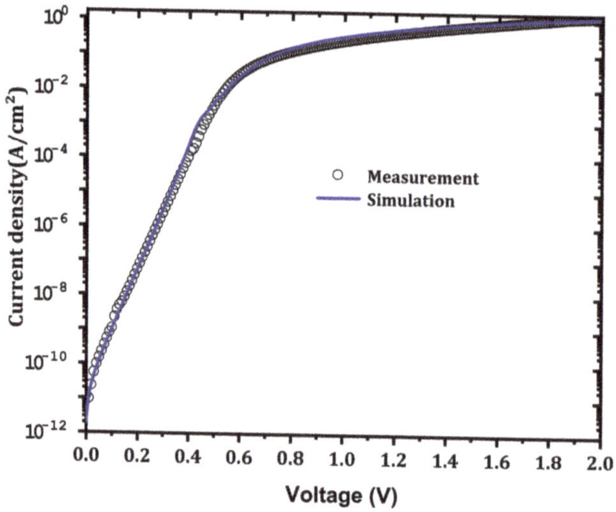

Figure 11. The best comparison between the simulation and the experimental results.

Table 3. Output parameters from simulation and measurement.

Parameter	n	ϕ_B (eV)	R_s (Ω cm^2)	R_{on} (Ω cm^2)	J_s (A/cm^2)
Simulation	1.02	1.25	1.78	1.01	1.72×10^{-12}
Measurement	1.03	1.29	1.91	1.04	1.11×10^{-11}

4.6. The Effect of 255 nm Wavelength Illumination on Forward Current

The simulated SBD J–V characteristics were successfully compared to measurements made at room temperature. This good agreement was achieved by modeling the effect of IL electron affinity, IZTO workfunction, and the concentrations of the IL traps. As shown in Table 3, a low dark saturation current was obtained, so this SBD is proposed as a high-performance solar-blind PD. The IZTO/β-Ga_2O_3 SBD was illuminated at 255 nm with a light intensity of 1mW/cm^2. To evaluate the performance of the IZTO/β-Ga_2O_3 solar-blind SBD under 255 nm, the forward photocurrent was extracted with the consideration of the previous IZTO workfunction, the IL electron affinity, and the concentrations of the interfacial traps. As presented in Figure 12, good agreement was demonstrated between the simulation results and the actual measurements. The solar-blind SBD exhibited a high rectifying characteristic after illumination at 255 nm in forward voltage (J_{Photon} / $J_{dark} = 3.70 \times 10^5$) under 0 V. The responsivity reached 0.64 (mA/W). The responsivity was estimated as follows [37]:

$$R_\lambda = \frac{J_{Photon} - J_{dark}}{P} \tag{10}$$

where J_{Photon}, J_{dark}, and P are the photocurrent at a given voltage, the dark current, and the power density, respectively. With zero bias voltage, the photodetector had good responsivity under 255 nm light illumination with intensity of 1 (mW/cm^2). In addition, a decrease in ϕ_B from 1.25 eV in dark to 1.18 after illumination was observed.

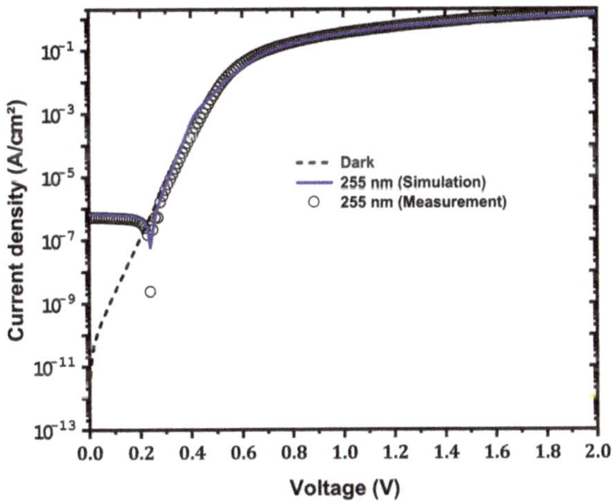

Figure 12. Comparison between simulation and measurement (under incident light power density of 1 mW/cm^2) of J-V characteristics of IZTO/β-Ga_2O_3 at 255 nm wavelength.

In this solar-blind PD, the built-in electric field in the depletion region of the β-Ga_2O_3 and IZTO interface was enough to separate the photogenerated electron–hole pairs toward corresponding electrodes. In this structure, a high built-in potential is related to the high ϕ_B. In addition, the effect of light power density on simulated J-V characteristics was studied. As presented in Figure 13, with increasing power density, the photocurrent increased, and a responsivity of 17.80 mA/W under 10 mW/cm^2 was achieved. This increase in photocurrent is related to the increase in photo-excited, separated, and collected carriers [38]. This result agrees with the result obtained by Wu et al. [39].

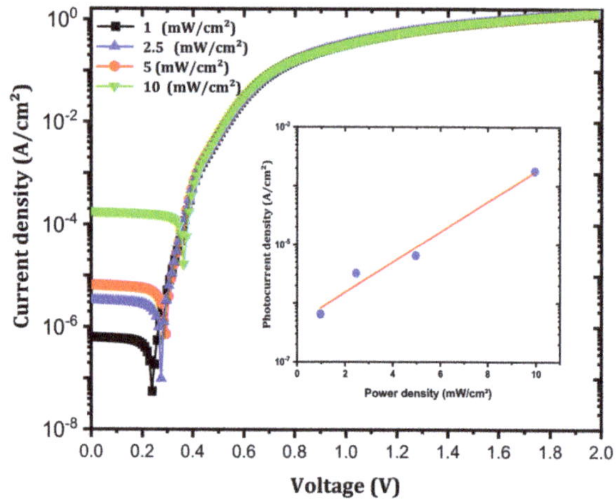

Figure 13. Effect of light power density on the simulated J–V characteristics. The inset is the variation of photocurrent with light power density at 0 V.

5. Conclusions

The J–V characteristics of an IZTO/β-Ga_2O_3 Schottky barrier diode was simulated by SILVACO-Atlas and compared with measurements. The effects of the IZTO workfunction and the interfacial layer electron affinity and trap concentrations were studied for further agreement with measurements made in the dark. Then, we demonstrated the IZTO/β-Ga_2O_3 self-powered Schottky photodetector with a high photo-to-dark ratio and responsivity of 3.70×10^5 and 0.64 (mA/W) under 255 nm illumination with 1 mW/cm^2 for 0 V, respectively. Finally, with increasing power density, the photocurrent increased, and a 17.80 mA/W responsivity under 10 mW/cm^2 was obtained.

Author Contributions: Software, writing—original draft preparation, M.L. (Madani Labed); conceptualization, visualization, H.K.; visualization, J.H.P.; visualization, M.L. (Mohamed Labed); methodology A.M.; writing—review and editing N.S.; writing—review and editing Y.S.R. All authors have read and agreed to the published version of the manuscript.

Funding: This paper was supported by the Korea Institute for Advancement of Technology (KIAT) grant funded by the Ministry of Trade, Industry & Energy (MOTIE, Korea). (P0012451, The Competency Development Program for Industry Specialist) and was also supported by the Technology Innovation Program—(20016102, Development of 1.2 kV Gallium oxide power semiconductor devices technology) funded by MOTIE, Korea.

Institutional Review Board Statement: Not applicable.

Informed Consent Statement: Not applicable.

Data Availability Statement: Data presented in this article is available on request from the corresponding author.

Acknowledgments: Madani Labed and Hojoong Kim contributed equally to this work.

Conflicts of Interest: The authors declare that they have no known competing financial interests or personal relationships that could have appeared to influence the work reported in this paper.

References

1. Galazka, Z. β-Ga_2O_3 for wide-bandgap electronics and optoelectronics. *Semicond. Sci. Technol.* **2018**, *33*, 113001. [CrossRef]
2. Labed, M.; Sengouga, N.; Labed, M.; Meftah, A.; Kyoung, S.; Kim, H.; Rim, Y.S. Modeling a Ni/β-Ga_2O_3 Schottky barrier diode deposited by confined magnetic-field-based sputtering. *J. Phys. D Appl. Phys.* **2021**, *54*, 115102. [CrossRef]
3. Labed, M.; Sengouga, N.; Labed, M.; Meftah, A.; Kyoung, S.; Kim, H.; Rim, Y.S. Modeling and analyzing temperature-dependent parameters of Ni/β-Ga_2O_3 Schottky barrier diode deposited by confined magnetic field-based sputtering. *Semicond. Sci. Technol.* **2021**, *36*, 35020. [CrossRef]
4. Labed, M.; Sengouga, N.; Meftah, A.; Labed, M.; Kyoung, S.; Kim, H.; Rim, Y.S. Leakage Current Modelling and Optimization of β-Ga_2O_3 Schottky Barrier Diode with Ni Contact under High Reverse Voltage. *ECS J. Solid State Sci. Technol.* **2020**, *9*, 125001. [CrossRef]
5. Pearton, S.J.; Ren, F.; Tadjer, M.; Kim, J. Perspective: Ga_2O_3 for ultra-high power rectifiers and MOSFETS. *J. Appl. Phys.* **2018**, *124*, 220901. [CrossRef]
6. Thomas, S.R.; Adamopoulos, G.; Lin, Y.-H.; Faber, H.; Sygellou, L.; Stratakis, E.; Pliatsikas, N.; Patsalas, P.A.; Anthopoulos, T.D. High electron mobility thin-film transistors based on Ga_2O_3 grown by atmospheric ultrasonic spray pyrolysis at low temperatures. *Appl. Phys. Lett.* **2014**, *105*, 092105. [CrossRef]
7. Grillo, A.; Barrat, J.; Galazka, Z.; Passacantando, M.; Giubileo, F.; Iemmo, L.; Luongo, G.; Urban, F.; Dubourdieu, C.; Di Bartolomeo, A. High field-emission current density from β-Ga_2O_3 nanopillars. *Appl. Phys. Lett.* **2019**, *114*, 193101. [CrossRef]
8. Jian, G.; He, Q.; Mu, W.; Fu, B.; Dong, H.; Qin, Y.; Zhang, Y.; Xue, H.; Long, S.; Jia, Z.; et al. Characterization of the inhomogeneous barrier distribution in a Pt/(100)β-Ga_2O_3 Schottky diode via its temperature-dependent electrical properties. *AIP Adv.* **2018**, *8*, 015316. [CrossRef]
9. Labed, M.; Park, J.H.; Meftah, A.; Sengouga, N.; Hong, J.Y.; Jung, Y.-K.; Rim, Y.S. Low Temperature Modeling of Ni/β-Ga_2O_3 Schottky Barrier Diode Interface. *ACS Appl. Electron. Mater.* **2021**, *3*, 3667–3673. [CrossRef]
10. Peng, B.; Yuan, L.; Zhang, H.; Cheng, H.; Zhang, S.; Zhang, Y.; Zhang, Y.; Jia, R. Fast-response self-powered solar-blind photodetector based on Pt/β-Ga_2O_3 Schottky barrier diodes. *Optik* **2021**, *245*, 167715. [CrossRef]
11. Kim, H.; Seok, H.-J.; Park, J.H.; Chung, K.B.; Kyoung, S.; Kim, H.-K.; Rim, Y.S. Fully Transparent InZnSnO/β-Ga_2O_3/InSnO Solar-Blind Photodetectors with High Schottky Barrier Height and Low-Defect Interfaces. *J. Alloys Compd.* **2021**, *890*, 161931. [CrossRef]
12. Guo, D.; Liu, H.; Li, P.; Wu, Z.; Wang, S.; Cui, C.; Li, C.; Tang, W. Zero-Power-Consumption Solar-Blind Photodetector Based on β-Ga_2O_3/NSTO Heterojunction. *ACS Appl. Mater. Interfaces* **2017**, *9*, 1619–1628. [CrossRef] [PubMed]
13. Qin, Y.; Li, L.; Zhao, X.; Tompa, G.S.; Dong, H.; Jian, G.; He, Q.; Tan, P.; Hou, X.; Zhang, Z.; et al. Metal–Semiconductor–Metal ε-Ga_2O_3 Solar-Blind Photodetectors with a Record-High Responsivity Rejection Ratio and Their Gain Mechanism. *ACS Photonics* **2020**, *7*, 812–820. [CrossRef]
14. Zhuo, R.; Wu, D.; Wang, Y.; Wu, E.; Jia, C.; Shi, Z.; Xu, T.; Tian, Y.; Li, X. A self-powered solar-blind photodetector based on a MoS_2/β-Ga_2O_3 heterojunction. *J. Mater. Chem. C* **2018**, *6*, 10982–10986. [CrossRef]
15. Liu, Z.; Wang, X.; Liu, Y.; Guo, D.; Li, S.; Yan, Z.; Tan, C.-K.; Li, W.; Li, P.; Tang, W. A high-performance ultraviolet solar-blind photodetector based on a β-Ga_2O_3 Schottky photodiode. *J. Mater. Chem. C* **2019**, *7*, 13920–13929. [CrossRef]
16. Zou, Y.; Zhang, Y.; Hu, Y.; Gu, H. Ultraviolet Detectors Based on Wide Bandgap Semiconductor Nanowire: A Review. *Sensors* **2018**, *18*, 2072. [CrossRef]
17. Chen, Y.; Zhang, K.; Yang, X.; Chen, X.; Sun, J.; Zhao, Q.; Li, K.; Shan, C. Solar-blind photodetectors based on MXenes–β-Ga_2O_3 Schottky junctions. *J. Phys. D Appl. Phys.* **2020**, *53*, 484001. [CrossRef]
18. Chen, X.; Liu, K.; Zhang, Z.; Wang, C.; Li, B.; Zhao, H.; Zhao, D.; Shen, D. Self-Powered Solar-Blind Photodetector with Fast Response Based on Au/β-Ga_2O_3 Nanowires Array Film Schottky Junction. *ACS Appl. Mater. Interfaces* **2016**, *8*, 4185–4191. [CrossRef]
19. Zhi, Y.; Liu, Z.; Chu, X.; Li, S.; Yan, Z.; Wang, X.; Huang, Y.; Wang, J.; Wu, Z.; Guo, D.; et al. Self-Powered β-Ga2O3 Solar-Blind Photodetector Based on the Planar Au/Ga_2O_3 Schottky Junction. *ECS J. Solid State Sci. Technol.* **2020**, *9*, 65011. [CrossRef]
20. Liu, Z.; Zhi, Y.; Zhang, S.; Li, S.; Yan, Z.; Gao, A.; Zhang, S.; Guo, D.; Wang, J.; Wu, Z.; et al. Ultrahigh-performance planar β-Ga_2O_3 solar-blind Schottky photodiode detectors. *Sci. China Technol. Sci.* **2021**, *64*, 59–64. [CrossRef]
21. Cui, S.; Mei, Z.; Zhang, Y.; Liang, H.; Du, X. Room-Temperature Fabricated Amorphous Ga_2O_3 High-Response-Speed Solar-Blind Photodetector on Rigid and Flexible Substrates. *Adv. Opt. Mater.* **2017**, *5*, 1700454. [CrossRef]
22. Labed, M.; Sengouga, N.; Meftah, A.; Meftah, A.; Rim, Y.S. Study on the improvement of the open-circuit voltage of NiOx/Si heterojunction solar cell. *Opt. Mater.* **2021**, *120*, 111453. [CrossRef]
23. Polyakov, A.Y.; Lee, I.-H.; Smirnov, N.B.; Yakimov, E.B.; Shchemerov, I.V.; Chernykh, A.V.; Kochkova, A.I.; Vasilev, A.A.; Carey, P.H.; Ren, F.; et al. Defects at the surface of β-Ga_2O_3 produced by Ar plasma exposure. *APL Mater.* **2019**, *7*, 061102. [CrossRef]
24. Labed, M.; Sengouga, N.; Rim, Y.S. Control of Ni/β-Ga_2O_3 Vertical Schottky Diode Output Parameters at Forward Bias by Insertion of a Graphene Layer. *Nanomaterials* **2022**, *12*, 827. [CrossRef]
25. Sze, S.M.; Ng, K.K. Physics and Properties of Semiconductors—A Review. *Phys. Semicond. Devices* **2006**, *3*, 5–75.
26. Carey, P.H.; Yang, J.; Ren, F.; Hays, D.C.; Pearton, S.J.; Kuramata, A.; Kravchenko, I.I. Improvement of Ohmic contacts on Ga_2O_3 through use of ITO-interlayers. *J. Vac. Sci. Technol. B* **2017**, *35*, 61201. [CrossRef]

27. Oshima, T.; Wakabayashi, R.; Hattori, M.; Hashiguchi, A.; Kawano, N.; Sasaki, K.; Masui, T.; Kuramata, A.; Yamakoshi, S.; Yoshimatsu, K.; et al. Formation of indium–tin oxide ohmic contacts for β-Ga_2O_3. *Jpn. J. Appl. Phys.* **2016**, *55*, 1202B7. [CrossRef]
28. Li, K.-D.; Chen, P.-W.; Chang, K.-S.; Hsu, S.-C.; Jan, D.-J. Indium-Zinc-Tin-Oxide Film Prepared by Reactive Magnetron Sputtering for Electrochromic Applications. *Materials* **2018**, *11*, 2221. [CrossRef]
29. Hakkoum, H.; Tibermacine, T.; Sengouga, N.; Belahssen, O.; Ghougali, M.; Benhaya, A.; Moumen, A.; Comini, E. Effect of the source solution quantity on optical characteristics of ZnO and NiO thin films grown by spray pyrolysis for the design NiO/ZnO photodetectors. *Opt. Mater.* **2020**, *108*, 110434. [CrossRef]
30. Hassanien, A.S.; Akl, A.A. Influence of composition on optical and dispersion parameters of thermally evaporated non-crystalline $Cd_{50}S_{50-x}Se_x$ thin films. *J. Alloys Compd.* **2015**, *648*, 280–290. [CrossRef]
31. Yang, W.-C.; Rodriguez, B.J.; Gruverman, A.; Nemanich, R.J. Polarization-dependent electron affinity of $LiNbO_3$ surfaces. *Appl. Phys. Lett.* **2004**, *85*, 2316–2318. [CrossRef]
32. Mohamed, M.; Irmscher, K.; Janowitz, C.; Galazka, Z.; Manzke, R.; Fornari, R. Schottky barrier height of Au on the transparent semiconducting oxide β-Ga_2O_3. *Appl. Phys. Lett.* **2012**, *101*, 132106. [CrossRef]
33. Lee, H.Y.; Lichtenwalner, D.J.; Jur, J.S.; Kingon, A.I. Investigation of Conducting Oxide and Metal Electrode Work Functions on Lanthanum Silicate High-k Dielectric. *ECS Trans.* **2019**, *11*, 607–612. [CrossRef]
34. Choi, K.-H.; Nam, H.-J.; Jeong, J.-A.; Cho, S.-W.; Kim, H.-K.; Kang, J.-W.; Kim, D.-G.; Cho, W.-J. Highly flexible and transparent $InZnSnO_x$/Ag/$InZnSnO_x$ multilayer electrode for flexible organic light emitting diodes. *Appl. Phys. Lett.* **2008**, *92*, 223302. [CrossRef]
35. Buchholz, D.B.; Proffit, D.E.; Wisser, M.D.; Mason, T.O.; Chang, R.P.H. Electrical and band-gap properties of amorphous zinc–indium–tin oxide thin films. *Prog. Nat. Sci. Mater. Int.* **2012**, *22*, 1–6. [CrossRef]
36. Sato, K.; Yasumura, Y. Study of forward I-V plot for Schottky diodes with high series resistance. *J. Appl. Phys.* **1985**, *58*, 3655–3657. [CrossRef]
37. Hu, Q.; Wang, P.; Yin, J.; Liu, Y.; Lv, B.; Zhu, J.-L.; Dong, Z.; Zhang, W.; Ma, W.; Sun, J. High-Responsivity Photodetector Based on a Suspended Monolayer Graphene/$RbAg_4I_5$ Composite Nanostructure. *ACS Appl. Mater. Interfaces* **2020**, *12*, 50763–50771. [CrossRef]
38. Periyanagounder, D.; Gnanasekar, P.; Varadhan, P.; He, J.-H.; Kulandaivel, J. High performance, self-powered photodetectors based on a graphene/silicon Schottky junction diode. *J. Mater. Chem. C* **2018**, *6*, 9545–9551. [CrossRef]
39. Wu, D.; Zhao, Z.; Lu, W.; Rogée, L.; Zeng, L.; Lin, P.; Shi, Z.; Tian, Y.; Li, X.; Tsang, Y.H. Highly sensitive solar-blind deep ultraviolet photodetector based on graphene/PtSe2/β-Ga_2O_3 2D/3D Schottky junction with ultrafast speed. *Nano Res.* **2021**, *14*, 1973–1979. [CrossRef]

nanomaterials

MDPI

Article

An Energy Harvester Coupled with a Triboelectric Mechanism and Electrostatic Mechanism for Biomechanical Energy Harvesting

Lei Zhai [1,†], Lingxiao Gao [1,*,†], Ziying Wang [1,*], Kejie Dai [2], Shuai Wu [1] and Xiaojing Mu [3,*]

1 School of Mechanical Engineering, Hebei University of Technology, Tianjin 300401, China; zlhbgd@163.com (L.Z.); wushuai@hebut.edu.cn (S.W.)
2 School of Electrical and Mechanical Engineering, Pingdingshan University, Pingdingshan 467000, China; dkj671@163.com
3 Key Laboratory of Optoelectronic Technology & Systems, Ministry of Education, International R&D Center of Micro-Nano Systems and New Materials Technology, Chongqing University, Chongqing 400044, China
* Correspondence: lingxiao.gao@hebut.edu.cn (L.G.); wangzy@hebut.edu.cn (Z.W.); mxjacj@cqu.edu.cn (X.M.)
† These authors contributed equally to this work.

Abstract: Energy-harvesting devices based on a single energy conversion mechanism generally have a low output and low conversion efficiency. To solve this problem, an energy harvester coupled with a triboelectric mechanism and electrostatic mechanism for biomechanical energy harvesting is presented. The output performances of the device coupled with a triboelectric mechanism and electrostatic mechanism were systematically studied through principle analysis, simulation, and experimental demonstration. Experiments showed that the output performance of the device was greatly improved by coupling the electrostatic induction mechanisms, and a sustainable and enhanced peak power of approximately 289 μW was produced when the external impedance was 100 MΩ, which gave over a 46-fold enhancement to the conventional single triboelectric conversion mechanism. Moreover, it showed higher resolution for motion states compared with the conventional triboelectric nanogenerator, and can precisely and constantly monitor and distinguish various motion states, including stepping, walking, running, and jumping. Furthermore, it can charge a capacitor of 10 μF to 3 V within 2 min and light up 16 LEDs. On this basis, a self-powered access control system, based on gait recognition, was successfully demonstrated. This work proposes a novel and cost-effective method for biomechanical energy harvesting, which provides a more convenient choice for human motion status monitoring and can be widely used in personnel identification systems.

Keywords: energy harvesting; triboelectric mechanism; electrostatic mechanism; human motion status monitoring

Citation: Zhai, L.; Gao, L.; Wang, Z.; Dai, K.; Wu, S.; Mu, X. An Energy Harvester Coupled with a Triboelectric Mechanism and Electrostatic Mechanism for Biomechanical Energy Harvesting. *Nanomaterials* **2022**, *12*, 933. https://doi.org/10.3390/nano12060933

Academic Editors: Qiongfeng Shi, Jianxiong Zhu and Antonio Gloria

Received: 18 January 2022
Accepted: 9 March 2022
Published: 11 March 2022

1. Introduction

To realize the intellectualization and miniaturization of wearable electronics, the requirements of their power devices have been raised, such as long life, small volume, high energy density, and easy integration [1–5]. Traditional chemical batteries cannot satisfy the requirements of the fast development of wearable electronics due to their fatal defects, such as their large size, non-flexible structure, limited lifetime, and potential hazards to environments [6]. Biomechanical energy is generated in our daily life all the time, including arm swinging, walking, running, and heart beating [7–10]. If this energy is converted into electricity, it can meet the power consumption for part of wearable electronics [11–17]. As an emerging energy conversion technology, triboelectric nanogenerators (TENG) have been widely used in the field of self-powered wearable electronics because of their low cost, environmental friendliness, and flexibility [18–20]. A TENG can provide an effective approach to converting such biomechanical energy into electrical energy to provide power for part of wearable electronics, thus constructing a self-powered human motion signal monitoring and recognition system [21–23].

However, there are two main problems for current devices to harvest biomechanical energy: (1) Most of the devices are based on a single conversion mechanism, leading to low energy conversion efficiency [24–29]; and (2) the energy is usually harvested from a single source. The movements of the human body are coordinated by many parts. For example, leg movements are accompanied with arm movements when walking. If the device is integrated onto the human leg, it is difficult to harvest the energy generated by arm movements. Therefore, in order to improve the efficiency of biomechanical energy harvesting, a variety of energy harvesting strategies should be combined to achieve efficiently cooperative work.

To solve this problem, we propose an energy harvester coupled with a triboelectric mechanism and electrostatic mechanism for biomechanical energy harvesting. By exposing the electrodes of the bracelet which is integrated onto the arm, the energy of the swinging arm and the energy generated by the separation and contact between the foot and the ground during movement can be obtained simultaneously. Traditional walking energy acquisition requires the integration of the insole with the power generation function, which has a short lifetime and will induce a certain level of discomfort during use. The scheme proposed in this paper does not need special processing of the sole and can realize the efficient acquisition of triboelectric energy between the sole and the ground when walking, which has the advantages of comfort, low cost, and long lifetime. Moreover, the device, when coupled with a triboelectric mechanism and electrostatic mechanism, showed higher resolution in the sensing of motion states, which provided a more convenient choice for human motion status monitoring and can be widely used in personnel identification systems.

2. Results and Discussion

2.1. Working Principle

An energy harvester coupled with a triboelectric mechanism and electrostatic mechanism for biomechanical energy harvesting, as illustrated in Figure 1, is proposed in this paper, which consists of two main parts: a wristband-type triboelectric nanogenerator (W–TENG) worn on the wrist, and an electrostatic nanogenerator (EENG) formed by the friction of the sole against the ground. The structure of the W–TENG, shown in the illustration, mainly consists of three parts: a 3D-printed resin ring, some cross-fingered copper foil electrodes, and a polytetrafluoroethylene (PTFE) ball. Fluorine has the highest electronegativity among almost all the elements, so the material of the ball was PTFE, which contains lots of fluorine [30]. Due to its strong electronic attraction, a PTFE ball can easily capture a large amount of triboelectric charges when rubbed against other materials. Meanwhile, is copper at a more positive direction than precious gold and silver in the triboelectric sequence [31]. Therefore, a PTFE ball and copper electrode will accumulate opposite charges after contact [32]. The whole device can be realized with a simple process and cheap materials which are suitable for mass production. One end of the forked-finger electrode was placed in contact with the human wrist. When the human body moved, the PTFE ball would roll on the cross-fingered electrodes in the 3D-printed resin ring with the swing of the arm, resulting in a triboelectric effect with the cross-fingered electrodes. The EENG was built by the shoe and the floor. In this structure, the sole of the shoe was not installed with any triboelectric materials. The sole and the ground produced a periodic contact—separation process with walking, and opposite charges were accumulated on each other's surfaces. The surface of the skin can generate electric charges by electrostatic induction because of the contact with the electric shoe. By exposing one end of the electrode of the W–TENG to the human body, the triboelectric charges on the skin are transmitted to the electrode of the W–TENG. The scheme proposed in this paper does not need to place any material in the sole, and can realize the efficient acquisition of triboelectric energy between the sole and the ground when walking, which has the advantages of comfort, low cost, and long life.

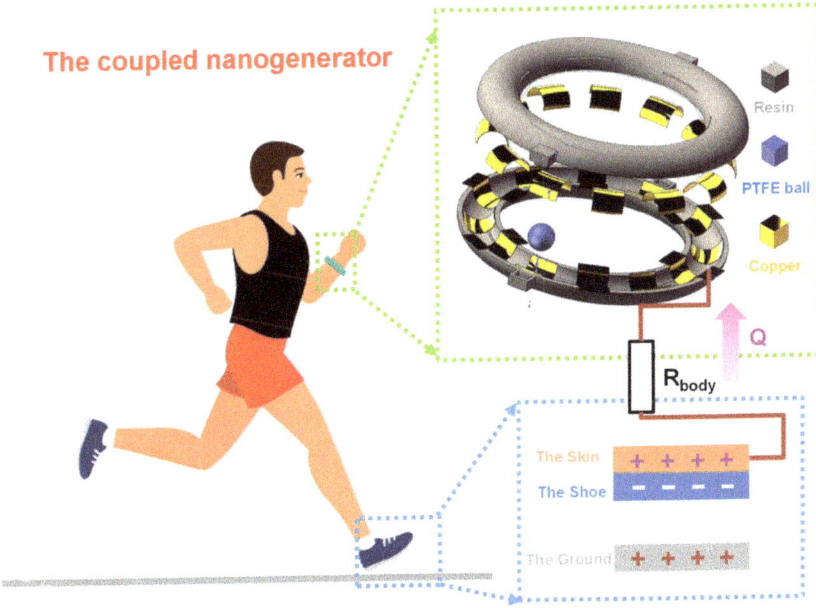

Figure 1. Schematic diagram of the energy harvester coupled with the triboelectric mechanism and electrostatic mechanism for biomechanical energy harvesting.

For the traditional triboelectric nanogenerator (W–TENG), its electrodes were generally shielded inside the shell, as shown in Figure 2a,b. The amount of charges accumulated on the equivalent capacitance $C_{W\text{-}TENG}$ was Q_1 after the PTFE ball contacted with the electrodes, and the open-circuit voltage can be expressed as [33]:

$$V_{OC} = R\frac{dQ_1}{dt} + \frac{Q_1}{C_{W-TENG}} \quad (1)$$

Due to the motion process of the human body, the sole produces a "contact—separation" process with the ground all the time, and the mechanical energy is converted to electric energy along with this process. The means of capturing this energy effectively is very important. In this work, a mechanism of triboelectric and electrostatic double-effect coupling, to achieve efficient collection of this energy, was proposed. By touching one end of the W–TENG electrode to the human body, the electric charges of the EENG was transmitted to the electrode of the W–TENG through the human body, and the coupled nanogenerator (C–ENG) was built. Indeed, the polarity of the triboelectric charge depends on the polarity of the triboelectric material. Therefore, the charge-transfer processes in two cases were analyzed respectively. When the material of the shoe had more electronegativity than the ground, the shoe accumulated negative charges when it made contact with the ground. Meanwhile, the ground accumulated positive charges, as shown in Figure 2d. When the foot was lifted, the surface of the skin of the human body generated positive charges through electrostatic induction. Due to the fact that the human skin was connected to one electrode (such as electrode one in Figure 2d) of the W–TENG, some of the negative charges were repelled all the way to electrode one. It created a potential difference between electrode one and electrode two of the W–TENG, resulting in a current from electrode two to electrode one in the external circuit. The other case was that the ground had more electronegativity than the material of the shoe, as shown in Figure 2e. As the shoe accumulated positive charges during its contact with the ground, the surface of the skin of the human body generated negative charges through electrostatic induction and some of the positive charges were repelled all the way to electrode one, resulting in a current from electrode one to electrode two in the external circuit. By comparing the two cases, only the direction of the current in the outer circuit changed with the change of the polarity of the material. It should be noted that the soles are usually made of chemical-resistant viton or polyvinyl chloride (PVC), and the floors are usually made of wood (marine-grade plywood) or wear-resistant garolite. Therefore, the soles of the shoes are usually more electronegative than the floors [32].

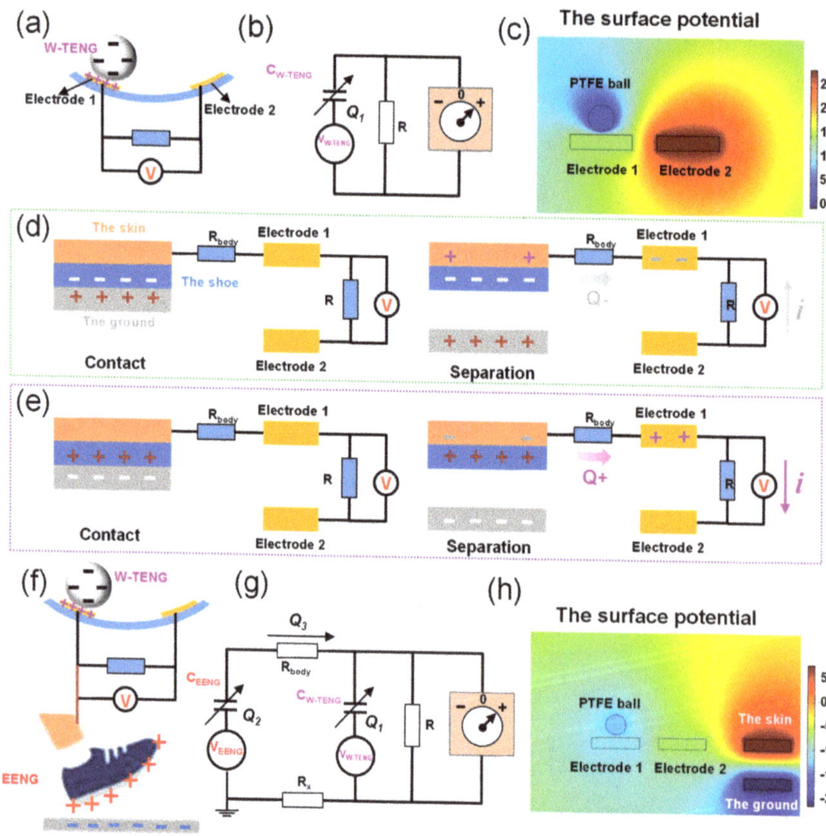

Figure 2. Analysis of the working principle: (**a**) the working principle of the W–TENG; (**b**) the circuit equivalent model of the W–TENG; (**c**) the simulation result of the output of W–TENG; (**d**,**e**) the analysis of charge-transfer direction for different materials; (**f**) the working principle of the C–ENG; (**g**) the circuit equivalent model of the C–ENG; and (**h**) the simulation result of the output of C–ENG.

The working principle of the C–ENG is shown in Figure 2f, and its circuit equivalent model is shown in Figure 2g. As the sole rubbed against the ground, the amount of charge accumulated on the equivalent capacitance (C_{EENG}) of the EENG was Q_2. Since the amount of triboelectric charge was positively correlated with the friction area, Q_2 was much larger than Q_1 because the sole had a larger area. When the electrode made contact with the human body, part of the charge on C_{EENG} flowed to C_{W-TENG}. Assuming that the transferred charge was Q_3, the open-circuit voltage of the device was:

$$V_{OC} = R\frac{d(Q_1 + Q_3)}{dt} + \frac{Q_1 + Q_3}{C_{W-TENG}} \quad (2)$$

Therefore, the output performance of the device is improved.

COMSOL 5.4 was employed in the simulation analysis, and the results are illustrated in Figure 2c,h. Figure 2c is the simulation result of the output of the W–TENG. In this figure, the circle represents the PTFE ball that accumulated negative charges, and the rectangles represent the adjacent electrodes that accumulated positive charges. Take the situation in Figure 2d as an example; the simulation result of the output of the C–ENG is shown in Figure 2h. Compared with Figure 2c, Figure 2h added two rectangles at the far right. The top rectangle represents the skin, which accumulated positive charges through electrostatic induction. The bottom rectangle represents the ground, accumulating negative charges.

2.2. Electrical Characterization

Firstly, in order to analyze the influence of the relationship between the diameter of the PTFE ball and the width of the electrode on the electrical performance of the C–ENG in the state of a human walking with swinging arms, PTFE balls with diameters of 6 and 8 mm were selected as test objects. The output voltages and currents of different electrode widths (4 mm, 6 mm, 9 mm, 11 mm, and 13 mm), with a PTFE ball with a diameter of 6 mm, were measured respectively, as illustrated in Figure 3a. The output voltages and currents of different electrode widths (6 mm, 8 mm, 10 mm, 12 mm, and 14 mm), with a PTFE ball with a diameter of 8 mm, are illustrated in Figure 3b. Figure 3c further analyzes the maximum voltage and current in Figure 3a. Its abscissa represents the ratio of electrode width (d) to PTFE ball diameter (D), and the ordinates on both sides represent the maximum voltage and current under different electrode widths. It can be seen that, when the ratio approached one, both the voltage and current reach their maximum values. The same result is obtained in Figure 3d. These results show that the best electrical properties were obtained when the electrode width was approximately the same as the PTFE ball diameter. After obtaining the above results, the electrical properties in jumping state were analyzed. The voltage and current values were measured when the electrode was not in contact with the human body (W–TENG) and when it was in contact with the human body (C–ENG) respectively. As seen in Figure 3e, when the electrode width was 6 mm and the diameter of the PTFE ball was 6 mm, the output voltages of the W–TENG and C−TENG were 42 V and 545 V respectively. It proves that the output voltage increased by, approximately, 12.98 times by coupling the mechanism of electrostatic. The maximum current correspondingly increased from 0.43 to 5.5 μA. Similarly, it can be seen from Figure 3f that when the electrode width was 8 mm and the diameter of the PTFE ball was 8 mm, the output voltage showed the same increasing trend. The results show that coupling the electrostatic mechanism can effectively improve the energy conversion efficiency of the nanogenerator, which is the same as in the theoretical analysis. In addition, all tests were conducted on the dry ground. This is because, in a dry environment, the sole and the ground can generate more triboelectric charges, which was conducive to verifying the coupling enhancement theory proposed in this paper.

In order to further study the influence of the diameter of the PTFE ball on the outputs, other sizes (7 mm, 9 mm, and 10 mm) were measured, under the condition that the size of the PTFE was equal to the width of the electrode, and the measured results were compared with those obtained previously at 6 mm and 8 mm, as illustrated in Figure 4a,b. As seen in the two figures, the diameter of the PTFE ball had no obvious effect on the outputs. This may be because, although the diameter of the ball changed, the total triboelectric area did not significantly change. Through the analysis of the measured experimental data, it can be concluded that the size combination of 7 mm is more suitable for subsequent experiments. As the signals generated in different motion states were different and the corresponding application occasions were also different, the signals in different motion states were recorded, such as walking without swinging arms, walking with swinging arms, running, jumping, and swinging arms only (as illustrated in the accompanying Supplementary Video S1). The voltage and current signals of the W–TENG (Figure 4c,d) and C–ENG (Figure 4e,f) were measured respectively. It can be seen from Figure 4c–f that both voltage and current show the maximum amplitude in the jumping state. The frequency of outputs measured in the running state are obviously higher than that in other states, but the amplitude of outputs decreased to a certain extent. This is because the triboelectric area of the sole with the ground was reduced in the running state. Through a comparison, it can be seen that the detection ability of the C–ENG for the five states is significantly better than the W–TENG. This is because the output performance of the device was greatly improved by coupling the electrostatic induction mechanism. The experiment proved that the C–ENG showed higher resolution for motion states compared with the conventional triboelectric nanogenerator and can precisely and constantly monitor and distinguish various motion states, including stepping, walking, running, and jumping.

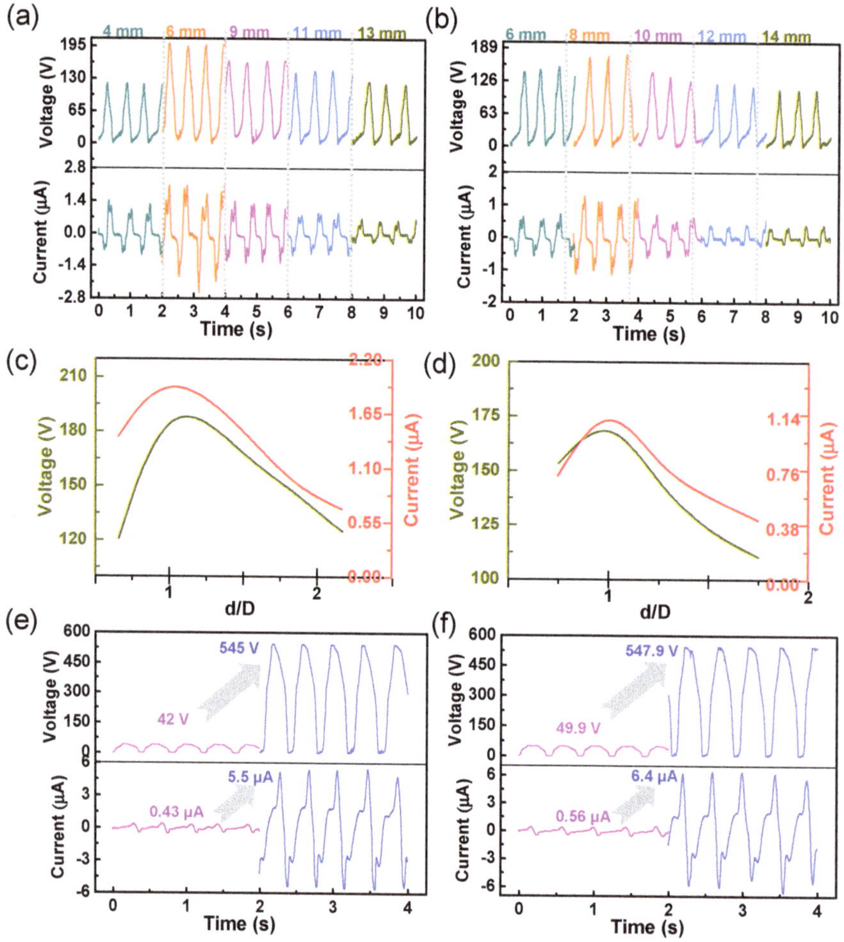

Figure 3. The effects of the coupling electrostatic energy and the relationship between PTFE ball diameter and electrode width on the output of the C–ENG: (**a**) the effect of cross-finger electrode width on the output of the C–ENG at the PTFE ball diameter of 6 mm; (**b**) the effect of cross-finger electrode width on the output of the C–ENG at the PTFE ball diameter of 8 mm; (**c**) a curve fitted according to Figure 3a; (**d**) a curve fitted according to Figure 3b; (**e**) a comparison of output characteristics between W–TENG and C–ENG at the PTFE ball diameter of 6 mm; and (**f**) a comparison of output characteristics between W–TENG and C–ENG at the PTFE ball diameter of 8 mm.

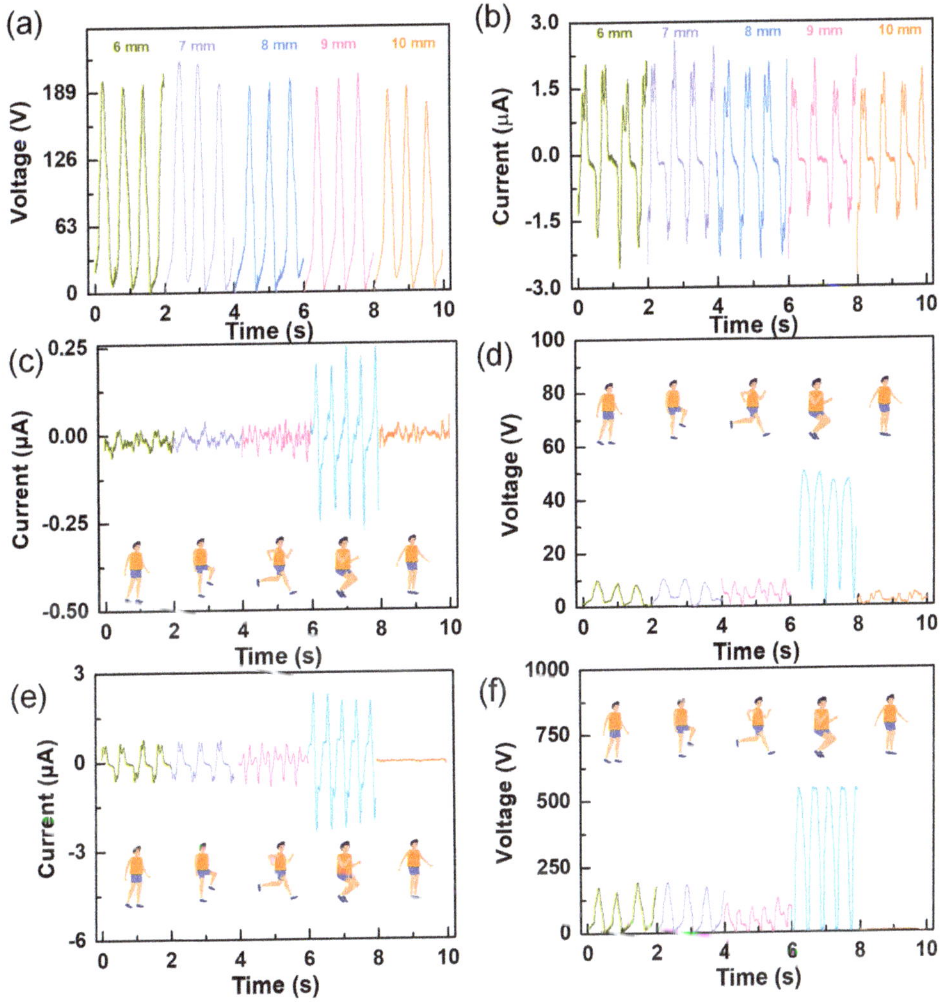

Figure 4. The influence of the diameter of the PTFE ball on the outputs and the outputs of the device under different motion modes: (**a**) the influence of the diameter of the PTFE ball on the output voltage; (**b**) the influence of the diameter of the PTFE ball on the output current; (**c**) the output voltages of W–TENG under different motion modes; (**d**) the output currents of W–TENG under different motion modes; (**e**) the output voltage of C–ENG under different motion modes; and (**f**) the output currents of C–ENG under different motion modes.

2.3. The Applications

According to the previous results, it is evident that the ouput voltages and currents of the C–ENG were greatly improved compared with those of the W–TENG. The enhancement in peak output power was further studied in walking condition. As illustrated in Figure 5a,b, the peak output power of the W–TENG and C–ENG were measured respectively, and the maximum values and corresponding impedance were marked. A sustainable and enhanced peak power of approximately 289 μW was produced for the C–ENG when the external impedance was 100 MΩ, which gave over a 46-fold enhancement to the conventional single triboelectric conversion mechanism (6.25 μW for W–TENG). Subsequently, the charging rates for different capacitors (1.0 μF, 2.2 μF, 4.7 μF and 10 μF) were studied to verify the excellent output of the C–ENG, as presented in Figure 5c. It is evident that the larger the capacitance, the longer the charging time, which is consistent with the theoretical basis. Furthermore, 16 LEDs were lit in a series to verify the excellent electrical output of the C–ENG, as illustrated in Figure 5d and Supplementary Video S2.

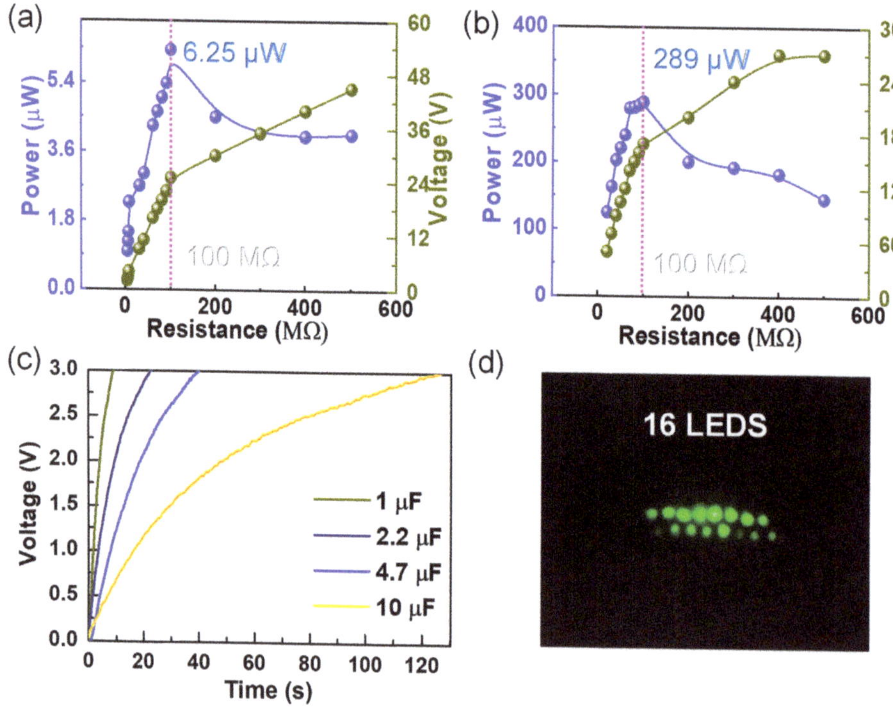

Figure 5. The power and charging characteristics of C–ENG: (**a**) the output power of the W–TENG; (**b**) the output power of the C–ENG; (**c**) the charge curves of the C–ENG for different capacitors; and (**d**) the experiments on lighting LED lights by the C–ENG.

To demonstrate that the C–ENG can serve as a sensing device for actual scenarios, a self-powered switching system was invented. The system was constructed of the coupled nanogenerator, an electrometer 6517B, a DAQ data acquisition card, and an online monitoring system, which is expressed in Figure 6a. To demonstrate this function, the C–ENG was worn on the wrist to record the voltage waveform output by different people swinging their arms on foot, while simultaneously measuring the maximum and minimum voltage values and calculating the corresponding peak to peak voltage values. The measurement process is illustrated in Figure 6b–e. When the system recognized that the gait information of a person was passable, the information was retrieved from the library and the access control was opened to allow the person to enter, as illustrated in Figure 6b,c, and Supplementary Video S3. In contrast, if the system recognized that the gait information of the personnel was not in the passable range, it called up the blocking command, so as to prevent the invasion of foreign personnel, as illustrated in Figure 6d,e, and Supplementary Video S4. It is worth noting that this part was only a general functional demonstration and the specific application of C–ENG in switching systems needs to be further improved.

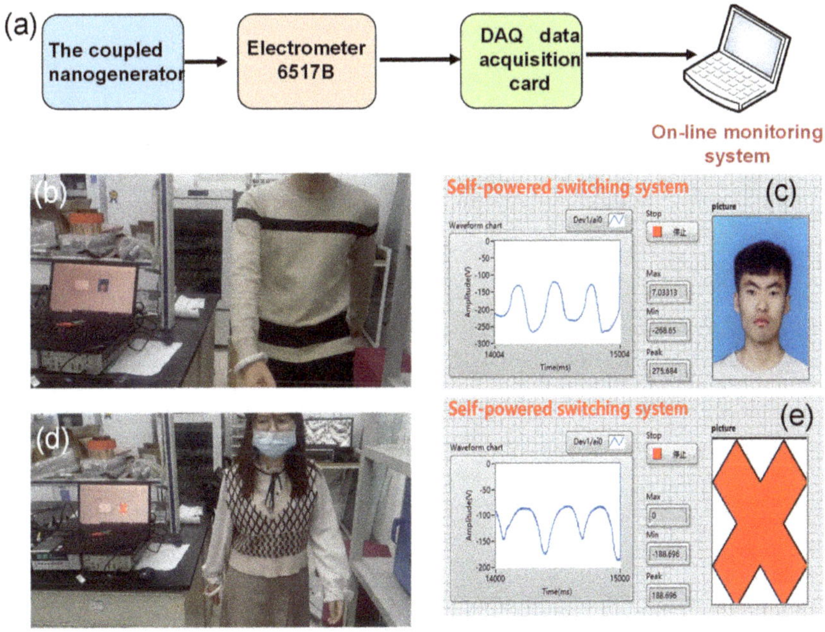

Figure 6. Self-powered switching system: (**a**) the block diagram of the self-powered switching system; (**b**,**c**) the demonstration results of passable personnel; and (**d**,**e**) the demonstration results of impassable personnel.

In order to verify the application prospect of the C–ENG in the field of self-powered wearable electronics, a self-powered heart rate testing system was built, as illustrated in Figure 7a,b, and Supplementary Video S5. Electric energy was generated by the walking state excitation device, and the generated electrical signal was stored in a 100 μF capacitor after rectification. It transpired that after about 10 min, the capacitor voltage reached 6 V, which is illustrated in Figure 7c. Afterwards, the heart rate test device was connected and the capacitor charging switch was closed. The voltage of the capacitor dropped sharply, as illustrated in Figure 7c. By observing the collected waveform, the single heartbeat can be clearly observed, including the diastolic (D) wave, tidal (T) wave, and percussion (P) wave. This proves that the C–ENG powered the pulse sensor successfully, and the normal working time of the device was two seconds, as illustrated in Figure 7d.

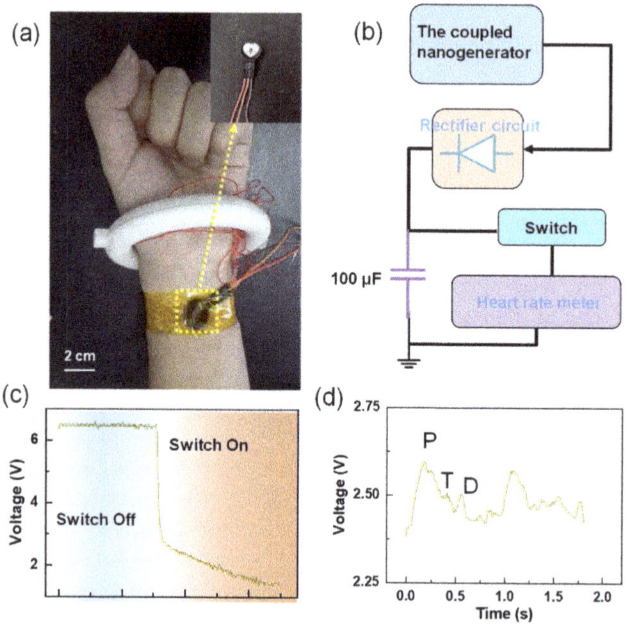

Figure 7. Self-powered heart rate testing system: (**a**) the electronic photo of system; (**b**) the block diagram of system composition; (**c**) the voltage changed at both ends of the capacitor before and after fast closing. After the switch was closed, the electric quantity in the capacitor began to power the heart rate sensor, and the voltage dropped sharply); and (**d**) pulse waveform collected.

3. Conclusions

In summary, a biomechanical energy harvesting device combining triboelectric and electrostatic mechanisms has been demonstrated. Aiming at the low output of energy acquisition devices based on a single conversion mechanism, a scheme of coupling electrostatic energy with triboelectric energy was proposed. The circuit-equivalent model of the coupled energy nanogenerator was constructed and the principle of output lifting was analyzed. Afterwards, the output characteristics of the coupled energy nanogenerator were demonstrated systematically through simulation and experiment. The results show that the output performance of the device was greatly improved by coupling the electrostatic induction mechanism. The best output performance was obtained when the electrode width and spacing were approximately the same as the PTFE ball diameter. The open circuit voltage increased by approximately 12.98 times, while the short circuit current increased by 12.79 times, when the human jumped, when the diameter of the PTFE ball was 6 mm. It is worth emphasizing that when people were walking, a sustainable and enhanced peak power of about 289 μW was produced when the external impedance was 100 MΩ, which gave over a 46-fold enhancement to the conventional single triboelectric conversion mechanism. Moreover, it showed higher resolution of motion states compared with the conventional triboelectric nanogenerator, and can precisely and constantly monitor and distinguish various motion states, including stepping, walking, running, and jumping. Furthermore, the charging rates and power supply features were studied to verify its excellent output, and the result show that it can charge a capacitor of 10 μF to 3 V within 2 min and light up 16 LEDs. After that, a self-powered access control system based on gait recognition was successfully demonstrated. This work provides a more convenient choice for human motion status monitoring, and provides diversified perspectives for personnel identification systems, simultaneously.

Supplementary Materials: The following are available online at https://www.mdpi.com/article/10.3390/nano12060933/s1: Video S1: the demonstration of gait recognition by the C–ENG; Video S2: the demonstration of lighting 16 LEDs by the C–ENG; Video S3: the demonstration of accessible personnel in access control systems; Video S4: the demonstration of inaccessible personnel in access control systems; and Video S5: the power supply process in a heart rate testing system.

Author Contributions: Conceptualization, L.G. and L.Z.; methodology, L.G. and X.M.; software, L.Z.; validation, L.G., L.Z. and Z.W.; formal analysis, K.D.; investigation, L.Z.; resources, L.G., Z.W. and X.M.; data curation, L.Z. and L.G.; writing—original draft preparation, L.G. and L.Z.; writing—review and editing, Z.W. and X.M.; visualization, K.D. and S.W.; supervision, L.G. and X.M.; project administration, X.M.; funding acquisition, L.G., Z.W. and X.M. All authors have read and agreed to the published version of the manuscript.

Funding: The National Key Research and Development Program of China: No. 2019YFB2004800; The General Program of National Natural Science Foundation of China: No.52075061; The Natural Science Foundation of Hebei Province: E2021202100, F2020202050 and E2021202008; The National Natural Science Foundation of China: No.52104189 and No.52005151.

Data Availability Statement: Request the corresponding author of this article.

Conflicts of Interest: The authors declare no conflict of interest.

References

1. Wei, Z.; Lin, S.; Li, Q.; Song, C.; Wang, F.; Tao, X.M. Fiber-based wearable electronics: A review of materials, fabrication, devices, and applications. *Adv. Mater.* **2014**, *26*, 5310.
2. Han, Y.Z.; Yi, F.; Jiang, C.; Dai, K.R.; Xu, Y.C.; Wang, X.F.; You, Z. Self-powered gait pattern-based identity recognition by a soft and stretchable triboelectric band. *Nano Energy* **2019**, *56*, 516–523. [CrossRef]
3. Zhang, Q.; Jin, T.; Cai, J.G.; Xu, L.; He, T.Y.Y.; Wang, T.H.; Tian, Y.Z.; Li, L.; Peng, Y.; Lee, C.K. Wearable triboelectric sensors enabled gait analysis and waist motion capture for iot-based smart healthcare applications. *Adv. Sci.* **2021**, *9*, 2103694. [CrossRef] [PubMed]
4. Liu, S.; Yuan, F.; Sang, M.; Zhou, J.Y.; Zhang, J.S.; Wang, S.; Li, J.S.; Xuan, S.H.; Gong, X.L. Functional sponge-based triboelectric nanogenerators with energy harvesting, oil–water separating and multi-mode sensing performance. *J. Mater. Chem. A* **2021**, *9*, 6913–6923. [CrossRef]
5. Lin, Z.M.; Wu, Z.Y.; Zhang, B.B.; Wang, Y.C.; Guo, H.Y.; Liu, G.L.; Chen, C.Y.; Chen, Y.L.; Yang, J.; Wang, Z.L. A Triboelectric nanogenerator-based smart insole for multifunctional gait monitoring. *Adv. Mater. Technol.* **2018**, *4*, 1800360. [CrossRef]
6. Wang, H.; Cheng, J., Wang, Z.; Ji, L.; Wang, Z.L. Triboelectric nanogenerators for human-health care. *Sci. Bull.* **2020**, *66*, 490–511. [CrossRef]
7. Maiti, S.; Karan, S.K.; Kim, J.K.; Khatua, B.B. Nature driven bio-piezoelectric/triboelectric nanogenerator as next-generation green energy harvester for smart and pollution free society. *Adv. Energy Mater.* **2019**, *9*, 1803027. [CrossRef]
8. Lim, J.; Choi, D.S.; Lee, G.Y.; Lee, H.J.; Sasikala, S.P.; Lee, K.E.; Kang, S.H.; Kim, S.O. Omnidirectional deformable energy textile for human joint movement compatible energy storage. *ACS Appl. Mater. Interfaces* **2017**, *9*, 41363–41370. [CrossRef]
9. Wu, H.Z.; Tatarenko, A.; Bichurin, M.I.; Wang, Y.J. A multiferroic module for biomechanical energy harvesting. *Nano Energy* **2021**, *83*, 105777. [CrossRef]
10. Gao, L.X.; Chen, X.; Lu, S.; Zhou, H.; Xie, W.B.; Chen, J.F.; Qi, M.K.; Yu, H.; Mu, X.J.; Wang, Z.L.; et al. Enhancing the output performance of triboelectric nanogenerator via grating-electrode enabled surface plasmon excitation. *Adv. Energy Mater.* **2019**, *9*, 1902725. [CrossRef]
11. Tcho, I.W.; Kim, W.G.; Choi, Y.K. A self-powered character recognition device based on a triboelectric nanogenerator. *Nano Energy* **2020**, *70*, 104534. [CrossRef]
12. Zhen, W.; Shen, Q.; Sun, X. Nanogenerators for self-powered gas sensing. *Nano-Micro Lett.* **2017**, *9*, 1–19.
13. Fu, H.L.; Mei, X.T.; Yurchenko, D.; Zhou, S.X.; Theodossiades, S.; Nakano, K.; Yeatamn, E.M. Rotational energy harvesting for self-powered sensing. *Joule* **2021**, *5*, 1074–1118. [CrossRef]
14. Chen, J.; Han, K.; Luo, J.J.; Xu, L.; Tang, W.; Wang, Z.L. Soft robots with self-powered configurational sensing. *Nano Energy* **2020**, *77*, 105171. [CrossRef]
15. Zhang, B.S.; Zhang, S.; Li, W.B.; Gao, Q.; Zhao, D.; Wang, Z.L.; Cheng, T.H. Self-powered sensing for smart agriculture by electromagnetic–triboelectric hybrid generator. *ACS Nano* **2021**, *15*, 20278–20286. [CrossRef]
16. Ma, M.Y.; Zhang, Z.; Liao, Q.L.; Yi, F.; Han, L.H.; Zhang, G.J.; Liu, S.; Liao, X.Q.; Zhang, Y. Self-powered artificial electronic skin for high-resolution pressure sensing. *Nano Energy* **2017**, *32*, 389–396. [CrossRef]
17. Askari, H.; Hashemi, E.; Khajepour, A.; Khamesee, M.B.; Wang, Z.L. Towards self-powered sensing using nanogenerators for automotive systems. *Nano Energy* **2018**, *53*, 1003–1019. [CrossRef]

18. Xie, W.B.; Gao, L.X.; Wu, L.K.; Chen, X.; Wang, F.Y.; Tong, D.Q.; Zhang, J.; Lan, J.Y.; He, X.B.; Mu, X.J.; et al. A non-resonant hybridized electromagnetic-triboelectric nanogenerator for irregular and ultralow frequency blue energy harvesting. *Research* **2021**, *1*, 5963293.
19. Islam, E.; Abdullah, A.M.; Chowdhury, A.R.; Tasnim, F.; Uddin, M.J. Electromagnetic-triboelectric-hybrid energy tile for biomechanical green energy harvesting. *Nano Energy* **2020**, *77*, 105250. [CrossRef]
20. Zou, Y.J.; Raveendran, V.; Chen, J. Wearable triboelectric nanogenerators for biomechanical energy harvesting. *Nano Energy* **2020**, *77*, 105303. [CrossRef]
21. Cheng, Y.; Gao, Y.Y.; Zhao, S.L.; Zhang, S.L.; Zhou, Y.H.; Deng, W.L.; Li, Z.W.; Jiang, G.; Jin, L.; Tian, G.; et al. A linear-to-rotary hybrid nanogenerator for high-performance wearable biomechanical energy harvesting. *Nano Energy* **2020**, *67*, 104235.
22. Shen, J.L.; Li, Z.L.; Yu, J.Y.; Ding, B. Humidity-resisting triboelectric nanogenerator for high performance biomechanical energy harvesting. *Nano Energy* **2017**, *40*, 282–288. [CrossRef]
23. Zhou, Z.H.; Weng, L.; Tat, T.; Libanori, A.; Lin, Z.M.; Ge, L.J.; Yang, J.; Chen, J. Smart insole for robust wearable biomechanical energy harvesting in harsh environments. *ACS Nano* **2020**, *14*, 14126–14133. [CrossRef] [PubMed]
24. Yang, Z.B.; Zhou, S.X.; Zu, J.; Inman, D. High-performance piezoelectric energy harvesters and their applications. *Joule* **2018**, *2*, 642–697. [CrossRef]
25. Zhang, S.L.; Bick, M.; Xiao, X.; Chen, G.R.; Nashalian, A.; Chen, J. Leveraging triboelectric nanogenerators for bioengineering. *Matter* **2021**, *4*, 845–887. [CrossRef]
26. Gao, L.X.; Lu, S.; Xie, W.B.; Chen, X.; Wu, L.K.; Wang, T.T.; Wang, A.B.; Yue, C.Q.; Tong, D.Q.; Lei, W.Q.; et al. A self-powered and self-functional tracking system based on triboelectric-electromagnetic hybridized blue energy harvesting module. *Nano Energy* **2020**, *72*, 104684. [CrossRef]
27. Ezzitouni, S.; Fernández-Yáez, P.; Sánchez, L.; Armas, O. Global energy balance in a diesel engine with a thermoelectric generator. *Appl. Energy* **2020**, *269*, 115139. [CrossRef]
28. Dragunov, V.P.; Ostertak, D.I.; Pelmenev, K.G.; Sinitskiy, R.E.; Dragunova, E.V. Electrostatic vibrational energy converter with two variable capacitors. *Sens. Actuators A Phys.* **2020**, *318*, 112501. [CrossRef]
29. Feser, J.P.; Ravichandran, J. More power to pyroelectrics. *Nat. Mater.* **2018**, *17*, 385–386. [CrossRef]
30. Garcia, C.; Trendafilova, I. Real-time diagnosis of small energy impacts using a triboelectric nanosensor. *Sens. Actuators A Phys.* **2019**, *291*, 196–203. [CrossRef]
31. Garcia, C.; Trendafilova, I.; de Villoria, R.G.; del Rio, J.S. Self-powered pressure sensor based on the triboelectric effect and its analysis using dynamic mechanical analysis. *Nano Energy* **2018**, *50*, 401–409. [CrossRef]
32. Zou, H.; Zhang, Y.; Guo, L.; Wang, P.; He, X.; Dai, G.; Zheng, H.; Chen, C.; Wang, A.C.; Xu, C.; et al. Quantifying the triboelectric series. *Nat. Commun.* **2019**, *10*, 1427. [CrossRef] [PubMed]
33. Niu, S.; Zhou, Y.S.; Wang, S.; Liu, Y.; Lin, L.; Bando, Y.; Wang, Z.L. Simulation method for optimizing the performance of an integrated triboelectric nanogenerator energy harvesting system. *Nano Energy* **2014**, *8*, 150–156. [CrossRef]

nanomaterials

MDPI

Review

Advances in Electrochemical Detection Electrodes for As(III)

Haibing Hu [1,*], Baozhu Xie [1], Yangtian Lu [1] and Jianxiong Zhu [2,*]

1 Academy of Opto-Electric Technology, Special Display and Imaging Technology Innovation Center of Anhui Province, National Engineering Laboratory of Special Display Technology, State Key Laboratory of Advanced Display Technology, Collaborative Innovation Center of Advanced Display Technology, Anhui Key Laboratory of Advanced Imaging and Display Technology, Opto-Electric Display Industry Innovation Center, Anhui Province Key Laboratory of Measuring Theory and Precision Instrument, School of Instrument Science and Optoelectronics Engineering, Hefei University of Technology, Hefei 230009, China; xie_bz@163.com (B.X.); hfut_lyt@163.com (Y.L.)

2 School of Mechanical Engineering, Southeast University, Nanjing 211189, China

* Correspondence: huhb@hfut.edu.cn (H.H.); mezhujx@seu.edu.cn (J.Z.); Tel.: +86-1811-097-2015 (H.H.)

Abstract: Arsenic is extremely abundant in the Earth's crust and is one of the most common environmental pollutants in nature. In the natural water environment and surface soil, arsenic exists mainly in the form of trivalent arsenite (As(III)) and pentavalent arsenate (As(V)) ions, and its toxicity can be a serious threat to human health. In order to manage the increasingly serious arsenic pollution in the living environment and maintain a healthy and beautiful ecosystem for human beings, it is urgent to conduct research on an efficient sensing method suitable for the detection of As(III) ions. Electrochemical sensing has the advantages of simple instrumentation, high sensitivity, good selectivity, portability, and the ability to be analyzed on site. This paper reviews various electrode systems developed in recent years based on nanomaterials such as noble metals, bimetals, other metals and their compounds, carbon nano, and biomolecules, with a focus on electrodes modified with noble metal and metal compound nanomaterials, and evaluates their performance for the detection of arsenic. They have great potential for achieving the rapid detection of arsenic due to their excellent sensitivity and strong interference immunity. In addition, this paper discusses the relatively rare application of silicon and its compounds as well as novel polymers in achieving arsenic detection, which provides new ideas for investigating novel nanomaterial sensing. We hope that this review will further advance the research progress of high-performance arsenic sensors based on novel nanomaterials.

Keywords: As(III) detection; electrochemical sensing; nanosensing; biosensing

Citation: Hu, H.; Xie, B.; Lu, Y.; Zhu, J. Advances in Electrochemical Detection Electrodes for As(III). *Nanomaterials* **2022**, *12*, 781. https://doi.org/10.3390/nano12050781

Academic Editor: Lyubov G. Bulusheva

Received: 27 January 2022
Accepted: 17 February 2022
Published: 25 February 2022

1. Introduction

In China and the rest of the world, environmental pollution has always been an urgent problem that threatens the living environment and health of human beings. A variety of forms of environmental pollution exist at the same time, including soil, air, radioactive elements, water pollution, and so on. Among the most direct and obvious causes of harm to human beings, the heavy metal pollution of water environments, the toxicity of heavy metal-like arsenic (As) ions, and the degree of difficulty in detecting and removing them constitutes one of the most important forms of environmental pollution. Arsenic is an element that is widely distributed in soil, minerals, the aquatic environment, and the atmosphere. In terms of arsenic abundance, it ranks 20th in the Earth's crust, 14th in seawater, and 12th in human systems when comparing all elements [1]. Widespread in nature, arsenic and its compounds are mobile in the environment and, when dissolved in the water species of arsenic, contamination will even enter the biosphere through the food chain. As early as the 1970s, in a study by the U.S. National Pesticide Monitoring Program targeting the detection of mercury, arsenic, lead, cadmium, and selenium residues in fish, the results show that more than 95% of the combined samples have detectable residues of all metals. This number is increasing year by year [2]. In natural water bodies, arsenic is usually present as arsenite (As(III)), arsenate (As(V)), monomethylarsenic acid (MMA), and dimethylarsenic acid (DMA). Inorganic forms of arsenite (As(III)) and arsenate (As(V)) have stronger circulation and are more likely to be enriched in the food chain, which is the main form of arsenic present in nature and has a great impact on human survival.

Heavy metals are a very important class of carcinogens, especially nickel, chromium, and arsenic, which are recognized as human carcinogens [3]. Radon and arsenic exposure were recognized as major risk factors in a cohort of Chinese tin miners with lung cancer [4]. Additionally, in China, Y J Lu et al. studied mineral dust deposition in the lungs of tin miners in Yunnan Province, and data obtained in lung tissue provided evidence for arsenic as a cause of lung cancer [5]. Exposure to inorganic arsenic in drinking water has the potential to cause changes in left ventricular geometry and cardiomegaly in adult males [6]. There are many ways for humans and animals to be exposed to arsenic (mainly through drinking water and the accumulation of arsenic in the food chain). Long-term exposure to arsenic ions may lead to a series of physiological diseases, including neurological, physiological, reproductive, kidney, liver, and even genetic conditions [7]. There is serious arsenic contamination of water and soil in many countries around the world, including China, and there was a large-scale arsenic ion poisoning incident in Bangladesh [8,9].

It is on account of these characteristics of arsenic that the detection of trace amounts of arsenic is very important. As early as the nineteenth century, John Bostock realized the importance of detecting traces of arsenic and observed different methods of doing so [10]. In 1901, S Delepine applied arsenic detection technology to life, detecting arsenic in beer and brewing materials [11]. In the following years, various techniques were applied to arsenic detection, and these techniques are divided into two main categories according to the instruments used: one is the use of traditional instrumentation techniques, including atomic absorption spectrometry, atomic emission spectrometry, inductively coupled plasma mass spectrometry, etc. The second is the use of new sensor methods, mainly biosensor methods, electrochemical detection, and the ultraviolet–visible spectroscopy (UV–Vis), colorimetric method, etc. In this paper, electrochemical detection methods will be introduced. Electrochemical detection has a series of advantages such as simple instrumentation, simple operation, high sensitivity, high selectivity, being easy to carry, and being easy to use for rapid detection. Electrochemical detection methods are an effective alternative to traditional ion detection methods.

2. Principles of Electrochemical Detection of As(III)

Probably one of the oldest measurement techniques, the electrochemical method, is an analytical method based on electrochemical principles and founded on the electrochemical properties of a substance in solution, which implies the transfer of charge between the electrode and the liquid or solid phase [12]. During the detection, the solution to be tested is usually used as part of a chemical cell, which reacts to the concentration of the substance being detected by means of the electrical parameters of the chemical cell. Compared with other arsenic detection methods, electrochemical detection is considered to be user-friendly and achieves a more desirable detection result, while the detection process is simple and cost effective. After a long period of development, electrochemical detection techniques to detect arsenic have become more mature, and there are many commonly methods based on various electrical signals used in the detection of heavy metals. Electrochemical techniques are divided into amperometric, voltammetric, potentiometric, impedance measurements, coulometric, and electrochemiluminescent techniques [13]. The general apparatus for electrochemical detection is shown in Figure 1 and usually consists of an electrolytic cell in which the heavy metal ions or other classes of heavy metal ions to be detected act as the electrolyte. A chemical reaction occurs in the cell, partly at the working electrode, resulting in a change in the electrical parameters that establish a link with the concentration of the target element for the purpose of quantitative analysis.

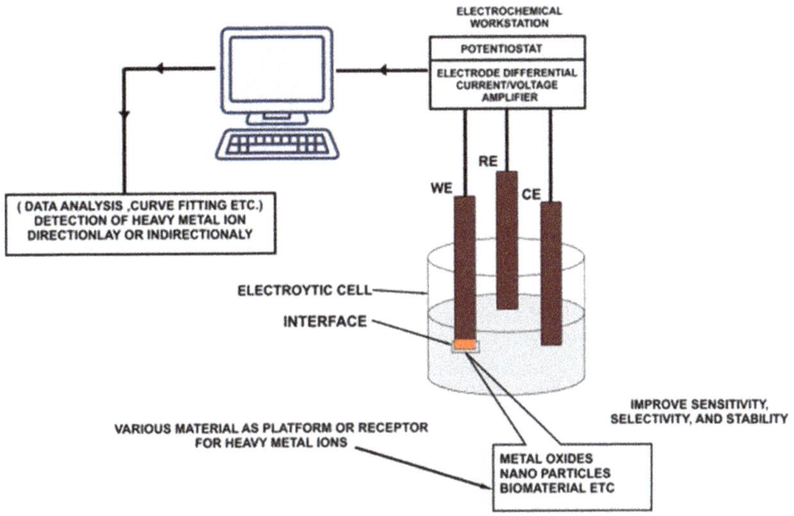

Figure 1. General apparatus for electrochemical detection of arsenic. (Reprinted with permission from [13], Copyright 2017, Elsevier B.V.)

Voltammetry is considered the most versatile of the arsenic ion assays and has been further advanced with the development of potential and current modulation techniques, such as differential pulse voltammetry (DPV), square-wave voltammetry (SWV), stripping chronopotentiometry (SCP), etc. Voltammetry is based on the voltage–current time relationship generated in a three-electrode cell; the position of the peak is reflective of the specific chemical, and the density of the peak is related to the concentration of the substance being detected. The combination of electrode modification and pre concentration techniques further increases the selectivity and sensitivity of voltammetry.

3. Electrochemical Detection Electrodes for As(III)

During decades of research, electrodes have often been modified in order to improve the sensitivity and selectivity of the electrochemical detection of ions. In the beginning, the excellent electrical properties of precious metals (e.g., gold and silver) were taken into account and a large number of electrodes made of precious metals were used for electrochemical detection. Later, there were considerations of cost constraints, reducing the amount of precious metals, and modifying the working electrodes with precious metals and their nanoparticles to achieve the same detection results. To further reduce costs, a range of materials including carbon nanomaterials, non-precious metal oxides, and bimetallic nanoparticles have been used for detection, again well below the WHO (World Health Organization) detection limit of 10 ppb. Additionally, in today's interdisciplinary world, there are many biotechnologies used for electrochemical detection, with examples of bioreceptors such as DNA or proteins being used for modified electrodes.

3.1. Precious Metal Electrodes and Precious-Metal-Modified Electrodes

The excellent electrical properties of precious metals with good conductivity make them the preferred choice for electrode materials and their modifications in electrochemical detection. They show distinct advantages over the conventional macroelectrodes, such as increased mass transport, decreased influence of the solution resistance, low detection limit, and better signal-to-noise ratio [14]. In addition, precious metal particles can easily be deposited onto the electrode surface by electrochemical means, thus modifying the electrode and improving its overall surface properties.

3.1.1. Gold Electrodes and Gold-Modified Electrodes

Gold electrodes in various forms and gold-modified materials have long been a popular topic in electrochemical detection. Agnese Giacomino et al. achieved a low detection limit of 0.060 ppb in the range of 1–15 ppb for the determination of As(III) by anodic dissolution voltammetry using a lateral gold electrode [15]. The electrochemical behavior of gold electrodes is closely related to their crystal orientation, and single-crystal Au(111) electrodes with clean, well-defined, and ordered surfaces can provide a more defined electrochemical behavior for As detection and are suitable for studying deposition mechanisms. Mohammad Rezaur Rahman et al. reported a simple method for the fabrication of Au(111)-like poly-gold electrodes, achieving a low detection limit of 0.28 ppb by square-wave anodic solvation voltammetry (SWASV) [16]. In addition, there are different sizes and forms of gold electrodes, including solid-disc gold electrodes [17], thin-film gold electrodes on glass carbon electrodes [18], graphite electrodes [19], platinum electrodes [20], microfilament gold electrodes [21], etc. Lijuan Bu et al. proposed the first method to electrically generate H_2 to reduce As(III) and improve As(0) preconcentration on a gold disc electrode for the determination of As(III) by anodic stripping voltammetry, which improved the sensitivity of As(III) detection and achieved a detection limit of 1.0 nM or 0.075 ppb [22]. Gold wire microelectrodes are also a common class of gold electrodes and ChiZhou was able to detect As(III) down to 2.6 ppb in 0.5 M H_2SO_4 by square-wave anodic dissolution voltammetry using electrochemically etched gold wire microelectrodes [23]. S. Laschi et al. tested a disposable gold screen-printed working electrode for arsenic detection in aqueous solutions using square-wave anodic dissolution voltammetry (SWASV) and achieved a detection limit of 2.5 ppb after 60 s of deposition [24]. A self-made nanoporous gold microelectrode by Darío Xavier Orellana Jaramillo et al. had a high sensitivity of 29.75 μA $(\mu g\ L^{-1})^{-1}$ cm^{-2} and a low detection limit of 0.62 ppb [25]. More comparisons of different gold-based electrodes are given in Table 1.

Table 1. Comparison of gold electrodes.

Electrode	Method	Sensitivity (μA/ppb)	Linear Range(ppb)	LOD (ppb)	Reference
composite gold electrode	DPASV			0.32	[26]
gold side disk rotating electrode	DPASV		5–80		[27]
gold screen-printed electrode	SWASV	0.03	0–200	2.5	[24]
gold film	SCP [1]			0.022	[28]
gold nanoelectrode ensembles	SWASV	3.14		0.02	[29]
gold nanofilm	LSV [2]		0.2–375	0.04	[17]
Au(111)-like polycrystalline gold electrode	SWASV	0.097 μA ppb^{-1} cm^{-2}	0–1123.8	0.28	[16]
vibrating gold microwire electrode	DPASV	0.014 μA ppb^{-1} mm^{-1} V^{-1}	0.07–3.0	0.07	[21]
lateral gold electrode	ASV	9.15	0.1–15	0.06	[15]
MEA-modified Au electrode	DPASV	0.0366	0.2–300	0.02	[30]
Au-MEE	LSASV		1–10	0.09	[31]
Au disc electrode	ASV	18.69	0.75–299.68	0.075	[22]
nanoporous gold microelectrode	LSV	1.74×10^{-4}	1.50–14,984	1.50	[32]
two gold electrodes	ASV		0.0374–0.7492	0.0097	[33]
Au-wire electrode	ASCP [3]			0.42	[34]

Table 1. *Cont.*

Electrode	Method	Sensitivity (µA/ppb)	Linear Range(ppb)	LOD (ppb)	Reference
porous gold electrode	SWV		0.1–14	0.1	[35]
gold electrode	ASV		0−5000	850	[36]
nanoporous gold microelectrode	SWASV	29.75 µA ppb^{-1} cm^{-2}	2–30 and 10–200	0.62	[25]
gold wire microelectrode	ASV	6.8 µA ppb^{-1} cm^{-2}		2.6	[23]
arsenite-selective ionophore film-Au	ASV		10–100	1.10	[37]
gold nanotextured electrode	ASV	39.54 µA ppb^{-1} cm^{-2}	0.1–9	0.1	[38]
gold nanostar	SWSV [4]		2.5–764.2	0.8	[39]

[1] stripping chronopotentiometry; [2] linear sweep voltammetry; [3] anodic stripping chronopotentiometry; [4] square wave stripping voltammetry.

Additionally, due to the detection cost of using gold as an electrode, there has been increasing amounts of research on gold nanoparticle-modified conventional macroscopic electrodes, such as modified graphite electrodes [19], glassy carbon electrodes [40], screen-printed electrodes [41], gold electrodes [17], solid carbon paste electrodes [42] or boron-doped diamond electrodes [43], etc. Forsberg et al. tested three different materials (platinum, mercury, and gold) for modified electrodes [44], using anodic solvation voltammetry (ASV) and differential pulse anodic solvation voltammetry (DPASV), and found that the gold-modified electrodes were the most sensitive to the electrical signal generated by arsenic oxidation (which is one of the reasons why gold nanomaterials became the first material for modified electrodes). In addition to various forms of nanogold, other materials of composite-modified electrodes for the electrochemical detection of arsenic are a hot topic of research.

Supunnee Duangthong et al. developed a flow injection differential pulsed anodic dissolution voltammetry (FI-DPASV) method for the detection of arsenic, using a gold film-modified glassy carbon electrode as the working electrode and optimizing the parameters to achieve a low detection limit of 0.81 ppb in the linear interval 1.0–30 ppb for As(III) [45]. Syeda Sara Hassan et al. synthesized gold nanoflower structures by heating a mixture of ibuprofen and gold chloride ($HAuCl_4$) at a constant temperature for about 30 min. The SEM images are shown in Figure 2. Ibu-AuPNFs modified screen-printed electrodes, followed by Nafion as a binder and stabilizer, were used in the 0.1–1800 ppb range, observed for As(III) with a linear calibration plot with a lower limit of detection of 0.018 ppb [46]. Tran Ngoc Huan proposed a three-dimensional (3D) gold (Au) nanodendritic network porous structure constructed by a simple electrochemical synthesis method, the SEM images of which in different views are shown in Figure 3, allowing for the more sensitive detection of As(III) due to its larger surface area [47]. Dao Anh Quang successfully synthesized and stabilized gold nanorods (GNR) using cetyltrimethylammonium bromide (GNR-CTAB) and poly(sodium-4-styrenesulfonate) (GNR-PSS); the TEM image and the diameter and length distribution of the synthesized nanorods are shown in Figure 4. The GNR-modified glassy carbon electrode showed an excellent response with a limit of detection (LOD) of 0.72 ppb and a linear concentration of As(III) between 0.90 and 38.99 ppb [48]. Dingnan Lu et al. reported a square-wave anodic dissolution voltammetry (SWASV) method using a new gold nanostar-modified screen-printed electrode (AuNS/SPCE) as the working electrode; the TEM image of the gold nanostar is shown in Figure 5. Electrochemical impedance spectroscopy tests showed that the charge transfer resistance of AuNS/SPCE (0.8 kΩ) was significantly lower compared with bare SPCE (2.4 kΩ), achieving a sensitivity of 0.2213 µA/ppb in the linear detection interval of 0–100 ppb for As(III) detection [49].

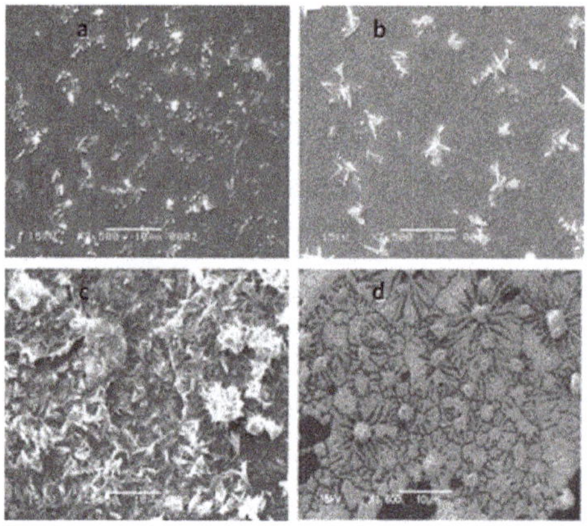

Figure 2. SEM images of Ibu-NSs at (**a**) 1, (**b**) 10, (**c**) 20, and (**d**) 30 min of heating time. (Reprinted with permission from [46], Copyright 2012, Elsevier B.V.)

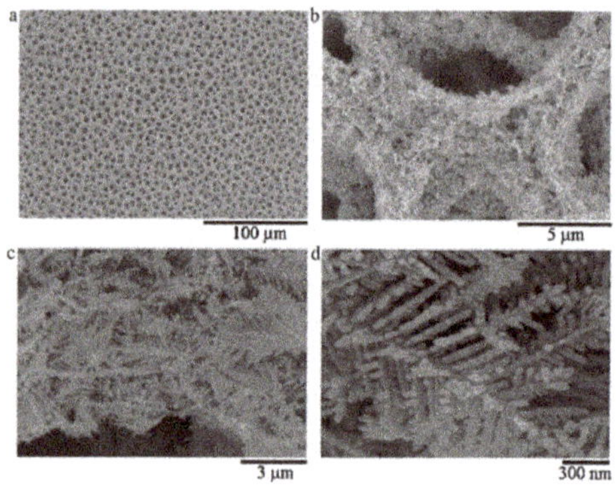

Figure 3. SEM images of porous structures of 3D Au nanodendritic networks with four different fields ((**a**) 100 µm, (**b**) 5 µm, (**c**) 3 µm, and (**d**) 300 nm) of view. (Reprinted with permission from [47], Copyright 2011, Elsevier B.V.)

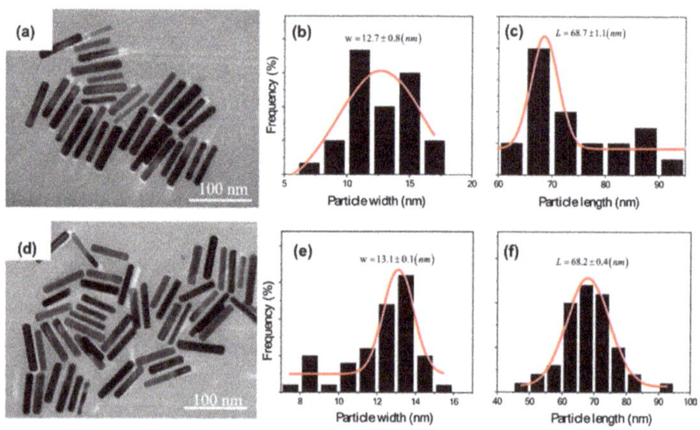

Figure 4. TEM images, diameter, and length distribution of GNR-CTAB (**a**–**c**) and GNR-PSS (**d**–**f**). (Reprinted with permission from [48], Copyright 2022, Springer Nature Switzerland AG.)

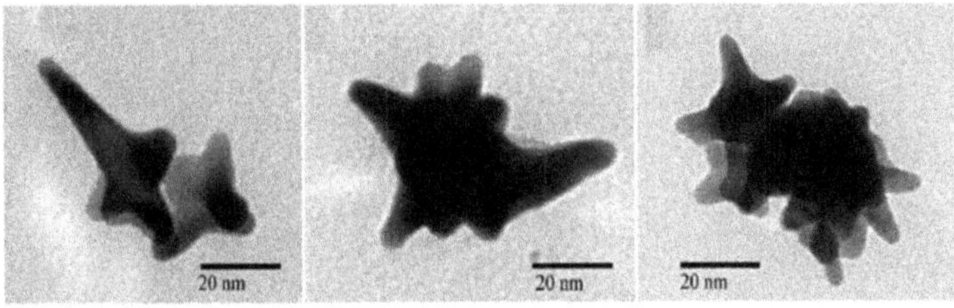

Figure 5. Representative TEM images of AuNS. (Reprinted with permission from [49], Copyright 2022, Springer Nature Switzerland AG.)

Among the many gold nanomaterials, gold nanoparticles and composites of gold nanoparticles on various materials are the absolute favorites in the field of electrochemical detection.

Anamarija Stanković et al. modified glassy carbon electrodes with gold nanoparticles and crystalline violet [50]. The electron impedance spectroscopy results show that the electron transfer resistance of the nanogold crystalline violet film was lower than that of the bare glassy carbon electrode, thus enhancing the electron transfer kinetics. The sensitivity of the modified electrode for the detection of As(III) was 5.6 A/μM cm^2, the detection limit was 0.8 μM, and it had a good linear response in the range of 2.0–22.0 μM. Connor Sullivan et al. used a screen-printed electrode modified with nanogold to detect As(III) in commercial apple juice by square-wave dissolution voltammetry. The sensitivity of the method was 0.1007 μA ppb^{-1} and the detection limit was 16.73 μg L^{-1}. The results of voltammetry were compared with those of graphite furnace atomic absorption spectrometry, with no systematic deviation and an R^2 of 0.939 [51]. In recent years, electro-membrane extraction (EME) and anodic stripping voltammetry (ASV) combined with gold nanoparticle-modified glassy carbon electrodes (AuNPs/GCE) have also been used to detect As (III) in water, with a detection limit of 0.18 μg L^{-1} for this method [52]. The application of electromagnetic radiation before electrochemical determination improves the selectivity and cleaning ability of the sample, which can increase the lifetime of the working electrode and reduce surface passivation.

A simple and easy method for synthesizing bentonite (bt) clay-supported gold nanoparticles (AuNPs) composites has been reported [53], and the test plots are shown in Figure 6. The AuNPs were successfully synthesized and doped into bt clay, as shown by spectroscopic, microscopic and electrochemical methods, and the synthesized Au bt material was used to modify the glassy carbon electrode (GC). The GC/Au bt electrode was used to detect As(III) in neutral solution by cyclic voltammetry. The GC/Au-bt electrode showed a wide linear range with good reproducibility and stability in As(III) solutions in the range of 1–1700 μM. The detection limit was 0.1 μM with high sensitivity. In addition, it had good selectivity for the determination of As(III) in the presence of Cu(II) and other interfering ions, providing an effective new route for the measurement of As(III) under neutral conditions. Qian Tang et al. prepared Au-PANI-Fe-CNFs composites by forming polyaniline (PANI) nanosheet arrays on Fe-CNFs substrates followed by the self-deposition of Au nanoparticles; the process is shown in Figure 7, using the composite-modified glassy carbon electrode as a sensing platform for As(III) measurements [54]. Polyaniline showed a uniform array structure on the surface of carbon fibers (CNFs), and the presence of Fe in CNFs promoted the formation of polyaniline nanosheets and the adsorption of As(III) in the subsequent sensing process. The detection of As(III) in water has good electrochemical performance, and the SWASV response plot is shown in Figure 8. The electrode has a wide linear range (5–400 ppb) and high sensitivity with a detection limit of 0.5 ppb, which provides a new route for the electrochemical analysis of arsenic in water.

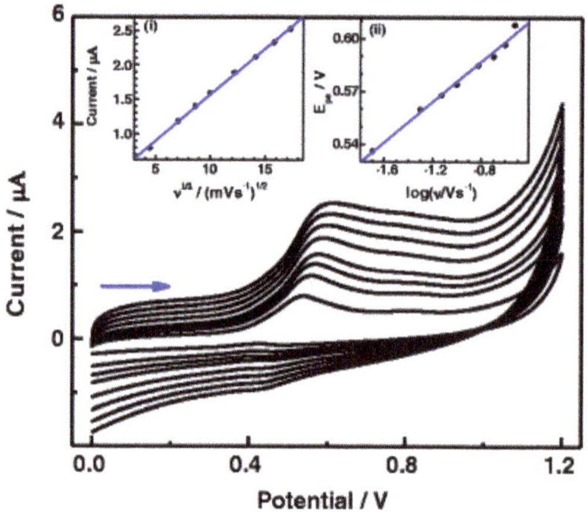

Figure 6. Cyclic voltammetric response of GC/Au-bt to 100.0 μM As(III) oxidation at different scan rates. Inset (i) shows the plot of the oxidation peak currents against the square root of scan rates and inset (ii) shows the plot of the oxidation peak potential against the log of scan rate. (Reprinted with permission from [53], Copyright 2016, Springer Nature).

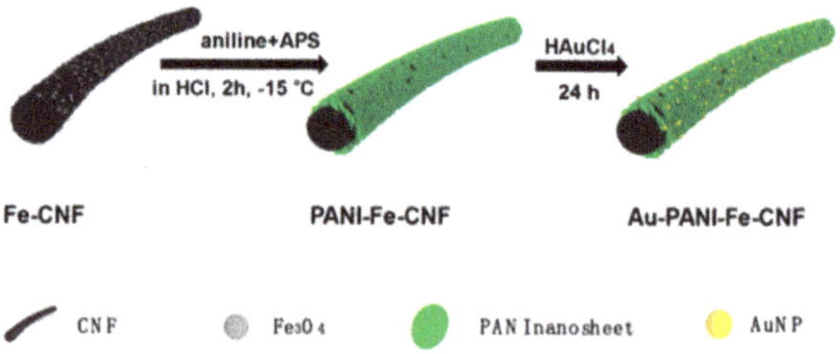

Figure 7. Schematic diagram of the preparation process of Au-PANI-Fe-CNFs. (Reprinted with permission from [54], Copyright 2020, Elsevier B.V.)

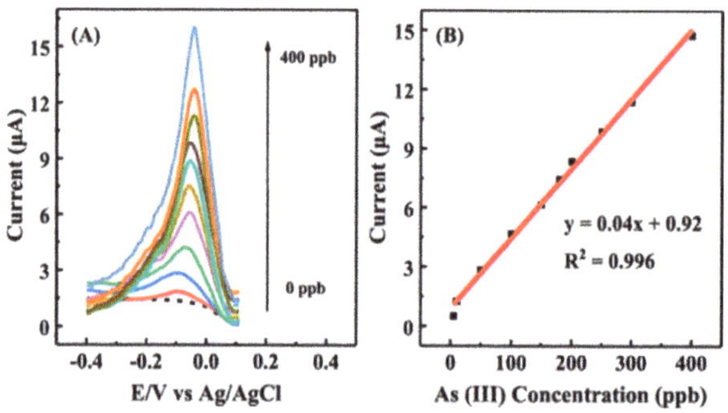

Figure 8. (**A**) SWASV response of different concentrations of As(III) on Au-PANI-Fe-CNFs/GCE. (**B**) Linear calibration curve of peak current versus As(III) concentration (0–400 ppb). (Reprinted with permission from [54], Copyright 2020, Elsevier B.V.)

Besides composites of gold nanomaterials with carbon materials, metal oxide nanomaterials and certain polymers have also received much attention. More pairs of gold nanomaterial-modified electrodes are shown in Table 2.

Table 2. Gold-trimmed electrodes.

Electrode	Method	Sensitivity (μA/ppb)	Linear Range (ppb)	LOD (ppb)	Reference
Au-ITO	LSV			5	[55]
AuNPs	ASV	0.400 $\mu A \cdot V\ ppb^{-1}$	0.5–15	0.25	[40]
AuNPs-PANI	SWV			0.4	[56]
Au-coated boron-doped diamond thin-film	DPASV		0.01–40	0.005	[57]
AuCNT	ASV	26.49		0.1	[58]
3DAu nanodendrite network porous structure	DPSV		0.1–70	0.1	[47]
AuNPs/Fe_3O_4	SWASV	13.55	0.01–1	0.00097	[59]
Nafion-Ibu-AuNSs	CV		0.1–1800	0.018	[46]
citrate stabilized AuNPs	SWV		0.05–1 and 1–15	0.025	[60]
PDDA-AuNPs	DPV	0.017	0–7492	4.36	[61]
NF (Au nano)	SWV	0.32	0.1–12.0	0.047	[62]
ERGO-AuNPs	ASV	0.16	0.75–374.6	0.20	[63]
AuNP-SPE	LSAdSV [1]	0.014	0.75–749.2		[64]
Au/Te	SWASV	6.35		0.0026	[65]
3D porous Au /TiO_2	SWASV	0.064	7.49–599.36	3.00	[66]
CB-AuNPs/SPE	ASV	0.63	2–30	0.4	[67]
MnO_x-AuNPs	LS-ASV	2.73 $\mu A\ ppb^{-1}\ cm^{-2}$	0.5–80	0.057	[68]
AuNP	DPASV		4–1498	0.9	[42]
TTCN-AuNPs	CV		0.0019–2.55	0.0006	[69]
gold film	FI-DPASV		1.0–30	0.81	[45]
AuNPs	EME [2]-ASV		0.5–10 and 10–600	0.18	[52]
AuNPs-C films	ASV	0.026	1–100	0.55	[70]
EG-AuNPs	SWASV			0.58	[71]
AuNPs-bt	CV		74.92–127,364	7.49	[53]
Au@Fe_3O_4-RTIL	SWASV	458.66 $\mu A\ ppb^{-1}\ cm^{-2}$ /86.89 $\mu A\ ppb^{-1}\ cm^{-2}$	0.1–1/1–10	0.0022	[72]
np-Au	SWASV	0.60	0.5–15	0.0315	[73]
$MnFe_2O_4$/Au hybrid nanospheres	SWASV	0.315		3.37	[74]
AuNpCμF	DPV			0.9	[75]
GCE-AuNPs	SWASV		10–12,000	0.15	[76]
AuNPs-PCWEs	AS-chronoamperometry	0.083		2.2	[77]
AuNPs-RGO	ASV	0.092	1–20	0.13	[78]
AuNPs-PpyNW	I-V	0.0029 and 0.0585	7.49–599.36 and 749.21–5244.51	23.97	[79]
Au/SiO_2	SWV		0.1–40	0.07	[80]
ZrF-8CAu	CV		5–700	1	[81]
3D-rGO/AuNPs	EIS		3.8×10^{-7}–3.0×10^{-4}	1.4×10^{-7}	[82]
SPE/CNF-CHIT@Au nano	FIA-ECD	0.2181	100–100,000	11.4	[83]

Table 2. *Cont.*

Electrode	Method	Sensitivity (μA/ppb)	Linear Range (ppb)	LOD (ppb)	Reference
AuNP/BDD	SWASV		100–1500	20	[84]
AuNPs/CeO_2-ZrO_2	SWASV	0.976	0.5–15	0.137	[85]
AuNPs-SPCE	ASV	0.11	30–150	8.9	[41]
Buckypaper modified by GNP	LSASV		0.75–750	0.75	[86]
AuNP-Au film	ASLSV	0.027 μA ppb^{-1} cm^{-2}	1–150	0.42	[87]
eAuNP-SPE	DPASV		0.5–20	0.22	[88]
gold film–plastic	DPSV		10–500	5	[89]
rGO-Au nano	SWASV		1.0–50.0	0.08	[90]
AuNPs/gC_3N_4	LSASV	3.07	0.375–74.921	0.22	[91]
3D NPG-ITO	DPASV	9.837	0.1–50	0.054	[92]
SiNPs/AuNPs	LSASV		10–100	5.6	[93]
MCPTH-AuNPs	CV			1	[94]
AuNS/SPCE	SWASV	0.2213	0–100	0.8	[49]
gold nanoparticles and crystal violet	DPV	0.075 μA ppb^{-1} cm^{-2}	149.84–1648.24	59.94	[50]
gold nanostar	SWV	0.101	0–100	2.9	[95]
Au-PANI-Fe-CNF	SWASV	0.04	5–400	0.5	[54]
quasi-hexagonal gold nano	DPASV		0.075–30	0.11	[96]
GNR	DPASV		0.90–38.99	0.72	[48]
AuNP-rLA-Lcyst	SWV-ASV	0.1	3–25	3	[97]
AuNPs-Co_3O_4	SWASV	12.1/3.7	0.1–1/1–20	0.09/0.79	[98]
GO/Fe_3O_4@ PMDA/AuNPs	SWASV		5–500	0.15	[99]
AuNPs	SWSV	0.1007		16.73	[51]
Au nano/Fe-MOF	SWASV	4.708	2–30	0.0085	[100]
GC-AuNP-ArOx	CV	46.05	0.75–749.21	0.37	[101]

[1] Linear sweep adsorptive stripping voltammetry; [2] electromembrane extraction.

3.1.2. Platinum Electrodes and Platinum-Modified Electrodes

In addition to gold, there are other precious metals used in the electrochemical detection of arsenic. Platinum wires are not only used as counter electrodes in many electrochemical detection systems, but there are also many working electrodes made of platinum or modified with platinum nanoparticles. Tung Son Vinh Nguyen et al. determined the As(III) ion concentration in water samples by the anodic solvation voltammetry (ASV) technique using a Nafion film-modified platinum electrode and obtained a wide linear range from 0 ppb to 40 ppb with a detection limit below 10 ppb [102]. He Xu, on the other hand, used a Pt nanotube array electrode, the SEM in cross-sectional view of which is shown in Figure 9, achieving a low detection limit of 0.1 ppb. Electrochemical experiments demonstrated that platinum-nanotube array electrodes (PtNTAEs) exhibited better performance for As(III) analysis than Pt nanoparticle-coated GCE (Pt nano/GCE) or Pt foil electrodes [103]. In addition, the modification of other electrodes with Pt nanoparticles and their composites is also a promising approach [104–108]; the detection limit is mostly below 10 ppb. Of interest is a Pt single-atom-anchored catalyst on MoS_2 (Pt_1 /MoS_2) developed by Pei-Hua L et al. to catalyze the determination of As(III). Pt_1/MoS_2 of 4% exhibited excellent stability and interference resistance, with a sensitivity of up to 3.31 μA/ppb for the detection of As(III) under near-neutral conditions, due to the Pt single atoms activated close to S atoms, greatly enhancing the catalytic activity of S atoms in the MoS_2 plane [109]. Dong-Dong Han et al. demonstrated a size-dependent effect of 2–5 nm Pt nanoparticles on the electrochemical behavior of arsenic (As(III)), with a decrease in sensitivity as the size of Pt nanoparticles increased from 2.3 nm to 5.5 nm [110].

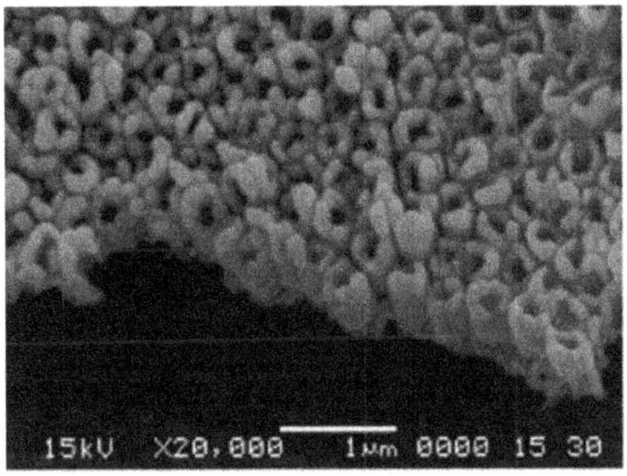

Figure 9. SEM image of cross-sectional view of PtNTAEs. (Reprinted with permission from [103], Copyright 2008, Elsevier B.V.)

3.1.3. Silver Electrodes and Silver-Modified Electrodes

Several papers have reported on the silver-based electrochemical sensing of arsenic. Among them, María del C. Aguirre used a silver wire electrode (SWE) for the electrochemical detection of arsenite with a minimum detection limit of 0.09 ppb, indicating that SWE can be used for the detection of trace arsenic in alkaline and neutral media [111]. Baudelaire Matangouo Sonkoue generated silver nanocolloids by the chemical reduction of silver salts using citrate in aqueous solution and used a gold electrode modified with silver nanoparticles as the working electrode in linear sweep voltammetry for the detection of arsenic ions. Under optimal conditions, calibration curves were plotted over a concentration range of 0.05–0.2 μM and the limit of detection was estimated to be 13.8 nM [112]. In addition, composites of silver nanomaterials with chitosan and graphene are also of interest for the electrochemical detection of arsenic. Silver nanoparticles (AgNPs) with chitosan (CT)-modified glassy carbon electrode (GCE) prepared by S. Prakash were used for the detection of As(III) by differential-pulse anodic dissolution voltammetry (DPASV), which has high sensitivity for the detection of As(III) in water due to the unique three-dimensional network and powerful adsorption capacity, and the designed nanostructured electrode has a wide linearity range (10–100 ppb), high sensitivity (0.309 μA ppb^{-1}) and a detection limit of 1.20 ppb (16.2 nM) [113]. Riyaz Ahmad Dar used cyclic voltammetry and anodic dissolution voltammetry measurements to evaluate the electrochemical properties of β-cyclodextrin-stabilized AgNPs-GO/GCE for As (III) detection with an approximately threefold increase in peak current compared with GO films alone, showing a wide linear range (13.33–375.19 nM) and high sensitivity (180.5 μA μM^{-1}), including a 0.24 nM detection limit [114]. Shao-Hua Wen et al. described the multimodal determination of arsenite (As(III)) in environmental samples by the stimulated response of multi-ligand functionalized silver nanoparticles (Ag NPs) in the electrochemical determination with GSH/DTT/Asn-Ag NPs as the signal probe of the redox electrochemical As(III) sensor. The As(III) concentration increased, the peak currents of Ag NPs in the DPV response curves were recorded, and the calibration curves showed good linearity of peak current intensity with As(III) concentration in the range of 0.01–40 ppb, with detection limits as low as 5.2 ppt [115].

3.1.4. Other Precious-Metal-Modified Electrodes

Agustiany studied the preparation of stable iridium-modified boron-doped diamond electrodes by electrodeposition, and As(III) was detected by cyclic voltammetry [116]. The electrode showed a linear curve in the concentration range of 1–100 μM with a detection limit of 4.64 μM and good stability and reproducibility, with a relative standard deviation of 2.6% under the optimal conditions of a pH 3 phosphate buffer and a scanning speed of 50 mV/s. In addition, the electrode showed good linearity (R^2 = 0.998) and sensitivity for the measurement of As(III) in both tap water and lake water samples. The test plots are shown in Figure 10. Erfan Mafakheri electrodeposited iridium oxide (IrO_2) nanotubes in polycarbonate (PC) stencils to form IrO_2 nanotubes with a uniform diameter of 110 ± 10 nm, an estimated length of 1–3 μm, and IrO_2 nanotube-modified glassy carbon electrodes for the detection of As(III) [117].

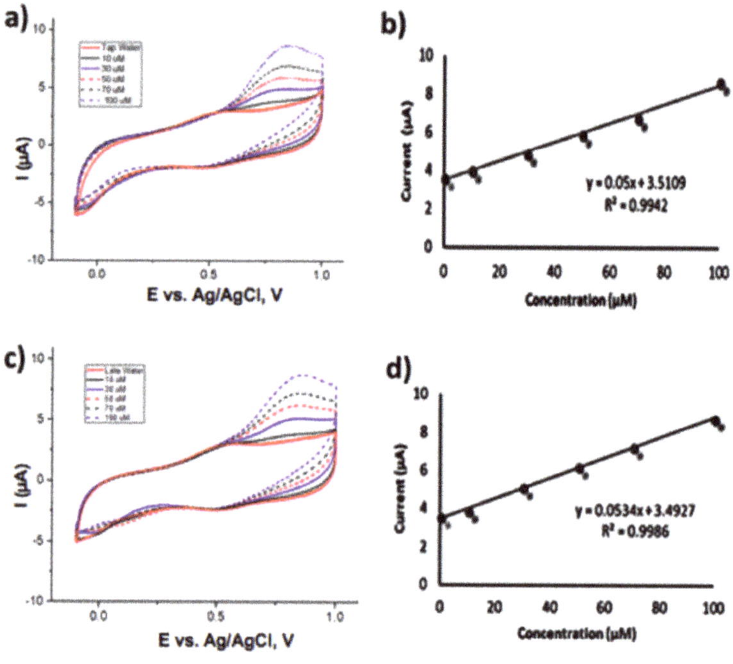

Figure 10. Cyclic voltammetry responses of different spike concentrations of arsenic (III) in (**a**) tap water and (**c**) lake water samples; scan rate 50 mV/s at Ir-BDD prepared using complete step deposition. (**b**) and (**d**) depict the dependence of current responses on arsenic (III) concentrations. (Reprinted with permission from [116], Copyright 2020, Elsevier B.V.)

R Gupta et al. deposited metallic ruthenium nanoparticles (RuNPs) on a glassy carbon electrode (GC) with a modified electrode with arsenite selectivity, schematically shown in Figure 11, for the detection of arsenite in water [118]. The differential pulse voltammetry (DPV) based on RuNPs/GC can determine the concentration of arsenite within minutes with a detection limit of 0.1 ppb, a reproducibility of 5.4%, and a sensitivity of 2.38 nA ppb^{-1}. RuNPs can be anchored on different chemical platforms such as graphene, electrodes, Fe_3O_4, etc. to design robust and reusable electrochemical sensors for the detection of arsenite in various aqueous solutions.

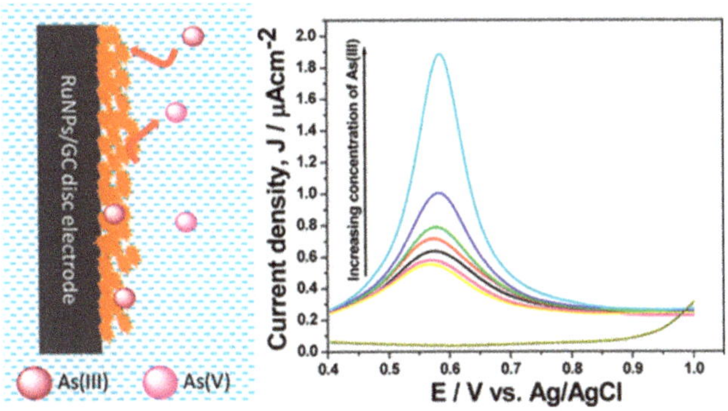

Figure 11. Schematic diagram of RuNPs/GC-based As electrochemical detection. (Reprinted with permission from [118], Copyright 2016, American Chemical Society.)

Sthitaprajna Dash investigated the electrodeposition of nanodendritic Pd on poly(3,4-ethylenedioxythiophene) (PEDOT)-modified Pd nanodendritic electrodes for the electroanalysis of As(III) in 1 M HCl solution. A wide detection range of up to 10 μM and a low detection limit of 7 nM (0.52 ppb) can be achieved with a pre-deposition time of 120 s under optimal conditions [119]. Md. Mahbubul Alam electrochemically immobilized Pd nanoparticles on Pt surface in the presence of sodium dodecyl sulfate (SDS) molecules. and its FE-SEM image is shown in Figure 12. The LOD of As(III) was determined to be 0.2 ppb using a Pt-Pd_{sds} sensor [120]. In addition, other noble-metal-modified electrode comparisons are presented in Table 3.

Figure 12. FE-SEM image of Pd particles deposited on Pt surface. (Reprinted with permission from [120], Copyright 2018, The Royal Society of Chemistry.)

Table 3. Other precious-metal-modified electrodes.

Electrode	Method	Sensitivity (μA/ppb)	Linear Range (ppb)	LOD (ppb)	Reference
Pt nano	ASV	2.94×10^{-3}	74.92–3746.08	2.1	[121]
Pt nano	SWV		0–100	0.5	[107]
PtNTAEs	LSV	0.011	749.21–14,984.32	0.1	[103]
Pt nano/CNTs	LSV	9.34×10^{-3}	374.61–74,921.6	0.12	[104]
Pt nano	CV			5.68	[122]

Table 3. *Cont.*

Electrode	Method	Sensitivity (μA/ppb)	Linear Range (ppb)	LOD (ppb)	Reference
Gr-nPt	SWASV		0.75–7.49	0.082	[106]
Pt nano (2.3 nm)	SWASV	0.356	0–1000		[110]
Nafion/Pt	ASV	0.036	0–40	<10	[102]
Pt nano	ASV		0–100	16.50	[105]
Pt_1 /MoS_2	SWASV	3.31	0.5-8	0.05	[109]
Pt nano	ASV	6.3×10^{-7} C μM^{-1}	3.75–74.92	4	[123]
SWE	LSASV			0.09	[111]
AgNPs/CT	DPASV		10–100	1.20	[113]
AgNPs-GO	ASV	2.41	1–28.11	0.018	[114]
Ag-GCE	SWASV	0.98	10–60	4.2	[124]
Ag-SPCE	SWASV	0.6	10–80	8.4	[124]
AgNPs	DPASV		3.75–14.98	1.03	[112]
GSH/DTT/Asn-Ag NPs	DPV		0.01–40	5.2×10^{-3}	[115]
IrOx-BDD	chronoamperometry	0.056	1.50–3746.08	0.15	[125]
Ir-BDD	CV	1.24×10^{-3} μA ppb^{-1} cm^{-2}		1.5	[126]
IrO_2 nanotubes	DPV		0–5993.73	7.49	[117]
Au-IrM	SWASV	$3.19 \times 10^{-4}/2.64 \times 10^{-3}$	0.75–3.75/0.07–0.75	0.037	[127]
Ir-BDD	CV	7.47×10^{-4} μA ppb^{-1} cm^{-2}	74.92–7492.16	347.63	[116]
Ru NPs	DPV	2.38×10^{-3}		0.1	[118]
$[Ru(bpy)_3]^{2+}$-GO	DPV	0.32	7.49–14.98	0.015	[128]
Ru(II)-tris(bipy)-GO	CV	1.42 μA ppb^{-1} cm^{-2}	3.75–59.93	2.25	[129]
Pd-PEDOT	DPASV	19.78 μA ppb^{-1} cm^{-2}	0–749.21	0.52	[119]
Pt–Pd_{sds}	SWV		74.92–16,857.36	0.2	[120]

3.2. Bimetallic Particle-Modified Electrodes

Precious metal nanomaterials are good for electrode modification, but the cost of modifying electrodes with gold is very high, so people choose to use gold composites for detection, in addition to deriving other bimetallic particles, in order to ensure the detection sensitivity and, at the same time, control the cost of detection. The bimetallic nanomaterials containing gold are Au-Pd, Au-Pt, Au-Ag, Au-Cu, etc. In addition, Fe has an excellent performance in arsenic ion detection, so it has also been studied in Fe-based bimetallic particle-modified electrodes, where bimetallic FePt, FeAu, FePd, and AuPt nanoparticles (NPs) are electrochemically deposited on Si(100) substrates and their electrochemical properties are investigated for As(III) detection. Trace amounts of As(III) can be determined by anodic stripping voltammetry at neutral pH. The synergistic effect with Fe alloying leads to the better performance of Fe precious metal NPs (Au, Pt and Pd) than pristine precious metal NPs (without Fe alloying). Detection limits and linear ranges were obtained for FePt, FeAu, and FePd NPs. The best performance was obtained for FePt NPs with a detection limit of 0.8 ppb and a sensitivity of 0.42 μA ppb^{-1}. The selectivity of the sensor was also tested in the presence of large amounts of Cu(II), the most detrimental interfering ion for As detection. Thus, bimetallic NPs are expected to be an effective and high-performance electrochemical sensor for the detection of ultra-trace amounts of arsenic [130]. More electrodes modified with bimetallic materials and their detection performance are summarized in Table 4.

Table 4. Bimetallic modified electrodes.

Electrode	Method	Sensitivity (µA/ppb)	Linear Range (ppb)	LOD (ppb)	Reference
Au-Pd NPs	SWV		1–25	0.25	[131]
Au-Pd NPs	SWASV	3.9		0.024	[132]
Au/Te crystals	SWASV	6.35	0.1–10	0.0026	[65]
Au-Pt NPs	LSASV		0.37–224.76	0.28	[133]
Au-Cu	SWASV	1.63 µA ppb^{-1} cm^{-2}		2.09	[134]
Au-PtNPs/PANI	SWASV	0.23	2.47–14.98	1.48	[135]
C-AuNPs	SWV		0.5–100	0.092	[136]
Ag-Au	CV/DPV		0.01–10	0.003×10^{-3}	[137]
Fe_3O_4-Au	SWASV	122	1–100	0.22	[138]
FePt	SWV	0.42	1–5	0.8	[130]
Pt-Fe	ASV	0.064		0.75	[139]

3.2.1. Gold–Platinum Bimetallic Modified Electrodes

A novel and easy-to-use nanohybrid platform suitable for the electrochemical detection of As(III) was prepared based on gold and platinum bimetallic nanoparticles (Au-Pt NPs) and the conducting polymer polyaniline [135]. Good detection limits were obtained by square-wave anodic dissolution voltammetry using modified screen-printed electrodes. The SWASV of 0–15 ppb (As(III) concentration) was recorded (Figure 13a) and this sensor was found to have good linearity in the range of 33–200 nM concentration of As(III) ions, with an LOD up to 19.7 nM, as shown in its calibration plot (Figure 13b).

3.2.2. Gold–Copper Bimetallic Modified Electrode

Recently, a simple hydrothermal method for the preparation of Au and Cu bimetallic nanoparticles of different compositions has been proposed [134]. The electrochemical performance of Au-Cu bimetallic nanoparticles in the determination of As(III) particle concentration was investigated using the square-wave anodic dissolution voltammetry SWASV method, and the SWASV response for As(III) detection in different concentration ranges is shown in Figure 14, where the Cu content in Au-Cu bimetallic nanoparticles is crucial for the detection efficacy. Compared with gold nanoparticles and gold electrodes, Au-Cu bimetallic nanoparticles exhibited better electrochemical performance with a lower detection limit (2.09 ppb) and higher sensitivity (1.63 µA ppb^{-1} cm^{-2}). In addition, the Au-Cu bimetallic nanoparticles also exhibited superb anti-interference performance for the detection of As(III).

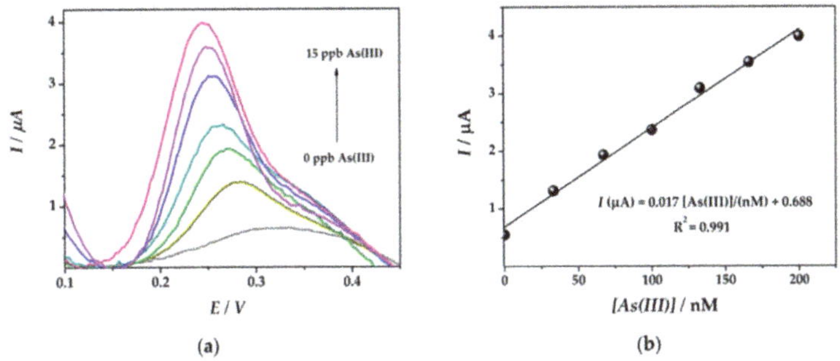

Figure 13. (**a**) SWASV curves obtained at the Au-PtNPs/SPCE platform. (**b**) Corresponding calibration plots (Reprinted with permission from [135]).

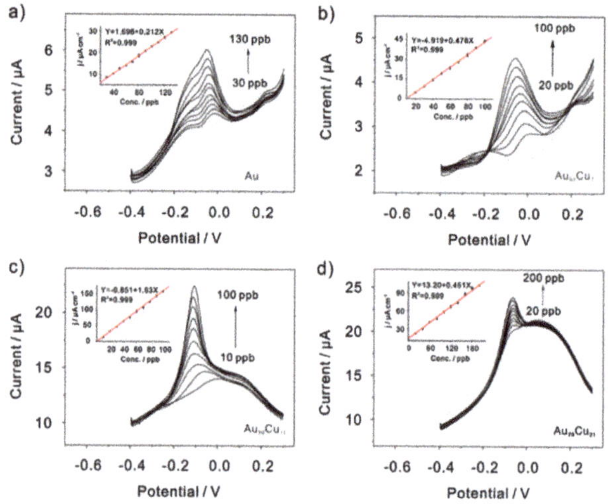

Figure 14. SWASV response of (a) Au, (b) $Au_{93}Cu_7$, (c) $Au_{89}Cu_{11}$, and (d) $Au_{79}Cu_{21}$ bimetallic nanoparticle-modified GCE for As(III) detection at different concentration ranges. (Reprinted with permission from [134], Copyright 2016, Elsevier B.V.)

3.2.3. Silver–Gold Bimetallic Modified Electrode

Reetu Yadav reported a sensor with silver–gold alloy nanoparticles (i.e., silver and gold alloy nanoparticles) modified with glassy carbon electrode and loaded with aptamer [137]. The bimetallic nanoparticles have a large surface area for adhesion with the aptamer, and thus have a large number of binding sites. The detection method uses cyclic voltammetry and differential pulse voltammetry, the electrode is used for the determination of As^{3+} in actual water samples, and the curve is linear when the As^{3+} concentration is 0.01–10 μg/L, with a detection limit of 0.003×10^{-3} μg/L. The sensor has good repeatability, stability, and selectivity, and can be applied to the detection of arsenic ion concentration in real water samples.

3.3. *Other Metals and Their Compound-Modified Electrodes*

3.3.1. Fe and Its Compound-Modified Electrodes

In the three-electrode-based electrochemical detection, the adsorption ability of the electrode surface on the target element plays a crucial role in the electrochemical detection performance, and the oxide nanoparticles of Fe become a hot spot for electrochemical detection due to their high adsorption ability and magnetic properties [140]. Pooja Devi et al. reported a chemically reduced rGO/Fe_3O_4 nanocomposite-modified glassy carbon electrode, which achieved a low detection limit of 0.12 ppb by square-wave anodic dissolution voltammetry [141]. Haibing Hu's team prepared Fe_3O_4 nanomaterials using a co-precipitation method, resulting in a Fe_3O_4 particle size of about 20 nm, which was then compounded with reduced graphene oxide. Under the optimized experimental conditions, the Fe_3O_4-rGO-modified glassy carbon electrode showed higher sensitivity (2.15 μA/ppb) for arsenic and achieved low detection limits [142]. The rGO/Fe_3O_4 nanocomposite has been shown to be a potential electrochemical and bioelectrochemical sensing material for the simultaneous detection of ascorbic acid, dopamine and uric acid, as well as for the electrocatalytic determination of nitrite. Akajionu Benjamin Chimezie et al. used differential pulsed anodic solvation voltammetry (DPASV) on a screen-printed electrode modified with reduced oxidation. An electrochemical sensor for the determination of As(III) in water resources was developed on the surface of graphene-magnetic nanocomposite (rGO-Fe_3O_4/SPEC) using differential pulsed anodic dissolution voltammetry (DPASV) [143]. The schematic diagram is shown in Figure 15. The electrode has a detection limit of 0.1 μg/L for As(III) in drinking water. The sensor has a wide operating range (2–300 μg L^{-1}), good repeatability, reproducibility and stability, and is virtually unaffected by common interfering ions.

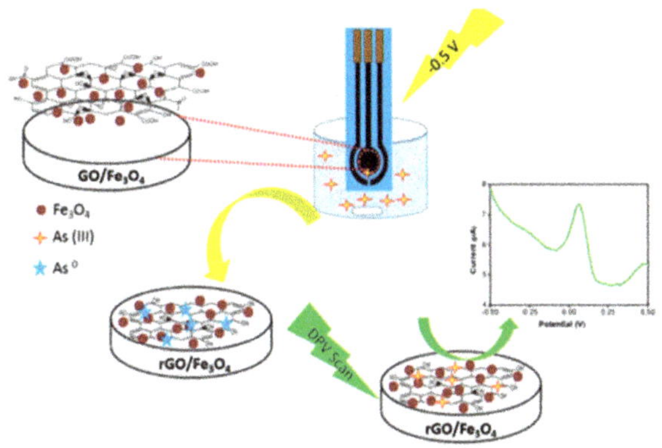

Figure 15. Schematic diagram of rGO-Fe_3O_4/SPEC-based electrochemical analysis of arsenic ions. (Reprinted with permission from [143], Copyright 2017, Elsevier B.V.)

Hong Cui et al. modified the glassy carbon electrode with a composite of magnetic Fe_3O_4 nanoparticles and gold nanoparticles, and the detection limit was 0.00097 ppb under optimal conditions [59]. Chao Gao et al. proposed an Fe_3O_4-RTIL (room temperature ionic liquid) composite-modified screen-printed carbon electrode (SPCE) in order to achieve ultra-low detection limits while reducing the dependence on precious metal gold, and obtained an ultra-low detection limit of 8×10^{-4} ppb by square-wave anodic solvation voltammetry (SWASV) while achieving a high sensitivity of 4.91 μA ppb^{-1} [144]. Other ferrites also have great potential for electrochemical detection. Shaofeng Zhou et al. reported Au nanoparticle-decorated mesoporous $MnFe_2O_4$ nanocrystal clusters for the detection of As(III) in water samples by square-wave anodic dissolution voltammetry (SWASV) with good reproducibility, stability and reproducibility, as well as interference resistance [74]. In a recent study, Hong-QiHuang et al. proposed a successful electrochemical sensor driven by noble metal-free layered porous Fe_3O_4/Co_3S_4 nanosheets for As(III) analysis, and obtained a considerable sensitivity of 4.359 μA/ppb for the electrochemical detection of As(III) in 0.1 M HAc-NaAc (pH 6.0) by square-wave anodic dissolution voltammetry (SWASV). The improved electrochemical performance of As(III) is attributed to its nanoporous structure, the presence of oxygen vacancies and the strong synergistic coupling effect between Fe_3O_4 and Co_3S_4 species [145].

3.3.2. Manganese and Cerium Oxide-Modified Electrodes

Due to the synergistic effect, Mn_2O_3/CeO_2 nanocubes have a high adsorption capacity for As(III), so its detection sensitivity is higher than any kind of oxide. Combined with the sensing properties of gold (Au) for As(III), a sensing material based on Mn_2O_3/CeO_2 nanocubes modified with gold electrode was fabricated [146], as shown in Figure 16. Under the optimized conditions, the sensitivity of the sensor was 0.0414 mA ppb^{-1} and the limit of detection (LOD) was 3.35 ppb with good stability and reproducibility, and the electrode had good selectivity for the presence of common interfering ions.

Manganese oxide (MnO_2) can be used as an active electrode material due to its good redox properties, porosity, low cost, and large specific surface area. In addition, the electrocatalytic properties of the composites can be further improved through synergistic effects by immobilizing metal oxides on the surface of polyhydroxytyramine (POT) and graphene oxide (rGO) composites. Sathish Kumar Ponnaiah et al. used a novel manganese dioxide/polyhydroxytyramine/reduced graphene oxide nanocomposite (MnO_2/POT/rGO/GCE) to fabricate sensing electrodes with a wide linear range (0.01–0.900 ppb) and minimum detection limit (42.0 ppt), and excellent selectivity, stability and reproducibility [147]. The process schematic is shown in Figure 17.

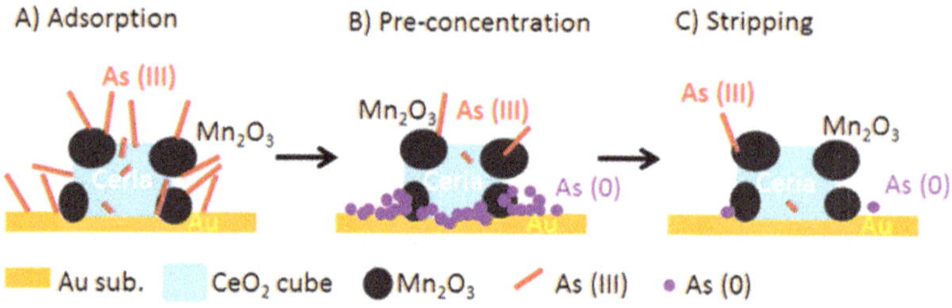

Figure 16. Schematic diagram of electrochemical analysis of arsenic on Mn_2O_3/CeO_2/gold electrode. (**A**) Adsorption, (**B**) pre-concentration, (**C**) stripping. (Reprinted with permission from [146], Copyright 2018, Wiley-VCH Verlag GmbH & Co. KGaA, Weinheim.)

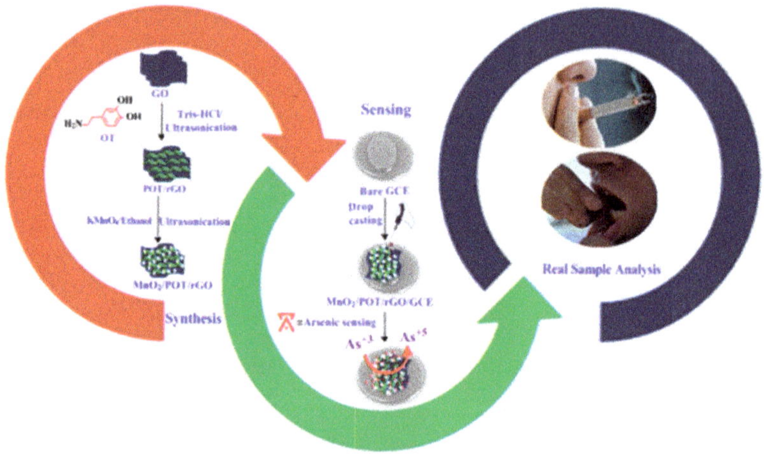

Figure 17. Synthesis of MnO_2/POT/rGO nanocomposites and arsenic detection process. (Reprinted with permission from [147], Copyright 2020, Springer Nature Switzerland AG.)

3.3.3. Cobalt Oxide-Modified Electrodes

An electrochemical sensor based on cobalt oxide nanoparticles was developed by Abdollah Salimi et al. Cobalt oxide (CoOx) nanoparticles were prepared from an aqueous buffer solution of $CoCl_2$ using cyclic voltammetry and deposited on the surface of a glassy carbon electrode. Then, As(III) was detected by cyclic voltammetry, and a detection limit of 11 nM was achieved. The authors concluded that immobilizing cobalt oxide nanoparticles on the surface of GC electrode seems to be an efficient method to develop a new class of sensitive, stable and reproducible electrochemical sensors for As(III) [148]. Chun-Yang Li combined the excellent catalytic properties of AuNPs with the high adsorption capacity of Co_3O_4 nanomaterials to construct an ultra-sensitive electrochemical sensor for electrochemical analysis of As(III) by homogeneously assembling gold nanoparticles on porous cobalt oxide (Co_3O_4) microsheets to form nanocomposites [98]. The experimental results show that the AuNPs/ Co_3O_4 nanocomposite-modified SPCE achieved an ultra-high sensitivity of 12.1 ± 0.2 μA ppb^{-1} and a detection limit of 0.09 ppb for As(III) using the SWASV method. This excellent electrochemical performance was attributed to the high adsorption capacity of the porous Co_3O_4 microporous sheet and AuNPs for the favorable electrocatalysis of As(III) reduction. In addition, the method also exhibits good anti-interference performance in the presence of other metal ions (Cu(II), Pb(II), Cd(II), etc.) with good stability and reproducibility. Most importantly, the electrochemical sensor has been successfully applied to the electroanalysis of As(III) in water and human serum samples, which provides a new approach to design sensitive and stable electrochemical sensors. The schematic diagram of the electrochemical analysis is shown in Figure 18 below.

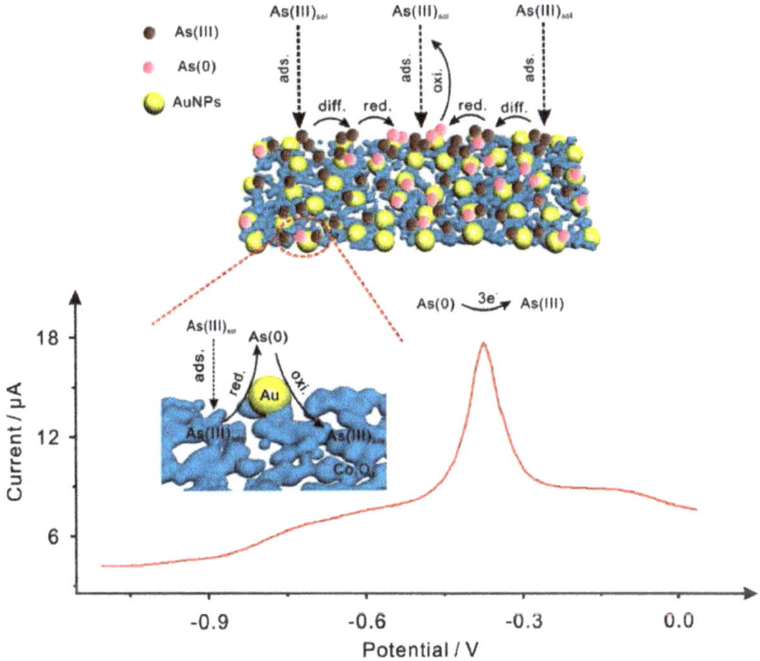

Figure 18. Electrochemical analysis of arsenic ions in solution based on AuNPs/Co_3O_4/SPCE. (Reprinted with permission from [98], Copyright 2020, Elsevier Ltd.)

3.3.4. Tin Oxide-Modified Electrodes

Tin oxide (SnO_2) has become the material of choice for arsenic (As^{3+}) redox sensing due to its high catalytic activity, environmental performance, wide band gap (3.64 eV) and high specific surface area (nanoscale) [149]. Tian-JiaJiang et al. reported an ultrathin SnO_2 nanosheet for the modification of gold electrodes, resulting in enhanced adsorption capacity on the gold electrode surface [150]. Gaurav Bhanjana et al. synthesized SnO_2 nanopins (particle size 60–80 nm) by the chemical precipitation method, characterized their elemental, topological, morphological and structural features, and then coated these nanopins on the surface of pencil cores (containing graphite/carbon(C)) to serve as working electrodes to prepare nanomaterial sensors for the detection of arsenic ions [151]. By the electrochemical determination of arsenic in real samples, the sensor has a detection limit of 10 ppb, a linear range of 50–500 ppb, and a sensitivity of 28.13 μA ppb^{-1} cm^{-2}. The experimental results provide a feasible method for the field detection of As^{3+} in environmental samples such as food, beverages, industrial samples, and wastewater. The modified electrode current response relationship under certain conditions is shown in Figures 19 and 20.

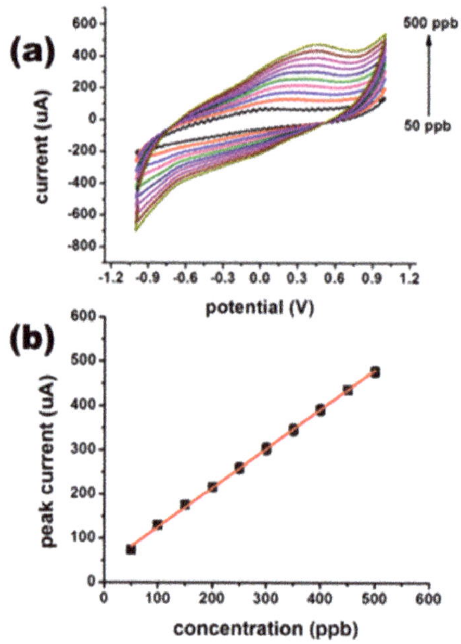

Figure 19. (**a**) Current response to different concentrations of arsenic; (**b**) oxidation peak current and concentration linear calibration curve. (Reprinted with permission from [151], Copyright 2018, Elsevier B.V.)

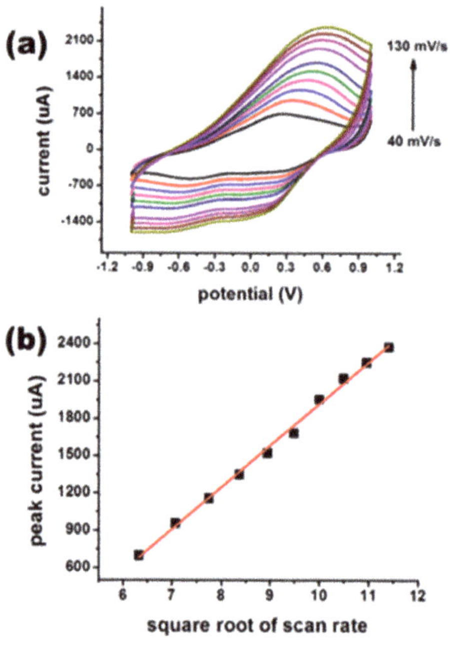

Figure 20. (**a**) Current response at different scan rates; (**b**) oxidation peak current and scan rate linear calibration curve (Reprinted with permission from [151], Copyright 2018, Elsevier B.V.)

3.3.5. Strontium Compound-Modified Electrodes

A. Karthika et al. reported a glassy carbon electrode modified with strontium titanium trioxide ($SrTiO_3$) and β-cyclodextrin (β-CD)-based nanocomposite, for the determination of toxic As(III) ions in water and serum samples [152]. The prepared $SrTiO_3$/β-CD nanocomposite-modified glassy carbon electrode has a high specific surface area and a sensitive electrochemical response. After testing, the oxidation peak current of As(III) increased linearly with the concentration in the concentration range of 10–140 μM of As(III) particles, and the detection limit was 0.02 μM. The electrode is stable, sensitive and reproducible for the detection of As(III) in water and serum.

3.3.6. Bismuth Compound-Modified Electrode

Thabile Ndlovu electrodeposited a bismuth film onto an exfoliated graphite (EG) electrode at a potential of −600 mV. The modification of EG resulted in an increase in the electroactive surface area of the electrode, and square-wave anodic dissolution voltammetry using the modified electrode (EG-Bi) in As(III) solution was able to detect 5 ppb of As(III) and was insensitive to many interfering cations except Cu(II) [153]. Potlako J. Mafa also electrodeposited bismuth nanoparticles onto graphite electrodes and used square-wave anodic solvation voltammetry (SWASV) to co-detect heavy metal ions in water samples with a detection limit of 0.014 ppb for As(III) under optimized experimental conditions [154]. In addition to bismuth nanoparticles, Lignesh Durai reported a novel and facile hydrothermal synthesis of bismuth vanadate ($BiVO_4$) nanoflakes for the trace detection of arsenic in biological samples by the electrodeposition of a screen-printed carbon electrode (SPCE) coated with polyaniline (PANI), with the As(III) sensing mechanism as shown in Figure 21. The sensor can detect As^{3+} ions by the differential pulse dissolution voltammetry (DPASV) technique with a significantly low limit of detection (LOD) of 0.0072 ppb and a sensitivity of 6.06 μA ppb^{-1} cm^{-2} with a linear range of 0.01–300 ppb [155].

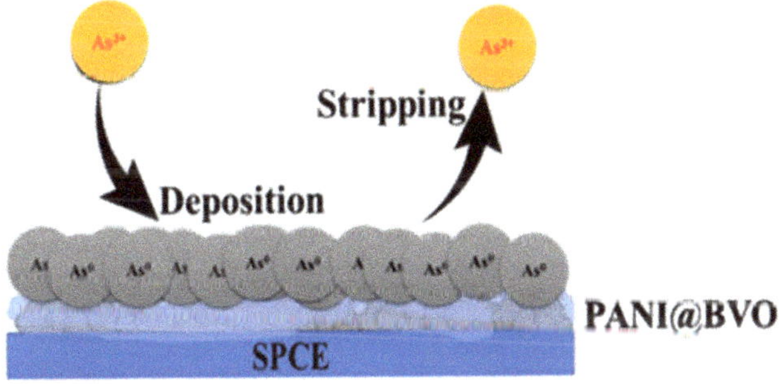

Figure 21. DPASV sensing mechanism of PANI@$BiVO_4$ /SPCE for As^{3+} ions. (Reprinted with permission from [155], Copyright 2020, Elsevier B.V.)

3.3.7. Zirconium Compound-Modified Electrodes

Gaurav Bhanjana used gold electrodes modified with zirconia nanocubes synthesized by a facile hydrothermal route, and electrochemical sensing of arsenic was achieved by cyclic voltammetry (CV) and chronoamperometry with a sensitivity of 550 nA cm^{-2} ppb^{-1} and a detection limit of 5 ppb (linear range of 5–60 ppb, response time below 2 s). The synthesized nanoparticles are nanocubes, and from the CV plots under different conditions (Figure 22) it can be seen that the peak oxidation current is more pronounced for zirconia nanocube-modified electrodes than for zirconia nanoparticle-modified electrodes, which are used as effective electrocatalysts in the direct redox sensing of arsenic [156]. In addition, zirconia composites were used for the electrochemical detection of arsenic. MengYang used AuNPs/CeO_2-ZrO_2 nanocomposite-modified glassy carbon electrodes (GCE) to fabricate a sensing interface for the sensitive and accurate analysis of As(III) in near groundwater pH values, and square-wave anodic dissolution voltammetry (SWASV) was used to determine As(III) in real water samples. Thanks to the strong adsorption capacity of CeO_2-ZrO_2, the electroanalytical sensitivity and theoretical detection limit of As(III) were 0.976 μA ppb^{-1} and 0.137 ppb, respectively, at the optimal parameters. In addition, the method has good anti-interference performance [85]. The performance analysis of different electrodes modified with metals other than noble metals and their compounds for the electrochemical detection of arsenic is given in Table 5.

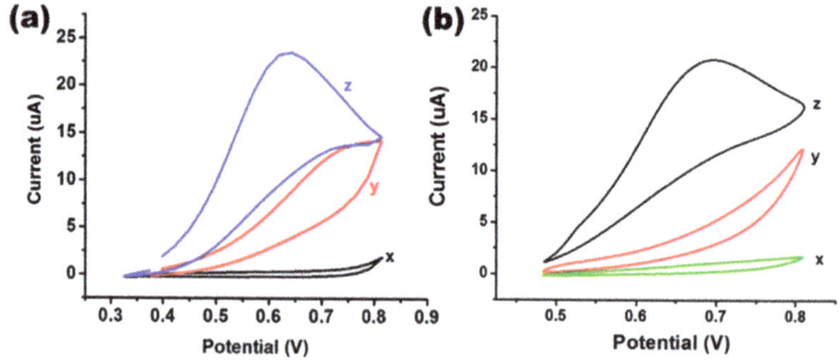

Figure 22. (**a**) CV curves of 50 ppm arsenic(III) ion standard under varying conditions for zirconia nanocubes, x: blank (black line), y: bare gold electrode (red line), and z: after fabrication (blue line); (**b**) CV curves of 50 ppm arsenic(III) ion standard under varying conditions for zirconia nanoparticles: x: blank (green line), y: bare gold electrode (red line), and z: after fabrication (black line). (Reprinted with permission from [156], Copyright 2016, Royal Society of Chemistry.)

Table 5. Other metal- and compound-modified electrodes.

Electrode	Method	Sensitivity (μA/ppb)	Linear Range (ppb)	LOD (ppb)	Reference
Au NPs/Fe_3O_4	SWV	13.55	0.01–1	0.00097	[59]
Fe-CB/PE	ASV		0.4–20	0.16	[157]
Fe_3O_4-RTIL	SWASV	4.91	1–10	0.0008	[144]
Chitosan-$Fe(OH)_3$	ASV	8.39	2–100	0.072	[158]
Au @Fe_3O_4-RTIL	SWASV	58.66 μA ppb^{-1} cm^{-2}	0.1–1	0.0022	[72]
rGO/Fe_3O_4		1.922	2–20	0.3	[159]
rGO-Fe_3O_4	DPASV		2–300	0.10	[143]
rGO/Fe_3O_4	SWASV	0.281		0.12	[141]
Fe-MOF @mFe_3O_4@mC	EIS		7.49×10^{-4}–0.75	5.04×10^{-4}	[160]
Fe Pc/Si-NP	DPASV	0.20		3.66	[161]
CoPc/Si-NP	DPASV	0.18		4.39	[161]
Au/Fe_3O_4	SWASV	9.43	0.1–10	0.0215	[162]
α-FeOOH	CV		0.75–1498.43	0.37	[163]
$Fe_2V_4O_{13}$–polypyrrole	DPASV		0–500	0.3	[164]
Fe_3O_4-Ag/Au HNSs-rGONs	CV	52	0.1–20	0.01	[165]
CN-wrapped IL-modified ZF-Ms (CN@ZF-Ms-IL)	SWASV	41.08	1–60	0.0006	[166]
Fe_3O_4–rGO	SWV	2.15	1–20	1.19	[142]

Table 5. *Cont.*

Electrode	Method	Sensitivity (μA/ppb)	Linear Range (ppb)	LOD (ppb)	Reference
GO/Fe_3O_4@PMDA/AuNPs	SWSV		0.5–750	0.15	[99]
Fe_3O_4-Au-IL	SWASV	122	1–100	0.22	[138]
Fe_3O_4/Co_3S_4	SWASV	4.359	1.0–10.0	0.691	[145]
MnO_x/AuNPs	LS-ASV	2.73 μA ppb^{-1} cm^{-2}	0.5–80	0.057	[68]
$MnFe_2O_4$ NCs	SWASV	0.295		1.95	[167]
rGO/MnO_2	SWASV	0.175	0.1–50	0.05	[168]
AuNPs/α-MnO_2	SWASV	0.828	1–10	0.019	[169]
$MnFe_2O_4$/Au hybrid nanospheres	SWASV	0.315		3.37	[74]
Mn_2O_3/CeO_2	SWASV	0.0414		3.35	[146]
Nafion/α-MnO@PDA	SWASV	0.13	10–150	3.2	[170]
MnO_2 /POT/rGO	DPV	0.00163	0.01–0.9	0.042	[147]
Bi-NPs	SWV			5	[153]
EG-Bi	SWASV			0.014	[154]
PANI@$BiVO_4$	DPASV	6.06 μA ppb^{-1} cm^{-2}	0.01–300	0.0072	[155]
F-doped CdO thin films	CV	5.747×10^{-3}	4.55–41	0.00455	[171]
CoOx	CV	1.49×10^{-3}		0.82	[148]
Co-rGO	ASV			0.31	[172]
AuNPs-Co_3O_4	SWASV	12.1/3.7	0.1–1/1–20	0.09/0.79	[98]
Fe_3O_4 /Co_3S_4	SWASV	4.359	1.0–10.0	0.691	[145]
$In_{0.38}Ga_{0.62}N$/Si(111)	SWV		10–50	9.27	[173]
Pt_1 /MoS_2	SWASV	3.31	0.5-8	0.05	[109]
PbO_2/rGO				0.75	[174]
Au-ITO	ASV			5	[175]
SnO_2 nanosheets	SWASV	0.058	5–300	4.6	[150]
Nafion/SnO_2 nanoneedles	CV	28.13 μA ppb^{-1} cm^{-2}	50–500	10	[151]
3D NPG-ITO	DPASV	9.837	0.1–50	0.054	[92]
Gemini-ITO	SWV		1–100	0.88	[176]
$SrTiO_3$ /β-CD	Amperometry	0.0053 μA μM cm^{-2}	749.21–10,489.02	1.50	[152]
CP-ThO_2 NP	DPASV	0.54	3–180	0.1	[177]
3D porous Au/TiO_2	SWASV	0.064	7.49–599.36	3.00	[66]
TiO_2-GSE	LSV	1.10	10–80	10	[178]
ZrO_2-nanocubes	CV		5–60	5	[156]
ZrF-8CAu	CV		5–700	1	[81]
Zr-G-PGE (As(V))	DPV	1.36	0.10–40.0	0.12	[179]
AuNPs/CeO_2-ZrO_2	SWASV	0.976	0.5–15	0.137	[85]

3.4. Carbon Nanomaterial-Modified Electrodes

In electrochemical sensing, carbon-based electrodes such as glassy carbon electrodes, screen-printed carbon electrodes and graphite–carbon paste electrodes are widely used for arsenic ion detection. In addition, due to the unique electronic properties of carbon-based nanomaterials, carbon nanomaterials have proven to be very suitable for the modification of working electrodes in the electrochemical detection of arsenic. Carbon nanomaterials include single-walled carbon nanotubes (SWNT), multi-walled carbon nanotubes (MWNT), graphene, nanodiamond, fullerene, and graphene quantum dots. The most widely studied carbon nano-derivatives in electrochemical sensing are carbon nanotubes and graphene [180]. The performance analysis of different carbon nanomaterials and their composite-modified electrodes for the electrochemical detection of arsenic is statistically presented in Table 6.

Table 6. Carbon material-modified electrode.

Electrode	Method	Sensitivity (μA/ppb)	Linear Range (ppb)	LOD (ppb)	Reference
ERGO-AuNPs	ASV	0.16	0.75–374.61	0.20	[63]
Gr-nPt	SWASV		0.75–7.5	0.082	[106]
Au/GO/Leucine/Nafion	CV	0.03 μA ppb^{-1} cm^{-2}		500	[181]
NH_2-GO	SWASV	130.631 μA ppb^{-1} cm^{-2}		0.162	[182]
3D-rGO/AuNPs	EIS		3.8×10^{-7}–3.0×10^{-4}	1.4×10^{-7}	[82]
$[Ru(bpy)_3]^{2+}$-GO	DPV		6.00–1123.83	1.57	[183]
Ru(II)-tris(bipy)-GO	CV	1.42 μA ppb^{-1} cm^{-2}	3.75–59.94	2.25	[129]
TTCA/rGO	SWASV		0–10	0.054	[184]
Gr/MOF	DPASV		0.2–25	0.06	[185]
RM-rGO	SWASV	2.49		0.07	[186]
SPGE	DPAV		0.0–5.0	0.28	[187]
Au-PANI-Fe-CNF	SWASV	0.04	5–400	0.5	[54]
SH-SWCNTs	LSV	1.33		0.008	[188]
DNA–SWCNT	LSV	0.17	0–33.6	0.05	[189]
Pt nano/CNTs	LSV	9.34×10^{-3}	374.61–74,921.6	0.12	[104]
CNTs/Leucine/Nafion	CV	0.27	0.37–149.84	0.12	[190]
CNTs/Nafion/Glutamine		1.33	0.075–37.835	2.72	[191]
CNTs-GNPs	LSASV	135		0.5	[192]
ssDNA/SWCNTs	DPV		0.5–10	0.5	[193]
Eu-MGO/Au@MWCNT	SWSV		0.99–100.0	0.27	[194]
Buckypaper modified by GNP	LSASV		0.75–750	0.75	[86]
CQDs/f-MWCNTs/GO	DPV		7.49×10^{-3}–0.82	0.037	[195]
MMWCNTs-D-NH_2	SWASV	0.5613	1.0–50.0	0.46	[196]

3.4.1. Carbon Nanotube-Based Detection of Arsenic Ions

Carbon nanotubes (CNT) are an excellent support structure due to their large effective detection surface, fast electron transfer rate compared with bulk carbon electrodes, high electrocatalytic activity and low electrode contamination, which can further help to improve electrochemical analysis performance by immobilizing other chemical species, such as metal NPs and organic molecules [14]. The main features of carbon nanotubes in electrochemical sensors are a fast response and low detection limits.

He Xu et al. chose to form a composite with carbon nanotubes for the electrochemical detection of arsenic using Pt nanoparticles, as shown by the TEM image of the composite (Figure 23), where Pt nanoparticles are clearly decorated on carbon nanotubes, reflecting a higher electroactive area than the Pt nanoparticle modification alone [104]. Yaxiong Liu et al.

proposed a single-layer low-resistance single-walled carbon nanotube-modified glassy carbon electrode for the electrochemical detection of arsenic with non-covalent SH groups sensitive to As(III), achieving an ultra-low detection limit of 0.008 ppb [188]. In addition to compounding with metal particles, carbon nanotubes can also be combined with biomolecules and with DNA. Yaxiong Liu's team developed a layer-by-layer assembly of DNA-functionalized single-walled carbon nanotubes that achieves a detection limit of 0.05 ppb in a near-physiological environment and can be reused multiple times [189]. Subramanian Nellaiappan et al., on the other hand, proposed a gold nanoparticle/carbon nanofiber/chitosan chemically modified carbon screen-printed electrode by simultaneously combining carbon nanofibers with metal particle biomolecules, achieving comparable results to inductively coupled plasma-emission spectroscopy [83].

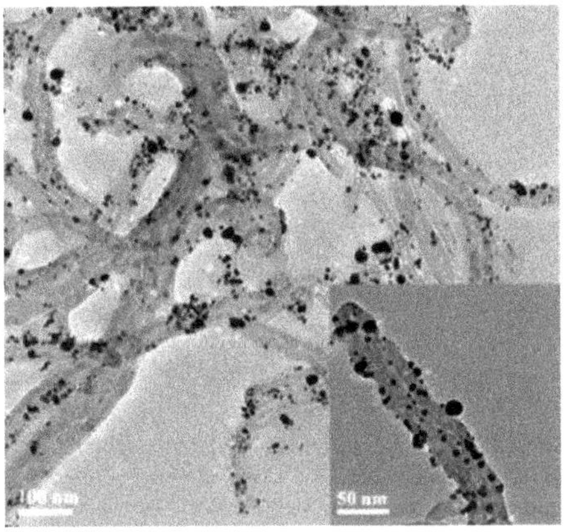

Figure 23. TEM photograph of Pt nano/CNTs synthesized by microwave irradiation; the inset is a higher magnification image of Pt/CNTs nanocomposite. (Reprinted with permission from [104], Copyright 2008, Elsevier B.V.)

3.4.2. Graphene Based Detection of Arsenic Ions

Graphene is a new material in which carbon atoms are tightly packed into a single two-dimensional honeycomb lattice structure. This material has excellent properties, with high strength and good toughness, and good electrical conductivity, and can be used directly as a modification material for electrodes. In addition, graphene has a characteristic that is very favorable for loading other materials, which is that it has many smooth folds, so there will be many graphene and other material composites to modify the electrode, so that the characteristics of the loading material can be more obvious.

Haibing Hu's team performed the electrochemical characterization of bare glassy carbon electrode, rGO, Fe_3O_4, and Fe_3O_4-rGO nanocomposite-modified glassy carbon electrode in a specific detection solution in the preparation of Fe_3O_4-rGO nanocomposite, and found that the electrochemical performance of glassy carbon electrode modified by rGO or Fe_3O_4 only was not as good as that of bare glassy carbon electrode, although it was better than that of bare glassy carbon electrode, while the electrochemical performance of glassy carbon electrode modified by both composites was very good, which indicated that the large surface area provided by rGO caused the Fe_3O_4 particles to adhere well to the surface [142]. In addition to metal oxides, graphene can also form composites with metals [185], multi-walled carbon nanotubes [195], precious metals [114], and biomolecules [197], among others.

Carbon nanoparticles and other carbon-based nanomaterials have been used for signal enhancement in electrochemical sensors and biosensors due to their advantageous specific surface area. An electrochemical sensor based on carbon nanoparticles (CNPs) and gold nanoparticles (AuNPs) comprising an immobilized platform for As(III) detection in water was reported in 2019 [136]. The carbon–gold nanoplatform was prepared by drop coating CNPs on a glassy carbon electrode (GCE), followed by the electrodeposition of AuNPs on the CNPs-modified electrode under certain conditions. The sensor has a detection limit of 0.092 ppb and exhibits insensitivity to the interference of Cd^{2+}, Cu^{2+}, and Hg^{2+}, providing an interference reduction method for the electrochemical detection of arsenic.

3.5. Biomolecule-Modified Electrodes

3.5.1. Arsenic Detection Based on DNA-Modified Electrodes

DNA in DNA-based biosensors provides biologically recognizable components with three modes of interaction [198], namely electrostatic interactions with negatively charged phosphates, binding interactions with minor and major grooves of the DNA double helix, and embedding between natural DNA stacked base pairs.

An advanced DNA biosensor was reported by J. Labuda et al. [199]. Using the Co(III) complex with 1,10-phenanthroline, $[Co(phen)_3]^{3+}$, as an electrochemical DNA marker and the Ru(II)complex with bipyridyne, $[Ru(bipy)_3]^{2+}$, as a DNA oxidation catalyst, calf thymus DNA (CT-DNA) immobilized on the surface of a screen-printed electrode (SPE) was placed in aqueous solutions of different concentrations of As(III), As(V), dimethylarsenic acid, phenylarsenic and p-arsenic acid. Although this system was reported to have a poor detection limit (75 mg/L), it showed a successful correlation between DNA-labeling signals and As(III) levels. Liu and Wei exploited the high electrical conductivity of carbon nanotubes (CNTs) to construct electrochemical biosensors and explored the concept of the direct oxidation of As(0) to As(III) on DNA-functionalized single-walled CNT-modified glassy carbon electrodes [189]. The developed biosensor was operationally stable over a wide pH range with a detection limit (S/N = 3) of 0.05 μg L^{-1} at pH 7.0. and demonstrated the ability to be reused 16 times. Shaohua Wen described a voltammetric method for the determination of arsenite (As(III)) based on the specific binding of As(III) to probe DNA (SBP DNA; single-stranded DNA) and the electrochemical indicator methylene blue (MB), the fabrication of which is schematically shown in Figure 24. Upon addition of As(III), it specifically binds to SBP DNA, which leads to conformational changes and the dissociation of SBP DNA from the electrode into the solution. As a result, the amount of MB remaining on the modified electrode is reduced, which decreases the peak MB current. Under optimized conditions, As(III) was quantified by measuring the DPV response of MB absorbed by the SBP/CP hybrid at the electrode surface, and the reduction peak current was linearly related to the logarithmic value of As(III) concentration, yielding a linear concentration range of 0.1–200 ppb and a detection limit as low as 75 ppt [200].

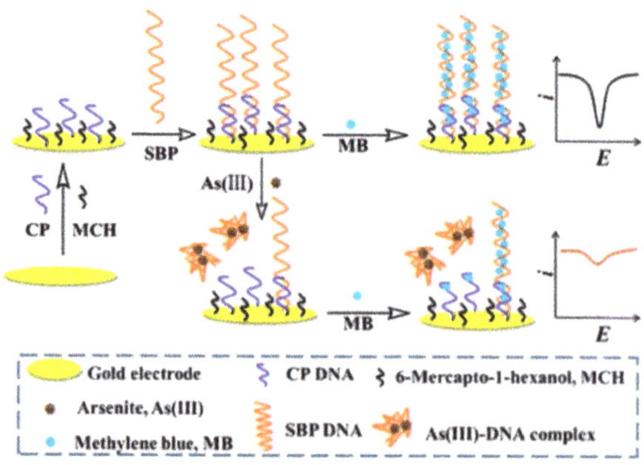

Figure 24. Schematic diagram of the fabrication of an electrochemical method for As(III) detection based on As(III)-induced DNA conformational changes and the electrochemical indicator MB. (Reprinted with permission from [200], Copyright 2017, Springer Nature Switzerland AG.)

3.5.2. Aptamer Sensors for Arsenic Detection

An aptamer-based biosensor is a small device that assembles one or more biomaterials/nanomaterials onto an electrode transducer, and electrochemical impedance spectroscopy (EIS), differential pulse voltammetry (DPV), etc., are often used with aptamer sensors for arsenic detection [201]. For example, Baghbaderani and Noorbakhsh constructed several aptamer sensors based on electrochemical signals for the determination of As(III) [202]. They designed an unlabeled impedance aptamer sensor for highly sensitive As(III) determination using a chitosan-Nafion(Chit-Naf) compound as an excellent conductive surface platform and a novel carbon nanotube based on the signal amplification process. The EIS experimental results show that the glassy carbon electrode (GCE) modified by Chit-Naf has higher electron transfer kinetics compared with bare GCE, GCE/Naf, and GCE/Chit electrodes, which provides great feasibility for an effective platform for biosensor design. In this work, based on a carbon nanotube–bovine serum albumin (CNT–BSA) hybrid system, they also used a signal amplification process to achieve an LOD of 74 pM. Lin Cui et al. designed an electrochemical aptamer sensor for the detection of As(III) based on gold-nanocoated screen-printed carbon electrodes (AuNPs/SPCE) [203], and the detection schematic is shown in Figure 25. By immobilizing the Ars-3 aptamer on AuNPs/SPCE, the Ars-3 aptamer is able to adsorb cations through electrostatic interactions with polydiallyldimethylammonium chloride (PDDA) and repel other cations. In the presence of arsenite, the Ars-3 conformation changes due to the formation of Ars-3/As(III) complexes, which reduces the adsorption of more positively charged electrochemically active indicator $[Ru(NH_3)_6]^{3+}$ on the surface of the PDDA adsorption electrode as a means to achieve coupling, thus enabling detection.

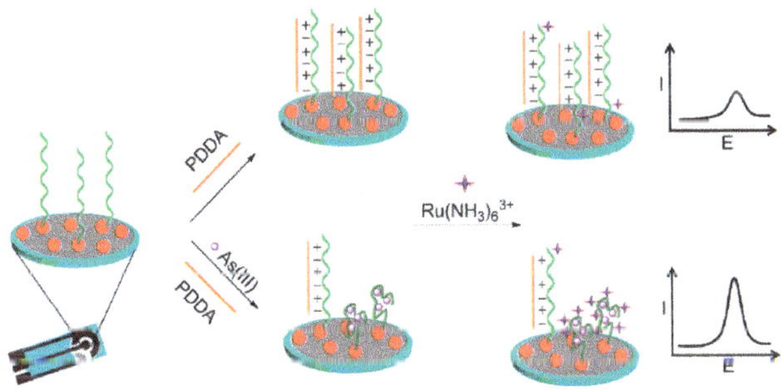

Figure 25. Schematic diagram of AuNPs/SPCE-based detection with Ars-3 aptamer. (Reprinted with permission from [203], Copyright 2016, Elsevier B.V.)

3.5.3. Arsenic Detection Based on Other Biomolecules

In addition to DNA with aptamers, certain proteins have been used as materials for the electrochemical detection of arsenic, and most protein-based arsenic detection is based on the inhibition phenomenon. Cytochrome-C (Cyt-C), an important component of the mitochondrial electron transport chain, is sensitive to all toxic compounds and is also used as a biorecognition element [204]. An electrochemical biosensor was constructed using Cyt-C, immobilized on a boron-doped diamond electrode. Square-wave voltammetry (SWV) and electrochemical impedance spectroscopy (EIS) were performed to investigate the interaction of Cyt-C with arsenic and cyanide. Subtractive normalized Fourier transform infrared spectroscopy (SNFTIR) was performed to confirm the effective protein adsorption onto the electrode. UV–vis studies of Cyt-C with the analytes confirmed the correct binding. The results indicate that their interaction was through the amino acids of the basic protein structure rather than through the heme portion of Cyt-C. Jae-Hoon Hwang et al. developed a novel As(III) sensor by depositing iron–chitosan complexes on screen-printed carbon electrodes using electrodeposition [205]. Mine wastewater and soil leachate were tested by square-wave anodic dissolution voltammetry. The detection limits of the Fe–chitosan-coated electrode were 1.12 ppb for mine wastewater and 1.01 ppb for soil leachate, both of which were significantly lower than the WHO requirements. The interference of Cu^{2+} ions had little effect on the detection, indicating that the chitosan-coated iron carbon could improve the stability. The sensor has high sensitivity and selectivity and provides a reliable level of detection of As(III) concentration in leachate from actual wastewater and contaminated sites. Suparna Saha et al. modified glassy carbon electrode with chitosan–$Fe(OH)_3$ composite and reducing agent L-cysteine [158], and its detection schematic is shown in Figure 26, under optimal optimized conditions, by differential pulse. The anodic dissolution voltammetry achieved a detection limit of 0.072 ppb in the linear interval of 2–100 ppb and avoided the interference of common co-existing ions. More examples of biomolecule-modified electrodes for the detection of arsenic are given in Table 7.

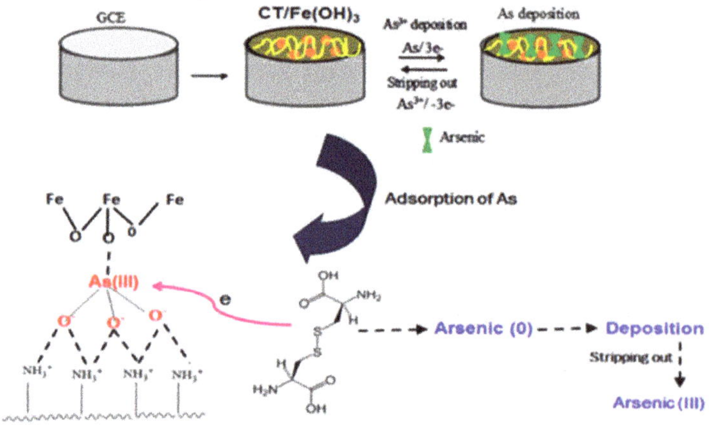

Figure 26. Schematic diagram of the scheme for the detection of arsenic ions by Fe–chitosan composites. (Reprinted with permission from [158], Copyright 2016, Elsevier B.V.)

Table 7. Biomolecule-modified electrodes.

Electrode	Method	Sensitivity (μA/ppb)	Linear Range (ppb)	LOD (ppb)	Reference
SBP DNA	DPV		0.1–200	0.075	[200]
$(GT)_{21}$-ssDNA and PB@GO	DPV		0.2–500	0.058	[206]
GH-APTES-Fe_3O_4 NP	SWV	1.92/0.12	13.3–65.8/117–241	1.6	[207]
SAMs	ASV		2–40	0.5	[208]
L-tryptophan	SWASV		7.49×10^{-3}–7.49	0.90×10^{-3}	[209]
AuNP-rLA-Lcyst	SWV-ASV	0.1	3–25	3	[97]
P. cruentum	DPASV		2.5–20	1.08	[210]
MTs	ASV		5–1000	13	[211]
AgNPs/CT	DPASV		10–100	1.20	[113]
Chitosan-$Fe(OH)_3$	ASV	8.39	2–100	0.072	[158]
SPE/CNF-CHIT@Au nano	FIA-ECD	0.2181	100–100,000	11.4	[83]
ACh-SPC	chronoamperometry			2689.63	[212]
HCR and RecJf exonuclease	EIS		0.1–500	0.02	[213]

Table 7. *Cont.*

Electrode	Method	Sensitivity (μA/ppb)	Linear Range (ppb)	LOD (ppb)	Reference
AF /AuNPs-SPCE	CV		0.500–1999.61	137.85	[214]
GC-AuNP-ArOx	CV	46.05	0.75–749.21	0.37	[101]
Escherichia coli	CV		0.94–3.75/3.75–30	0.8	[215]
Whole-Cell Biosensors	CV		0–100		[216]
E. coli	CV	0.122		1.5	[217]
Ars-3	DPV		0.015–7.49	0.011	[203]
3D-rGO/Au NPs/ssDNA	EIS		3.8×10^{-7}–3.0×10^{-4}	1.4×10^{-7}	[82]
ArsSApt	EIS		50–10,000	59.94	[218]
C-AuNPs	SWV		0.5–100	0.092	[136]
Ars-3/AuNPs-GO-MB	DPV		0.4–10,000	0.2	[197]

3.6. Others

3.6.1. Silicon and Its Compound-Modified Electrodes

Suhainie Ismail et al. developed an efficient electrochemical detection method for arsenite using linear scanning anodic solvation voltammetry (LSASV) based on silicon nanoparticles and gold nanoparticles (SiNPs/AuNPs/SPCE) modifying the screen-printed electrode surface [93]. The electrode showed good linearity in the concentration range of 10–100 ppb with a detection limit of 5.6 ppb. Multiple co-existing ions—Pb^{2+}, Ni^{2+}, Zn^{2+}, Hg^{2+} and Cu^{2+}—in the water samples did not interfere with the detection of arsenite. The method is highly sensitive and reproducible with a relative standard deviation of 4.52%, which is promising for application. In addition, they also tried to detect As(III) using silicon nanoparticles (SiNPs)-modified screen-printed electrodes (SPCE) and tested the electrochemical response of the electrode to arsenic using cyclic voltammetry (CV) and linear scanning anodic solvation voltammetry (LSASV) [219]. Under the optimized conditions, the peak anode current showed good linearity in the concentration range of 5–30 μg/L As(III) with a detection limit of 6.2 μg/L. This method can effectively detect As (III) in real water samples with low fabrication cost, good reproducibility and stability. In addition to silicon nanoparticles, Feng Sun prepared nano-Au/SiO_2 modified GCE by a one-step method. This nanohybrid material was used for the electrochemical detection of As (III). The calculated LOD was 0.07 μg/L, with a linear detection range of 0.1–40 μg/L [80].

3.6.2. Novel Polymer-Modified Electrodes

Mohammed M. Rahman synthesized a new class of thermally stable hybrid poly(arylene)(azomethanes) and copoly(arylene)(azomethanes) (PAAP) based on diarylidenecycloalkanes by solution polycondensation, combined with a conductive nafion (5%) coating agent-modified glassy carbon electrode for the detection of arsenic by the I-V method, which exhibits higher sensitivity and selectivity for As^{3+} ions. Based on the calibration curve, the sensitivity and detection limits were calculated as 2.714 μA μM^{-1} cm^{-2} and 6.8 ± 0.1 nM, respectively, and this novel method provides a new route for the electrochemical detection of arsenic ions [220]. Wuwei Ma et al. proposed an electrochemical sensor based on ion-imprinted polymers (IIPs) and nanoporous gold (NPG)-modified gold electrode (IIP/NPG/GE) for the determination of arsenic ions (As^{3+}) in different kinds of water, which was prepared by the electrodeposition of nanoporous gold on the gold electrode, and then a layer of IIPs with As^{3+} as the template ion was synthesized in situ on the NPG surface by electropolymerization. The IIPs/NPG/GE formation process is shown in Figure 27. The linear range of As^{3+} was obtained from 2.0×10^{-11} to 9.0×10^{-9} M by cyclic voltammetry, and the lower limit of detection was 7.1×10^{-12} M after the calibration curve [221].

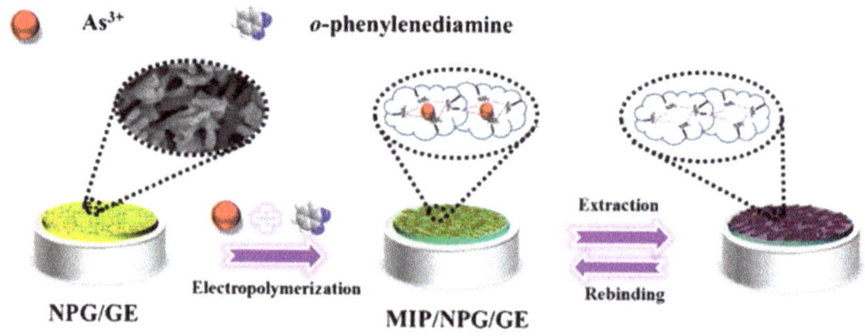

Figure 27. Schematic diagram of IIPs/NPG/GE formation process. (Reprinted with permission from [221], Copyright 2020, Springer Nature Switzerland AG.)

4. Conclusions

Arsenic contamination has seriously endangered the living environment and health of human beings, and achieving the efficient and reliable measurement of arsenic ions has gradually become a popular research area in the scientific community. The electrochemical detection method based on a nanomaterial-modified electrode has become a mainstream analytical method for measuring inorganic arsenic, with many advantages, such as simple operation, high sensitivity, good selectivity, low cost, and rapid portability. The working electrode is modified by nanomaterials so as to improve the performance of the electrochemical sensor, as described in this paper. The use of noble metal materials to modify the electrode can increase the mass transfer and reduce the effect of solution resistance; due to the synergistic effect, the bimetallic materials can ensure the detection performance while controlling the detection cost. Although noble metals show excellent performance in the electrochemical detection of arsenic, no noble metal nanomaterials are considered to be mainstream for the electrochemical detection of arsenic. The use of other metals and their compounds can achieve low cost, high sensitivity, and strong interference resistance; secondly, biomolecule-based electrochemical sensors for arsenic have better reproducibility and feasibility, and are increasingly being used in clinical diagnosis, food analysis, and environmental monitoring.

Most electrochemical studies are conducted under acidic conditions (i.e., acidic buffer solutions are used as solvents in sample preparation, along with the more common buffer solutions such as phosphate, acetic acid, hydrochloric acid, nitric acid buffers, etc.). Precious metal nanoparticles, bimetallic nanomaterials, metal oxide nanomaterials, and other modified electrodes show better performance and obtain better sensitivity in acidic environments. However, detection under ambient pH conditions has several advantages, such as avoiding unexpected changes in As morphology during acidification and simplifying the experimental procedures during field detection. In the operation of electrode preparation, deposition techniques, including electrodeposition and chemical deposition, are mostly used for metal nanoparticles, and drop casting is less frequently used. For metal oxide nanoparticles, the drop-casting method is mostly used directly. In real water samples, there may be interference from co-existing ions, such as Pb(II), Cu(II), Ni(II), Co(II), Cr(III), Zn(II), and NO_3^-. The electrodes in most of the previous studies showed high anti-interference and selectivity, and also performed well when the concentration of interfering ions was much higher than that of arsenic.

A method suitable for field analysis which can achieve a low detection limit (within 10 ppb) is urgently needed to detect arsenic in drinking water, and the use of a nanomaterial-modified electrode electrochemical system for this purpose represents a great opportunity. Research on new nanomaterials continues to make progress, such as reduced graphene oxide (rGO) and other metal oxide composites, which have been shown to have good detection performance, showing that metals and their compound nanomaterials for the detection of arsenic ions in water have a bright future.

Author Contributions: Conceptualization, H.H. and B.X.; validation, B.X. and Y.L.; formal analysis, H.H.; investigation, B.X. and Y.L.; resources, H.H.; data curation, B.X.; writing—original draft preparation, B.X. and Y.L.; writing—review and editing, J.Z.; project administration, H.H.; funding acquisition, H.H. All authors have read and agreed to the published version of the manuscript.

Funding: This research was funded by Natural Science Foundation of Anhui Province (1808085QE140) and the project of Research on Sensing Properties of Graphene-Based Composite Materials (W2020JSFW0662).

Institutional Review Board Statement: Not applicable.

Informed Consent Statement: Not applicable.

Data Availability Statement: Not applicable.

Acknowledgments: Thanks for the technical supports that are given by Liu Xingnan and Meng Fancheng.

Conflicts of Interest: The authors declare no conflict of interest.

References

1. Mandal, B.K.; Suzuki, K.T. Arsenic round the world: A review. *Talanta* **2002**, *58*, 201–235. [CrossRef]
2. Walsh, D.F.; Berger, B.L.; Bean, J.R. Mercury, arsenic, lead, cadmium, and selenium residues in fish, 1971–1973—National Pesticide Monitoring Program. *Pestic. Monit. J.* **1977**, *11*, 5–34. [PubMed]
3. Waalkes, M.P.; Coogan, T.P.; Barter, R.A. Toxicological principles of metal carcinogenesis with special emphasis on cadmium. *Crit. Rev. Toxicol.* **1992**, *22*, 175–201. [CrossRef] [PubMed]
4. Qiao, Y.L.; Taylor, P.R.; Yao, S.X.; Erozan, Y.S.; Luo, X.C.; Barrett, M.J.; Yan, Q.Y.; Giffen, C.A.; Huang, S.Q.; Maher, M.M.; et al. Risk factors and early detection of lung cancer in a cohort of Chinese tin miners. *Ann. Epidemiol.* **1997**, *7*, 533–541. [CrossRef]
5. Lu, Y.J.; Sun, S.Q. Study of deposition and translocation of mineral dust in lungs of tin miners in Yunnan by X-ray microanalysis. *Chin. J. Pathol.* **1992**, *21*, 103–105.
6. Kabir, R.; Sinha, P.; Mishra, S.; Ebenebe, O.V.; Taube, N.; Oeing, C.U.; Keceli, G.; Chen, R.; Paolocci, N.; Rule, A.; et al. Inorganic arsenic induces sex-dependent pathological hypertrophy in the heart. *Am. J. Physiol. Heart Circ. Physiol.* **2021**, *320*, H1321–H1336. [CrossRef]
7. Ozturk, M.; Metin, M.; Altay, V.; Bhat, R.A.; Ejaz, M.; Gul, A.; Unal, B.T.; Hasanuzzaman, M.; Nibir, L.; Nahar, K.; et al. Arsenic and Human Health: Genotoxicity, Epigenomic Effects, and Cancer Signaling. *Biol. Trace Elem. Res.* **2022**, *200*, 988–1001. [CrossRef]
8. Argos, M.; Kalra, T.; Rathouz, P.J.; Chen, Y.; Pierce, B.; Parvez, F.; Islam, T.; Ahmed, A.; Rakibuz-Zaman, M.; Hasan, R.; et al. Arsenic exposure from drinking water, and all-cause and chronic-disease mortalities in Bangladesh (HEALS): A prospective cohort study. *Lancet* **2010**, *376*, 252–258. [CrossRef]
9. Martinez, V.D.; Vucic, E.A.; Lam, S.; Lam, W.L. Arsenic and Lung Cancer in Never-Smokers: Lessons from Chile. *Am. J. Respir. Crit. Care Med.* **2012**, *185*, 1131–1132. [CrossRef]
10. Bostock, J. Observations on the Different Methods Recommended for Detecting Minute Portions of Arsenic. *Edinb Med. Surg. J.* **1809**, *5*, 166–174.
11. Delepine, S. The Detection of Arsenic in Beer and Brewing Material. *Br. Med. J.* **1901**, *1*, 81–84. [CrossRef] [PubMed]
12. Scozzari, A. Electrochemical sensing methods: A brief review. In Proceedings of the Conference on NATO-Advanced-Study-Institute on Sensor Systems for Biological Threats—Algal Toxins Case, Pisa, Italy, 30 September–11 October 2008; pp. 335–351.
13. Bansod, B.; Kumar, T.; Thakur, R.; Rana, S.; Singh, I. A review on various electrochemical techniques for heavy metal ions detection with different sensing platforms. *Biosens. Bioelectron.* **2017**, *94*, 443–455. [CrossRef] [PubMed]
14. Aragay, G.; Pons, J.; Merkoci, A. Recent Trends in Macro-, Micro-, and Nanomaterial-Based Tools and Strategies for Heavy-Metal Detection. *Chem. Rev.* **2011**, *111*, 3433–3458. [CrossRef] [PubMed]
15. Giacomino, A.; Abollino, O.; Lazzara, M.; Malandrino, M.; Mentasti, E. Determination of As(III) by anodic stripping voltammetry using a lateral gold electrode: Experimental conditions, electron transfer and monitoring of electrode surface. *Talanta* **2011**, *83*, 1428–1435. [CrossRef] [PubMed]
16. Rahman, M.R.; Okajima, T.; Ohsaka, T. Selective Detection of As(III) at the Au(III)-like Polycrystalline Gold Electrode. *Anal. Chem.* **2010**, *82*, 9169–9176. [CrossRef] [PubMed]
17. Diu, Y.; Zhao, W.; Xu, J.-J.; Chen, H.-Y. Electrochemical determination of arsenite in neutral media on reusable gold nanostructured films. *Talanta* **2009**, *79*, 243–248. [CrossRef]
18. Dugo, G.; La Pera, L.; Lo Turco, V.; Di Bella, G. Speciation of inorganic arsenic in alimentary and environmental aqueous samples by using derivative anodic stripping chronopotentiometry (dASCP). *Chemosphere* **2005**, *61*, 1093–1101. [CrossRef]
19. Viltchinskaia, E.A.; Zeigman, L.L.; Garcia, D.M.; Santos, P.F. Simultaneous determination of mercury and arsenic by anodic stripping voltammetry. *Electroanalysis* **1997**, *9*, 633–640. [CrossRef]
20. HuiliangHuang, K.; Dasgupta, P. A field-deployable instrument for the measurement and speciation of arsenic in potable water. *Anal. Chim. Acta* **1999**, *380*, 27–37. [CrossRef]

21. Alves, G.M.S.; Magalhaes, J.M.C.S.; Salauen, P.; van den Berg, C.M.G.; Soares, H.M.V.M. Simultaneous electrochemical determination of arsenic, copper, lead and mercury in unpolluted fresh waters using a vibrating gold microwire electrode. *Anal. Chim. Acta* **2011**, *703*, 1–7. [CrossRef]
22. Bu, L.; Gu, T.; Ma, Y.; Chen, C.; Tan, Y.; Xie, Q.; Yao, S. Enhanced Cathodic Preconcentration of As(0) at Au and Pt Electrodes for Anodic Stripping Voltammetry Analysis of As(III) and As(V). *J. Phys. Chem. C* **2015**, *119*, 11400–11409. [CrossRef]
23. Zhou, C.; Yang, M.; Li, S.-S.; Jiang, T.-J.; Liu, J.-H.; Huang, X.-J.; Chen, X. Electrochemically etched gold wire microelectrode for the determination of inorganic arsenic. *Electrochim. Acta* **2017**, *231*, 238–246. [CrossRef]
24. Laschi, S.; Bagni, G.; Palchetti, I.; Mascini, M. As(III) voltammetric detection by means of disposable screen-printed gold electrochemical sensors. *Anal. Lett.* **2007**, *40*, 3002–3013. [CrossRef]
25. Orellana Jaramillo, D.X.; Sukeri, A.; Saravia, L.P.H.; Espinoza-Montero, P.J.; Bertotti, M. Nanoporous Gold Microelectrode: A Novel Sensing Platform for Highly Sensitive and Selective Determination of Arsenic(III) using Anodic Stripping Voltammetry. *Electroanalysis* **2017**, *29*, 2316–2322. [CrossRef]
26. Navratil, T.; Kopanica, M.; Krista, J. Anodic stripping voltammetry for arsenic determination on composite gold electrode. *Chem. Anal.* **2003**, *48*, 265–272.
27. Prakash, R.; Srivastava, R.C.; Seth, P.K. Direct estimation of total arsenic using a novel metal side disk rotating electrode. *Electroanalysis* **2003**, *15*, 1410–1414. [CrossRef]
28. Vandenhecke, J.; Waeles, M.; Riso, R.D.; Le Corre, P. A stripping chronopotentiometric (SCP) method with a gold film electrode for determining inorganic arsenic species in seawater. *Anal. Bioanal. Chem.* **2007**, *388*, 929–937. [CrossRef]
29. Jena, B.K.; Raj, C.R. Gold nanoelectrode ensembles for the simultaneous electrochemical detection of ultratrace arsenic, mercury, and copper. *Anal. Chem.* **2008**, *80*, 4836–4844. [CrossRef]
30. Li, D.; Li, J.; Jia, X.; Han, Y.; Wang, E. Electrochemical determination of arsenic(III) on mercaptoethylamine modified Au electrode in neutral media. *Anal. Chim. Acta* **2012**, *733*, 23–27. [CrossRef]
31. Zakharova, E.A.; Noskova, G.N.; Antonova, S.G.; Kabakaev, A.S. Speciation of arsenic(III) and arsenic(V) by manganese-mediated stripping voltammetry at gold microelectrode ensemble in neutral and basic medium. *Int. J. Environ. Anal. Chem.* **2014**, *94*, 1478–1498. [CrossRef]
32. Jiang, J.; Holm, N.; O'Brien, K. Improved Anodic Stripping Voltammetric Detection of Arsenic(III) Using Nanoporous Gold Microelectrode. *ECS J. Solid State Sci. Technol.* **2015**, *4*, S3024–S3029. [CrossRef]
33. Korolczuk, M.; Ochab, M.; Rutyna, I. Determination of As(III) by anodic stripping voltammetry following double deposition and stripping steps at two gold working electrodes. *Talanta* **2015**, *144*, 517–521. [CrossRef] [PubMed]
34. Jedryczko, D.; Pohl, P.; Welna, M. Inorganic arsenic speciation in natural mineral drinking waters by flow-through anodic stripping chronopotentiometry. *Talanta* **2016**, *150*, 265–271. [CrossRef] [PubMed]
35. Kim, J.; Han, S.; Kim, Y. Electrochemical detection of arsenic(III) using porous gold via square wave voltammetry. *Korean J. Chem. Eng.* **2017**, *34*, 2096–2098. [CrossRef]
36. Lestarini, D.T.; Ivandini, T.A. Electrochemical detection of As^{3+} and As^{5+} by anodic stripping voltammetry at a gold electrode. In Proceedings of the 2nd International Symposium on Current Progress in Functional Materials (ISCPFM)—Functional Materials for Energy and Environment Applications, Bali, Indonesia, 8–9 November 2019.
37. Wang, L.; Lu, W.; Jing, G.; Cui, T. Trace Determination of Arsenite with an Ionophore-Coated Selective Micro Sensor. *IEEE Sens. J.* **2018**, *18*, 4364–4371. [CrossRef]
38. Babar, N.-U.-A.; Joya, K.S.; Tayyab, M.A.; Ashiq, M.N.; Sohail, M. Highly Sensitive and Selective Detection of Arsenic Using Electrogenerated Nanotextured Gold Assemblage. *ACS Omega* **2019**, *4*, 13645–13657. [CrossRef]
39. Dutta, S.; Strack, G.; Kurup, P. Gold nanostar electrodes for heavy metal detection. *Sens. Actuators B Chem.* **2019**, *281*, 383–391. [CrossRef]
40. Majid, E.; Hrapovic, S.; Liu, Y.L.; Male, K.B.; Luong, J.H.T. Electrochemical determination of arsenite using a gold nanoparticle modified glassy carbon electrode and flow analysis. *Anal. Chem.* **2006**, *78*, 762–769. [CrossRef]
41. Boonpeng, P.; Sooksamiti, P.; Lapanantnoppakhun, S.; Jakmunee, J. Determination of Inorganic Arsenic by Anodic Stripping Voltammetry with Gold Nanoparticles Modified Screen-printed Carbon Electrode and Anion Exchange Column Preconcentration. *Chiang Mai J. Sci.* **2019**, *46*, 106–117.
42. Sakira, A.K.; Some, I.T.; Ziemons, E.; Dejaegher, B.; Mertens, D.; Hubert, P.; Kauffmann, J.-M. Determination of Arsenic(III) at a Nanogold Modified Solid Carbon Paste Electrode. *Electroanalysis* **2015**, *27*, 309–316. [CrossRef]
43. Simm, A.O.; Banks, C.E.; Ward-Jones, S.; Davies, T.J.; Lawrence, N.S.; Jones, T.G.J.; Jiang, L.; Compton, R.G. Boron-doped diamond microdisc arrays: Electrochemical characterisation and their use as a substrate for the production of microelectrode arrays of diverse metals (Ag, Au, Cu) via electrodeposition. *Analyst* **2005**, *130*, 1303–1311. [CrossRef] [PubMed]
44. Forsberg, G.; O'Laughlin, J.W.; Megargle, R.G.; Koirtyihann, S.R. Determination of arsenic by anodic stripping voltammetry and differential pulse anodic stripping voltammetry. *Anal. Chem.* **1975**, *47*, 1586–1592. [CrossRef]
45. Duangthong, S.; Suwanin, A.; Chooto, P.; Innuphat, C. Flow injection-differential pulse anodic stripping voltammetry to measure As(III) and As(V) in natural water samples. *ScienceAsia* **2016**, *42*, 266–274. [CrossRef]
46. Hassan, S.S.; Sirajuddin; Solangi, A.R.; Kazi, T.G.; Kalhoro, M.S.; Junejo, Y.; Tagar, Z.A.; Kalwar, N.H. Nafion stabilized ibuprofen-gold nanostructures modified screen printed electrode as arsenic(III) sensor. *J. Electroanal. Chem.* **2012**, *682*, 77–82. [CrossRef]

47. Tran Ngoc, H.; Ganesh, T.; Kim, K.S.; Kim, S.; Han, S.-H.; Chung, H. A three-dimensional gold nanodendrite network porous structure and its application for an electrochemical sensing. *Biosens. Bioelectron.* **2011**, *27*, 183–186. [CrossRef]
48. Anh Quang, D.; Do Mai, N.; Tran Thanh Tam, T. Determination of arsenic(III) in water using gold nanorods-modified electrode. *J. Mater. Sci. Mater. Electron.* **2021**, *32*, 27962–27974. [CrossRef]
49. Lu, D.; Sullivan, C.; Brack, E.M.; Drew, C.P.; Kurup, P. Simultaneous voltammetric detection of cadmium(II), arsenic(III), and selenium(IV) using gold nanostar-modified screen-printed carbon electrodes and modified Britton-Robinson buffer. *Anal. Bioanal. Chem.* **2020**, *412*, 4113–4125. [CrossRef]
50. Stankovic, A.; Kajinic, Z.; Turkalj, J.V.; Romic, Z.; Sikiric, M.D.; Asserghine, A.; Nagy, G.; Medvidovic-Kosanovic, M. Voltammetric Determination of Arsenic with Modified Glassy Carbon Electrode. *Electroanalysis* **2020**, *32*, 1043–1051. [CrossRef]
51. Sullivan, C.; Lu, D.; Senecal, A.; Kurup, P. Voltammetric detection of Arsenic(III) using gold nanoparticles modified carbon screen printed electrodes: Application for facile and rapid analysis in commercial apple juice. *Food Chem.* **2021**, *352*, 129327. [CrossRef]
52. Kamyabi, M.A.; Aghaei, A. Electromembrane extraction coupled to square wave anodic stripping voltammetry for selective preconcentration and determination of trace levels of As(III) in water samples. *Electrochim. Acta* **2016**, *206*, 192–198. [CrossRef]
53. Rastogi, P.K.; Yadav, D.K.; Pandey, S.; Ganesan, V.; Sonkar, P.K.; Gupta, R. Synthesis and characterization of gold nanoparticles incorporated bentonite clay for electrocatalytic sensing of arsenic(III). *J. Chem. Sci.* **2016**, *128*, 349–356. [CrossRef]
54. Tang, Q.; Zhu, G.; Ge, Y.; Yang, J.; Huang, M.; Liu, J. AuNPs-polyaniline nanosheet array on carbon nanofiber for the determination of As(III). *J. Electroanal. Chem.* **2020**, *873*, 114381. [CrossRef]
55. Agir, S.K.; Kundu, M.; IEEE. Detection and Quantification of Arsenic in Water Using Electronic Tongue. In Proceedings of the 1st IEEE International Conference on Control, Measurement and Instrumentation (CMI), Jadavpur University, Kolkata, India, 8–10 January 2016; pp. 424–428.
56. Chowdhury, A.-N.; Ferdousi, S.; Islam, M.M.; Okajima, T.; Ohsaka, T. Arsenic detection by nanogold/conducting-polymer-modified glassy carbon electrodes. *J. Appl. Polym. Sci.* **2007**, *104*, 1306–1311. [CrossRef]
57. Song, Y.; Swain, G.M. Total inorganic arsenic detection in real water samples using anodic stripping voltammetry and a gold-coated diamond thin-film electrode. *Anal. Chim. Acta* **2007**, *593*, 7–12. [CrossRef] [PubMed]
58. Xiao, L.; Wildgoose, G.G.; Compton, R.G. Sensitive electrochemical detection of Arsenic(III) using gold nanoparticle modified carbon nanotubes via anodic stripping voltammetry. *Anal. Chim. Acta* **2008**, *620*, 44–49. [CrossRef] [PubMed]
59. Cui, H.; Yang, W.; Li, X.; Zhao, H.; Yuan, Z. An electrochemical sensor based on a magnetic Fe_3O_4 nanoparticles and gold nanoparticles modified electrode for sensitive determination of trace amounts of arsenic(III). *Anal. Methods* **2012**, *4*, 4176–4181. [CrossRef]
60. Lan, Y.; Luo, H.; Ren, X.; Wang, Y.; Wang, L. Glassy Carbon Electrode Modified with Citrate Stabilized Gold Nanoparticles for Sensitive Arsenic(III) Detection. *Anal. Lett.* **2012**, *45*, 1184–1196. [CrossRef]
61. Thotiyl, M.M.O.; Basit, H.; Sanchez, J.A.; Goyer, C.; Coche-Guerente, L.; Dumy, P.; Sampath, S.; Labbe, P.; Moutet, J.-C. Multilayer assemblies of polyelectrolyte-gold nanoparticles for the electrocatalytic oxidation and detection of arsenic(III). *J. Colloid Interface Sci.* **2012**, *383*, 130–139. [CrossRef]
62. Huang, J.-F.; Chen, H.-H. Gold-nanoparticle-embedded nafion composite modified on glassy carbon electrode for highly selective detection of arsenic(III). *Talanta* **2013**, *116*, 852–859. [CrossRef]
63. Liu, Y.; Huang, Z.; Xie, Q.; Sun, L.; Gu, T.; Li, Z.; Bu, L.; Yao, S.; Tu, X.; Luo, X.; et al. Electrodeposition of electroreduced graphene oxide-Au nanoparticles composite film at glassy carbon electrode for anodic stripping voltammetric analysis of trace arsenic(III). *Sens. Actuators B Chem.* **2013**, *188*, 894–901. [CrossRef]
64. Pifferi, V.; Ardizzone, S.; Cappelletti, G.; Falciola, L.; Meroni, D. Ultra-Traces Detection by Gold-Based Electrodes in As(III) Novel Photoremediation. *Electrocatalysis* **2013**, *4*, 306–311. [CrossRef]
65. Wang, D.; Zhao, Y.; Jin, H.; Zhuang, J.; Zhang, W.; Wang, S.; Wang, J. Synthesis of Au-Decorated Tripod-Shaped Te Hybrids for Applications in the Ultrasensitive Detection of Arsenic. *ACS Appl. Mater. Interfaces* **2013**, *5*, 5733–5740. [CrossRef] [PubMed]
66. Wu, J.; Yang, M.; Xiao, J.; Fu, X.; Jin, J.; Li, L.; Chang, W.; Xie, C. Gold Nanoparticle Dropped Titania Microsphere Hybrids as an Enhanced Sensitive Material for Stripping Voltammetry Determination of As(III). *J. Electrochem. Soc.* **2013**, *160*, B225–B230. [CrossRef]
67. Cinti, S.; Politi, S.; Moscone, D.; Palleschi, G.; Arduini, F. Stripping Analysis of As(III) by Means of Screen-Printed Electrodes Modified with Gold Nanoparticles and Carbon Black Nanocomposite. *Electroanalysis* **2014**, *26*, 931–939. [CrossRef]
68. Wu, S.; Zhao, Q.; Zhou, L.; Zhang, Z. Stripping Analysis of Trace Arsenic Based on the MnO_x/AuNPs Composite Film Modified Electrode in Alkaline Media. *Electroanalysis* **2014**, *26*, 1840–1849. [CrossRef]
69. Sanghavi, B.J.; Gadhari, N.S.; Kalambate, P.K.; Karna, S.P.; Srivastava, A.K. Potentiometric stripping analysis of arsenic using a graphene paste electrode modified with a thiacrown ether and gold nanoparticles. *Microchim. Acta* **2015**, *182*, 1473–1481. [CrossRef]
70. Kato, D.; Kamata, T.; Kato, D.; Yanagisawa, H.; Niwa, O. Au Nanoparticle-Embedded Carbon Films for Electrochemical As^{3+} Detection with High Sensitivity and Stability. *Anal. Chem.* **2016**, *88*, 2944–2951. [CrossRef]
71. Mafa, J.P.; Mabuba, N.; Arotiba, O.A. An Exfoliated Graphite Based Electrochemical Sensor for As(III) in Water. *Electroanalysis* **2016**, *28*, 1462–1469. [CrossRef]
72. Wei, J.; Li, S.-S.; Guo, Z.; Chen, X.; Liu, J.-H.; Huang, X.-J. Adsorbent Assisted in Situ Electrocatalysis: An Ultra-Sensitive Detection of As(III) in Water at Fe_3O_4 Nanosphere Densely Decorated with Au Nanoparticles. *Anal. Chem.* **2016**, *88*, 1154–1161. [CrossRef]

73. Yang, M.; Chen, X.; Liu, J.-H.; Huang, X.-J. Enhanced anti-interference on electrochemical detection of arsenite with nanoporous gold in mild condition. *Sens. Actuators B Chem.* **2016**, *234*, 404–411. [CrossRef]
74. Zhou, S.; Han, X.; Fan, H.; Liu, Y. Electrochemical Sensing toward Trace As(III) Based on Mesoporous $MnFe_2O_4$/Au Hybrid Nanospheres Modified Glass Carbon Electrode. *Sensors* **2016**, *16*, 935. [CrossRef]
75. Carrera, P.; Espinoza-Montero, P.J.; Fernandez, L.; Romero, H.; Alvarado, J. Electrochemical determination of arsenic in natural waters using carbon fiber ultra-microelectrodes modified with gold nanoparticles. *Talanta* **2017**, *166*, 198–206. [CrossRef] [PubMed]
76. Idris, A.O.; Mabuba, N.; Arotiba, O.A. Electrochemical co-detection of Arsenic and Selenium on a Glassy Carbon Electrode Modified with Gold Nanoparticles. *Int. J. Electrochem. Sci.* **2017**, *12*, 10–21. [CrossRef]
77. Nunez-Bajo, E.; Carmen Blanco-Lopez, M.; Costa-Garcia, A.; Teresa Fernandez-Abedule, M. Electrogeneration of Gold Nanoparticles on Porous-Carbon Paper-Based Electrodes and Application to Inorganic Arsenic Analysis in White Wines by Chronoamperometric Stripping. *Anal. Chem.* **2017**, *89*, 6415–6423. [CrossRef] [PubMed]
78. Sahoo, S.; Sahoo, P.K.; Satpati, A.K. Gold Nano Particle and Reduced Graphene Oxide Composite Modified Carbon Paste Electrode for the Ultra Trace Detection of Arsenic(III). *Electroanalysis* **2017**, *29*, 1400–1409. [CrossRef]
79. Salunke, R.S.; Kasar, C.K.; Bangar, M.A.; Chavan, P.G.; Shirale, D.J. Electrodeposition of gold nanoparticles decorated single polypyrrole nanowire for arsenic detection in potable water: A chemiresistive sensor device. *J. Mater. Sci. Mater. Electron.* **2017**, *28*, 14672–14677. [CrossRef]
80. Sun, F.; Fan, G. Electrochemical Detection of As(III) Ions in Hot Spring Water Based on a SiO_2-Au Nanocomposite. *Int. J. Electrochem. Sci.* **2017**, *12*, 8982–8990. [CrossRef]
81. Coneo Rodriguez, R.; Bruno, M.M.; Angelome, P.C. Au nanoparticles embedded in mesoporous ZrO_2 films: Multifunctional materials for electrochemical detection. *Sens. Actuators B Chem.* **2018**, *254*, 603–612. [CrossRef]
82. Ensafi, A.A.; Akbarian, F.; Heydari-Soureshjani, E.; Rezaei, B. A novel aptasensor based on 3D-reduced graphene oxide modified gold nanoparticles for determination of arsenite. *Biosens. Bioelectron.* **2018**, *122*, 25–31. [CrossRef]
83. Nellaiappan, S.; Pillai, K.C.; Kumar, A.S. Flow-injection analysis coupled with electrochemical detection of poisonous inorganic arsenic(III) species using a gold nanoparticle/carbon nanofiber/chitosan chemically modified carbon screen printed electrode in neutral pH solution. *Anal. Methods* **2018**, *10*, 799–808. [CrossRef]
84. Pungjunun, K.; Chaiyo, S.; Jantrahong, I.; Nantaphol, S.; Siangproh, W.; Chailapakul, O. Anodic stripping voltammetric determination of total arsenic using a gold nanoparticle-modified boron-doped diamond electrode on a paper-based device. *Microchim. Acta* **2018**, *185*, 324. [CrossRef] [PubMed]
85. Yang, M.; Li, P.-H.; Xu, W.-H.; Wei, Y.; Li, L.-N.; Huang, Y.-Y.; Sun, Y.-F.; Chen, X.; Liu, J.-H.; Huang, X.-J. Reliable electrochemical sensing arsenic(III) in nearly groundwater pH based on efficient adsorption and excellent electrocatalytic ability of AuNPs/CeO_2-ZrO_2 nanocomposite. *Sens. Actuators B Chem.* **2018**, *255*, 226–234. [CrossRef]
86. Buffa, A.; Mandler, D. Arsenic(III) detection in water by flow-through carbon nanotube membrane decorated by gold nanoparticles. *Electrochim. Acta* **2019**, *318*, 496–503. [CrossRef]
87. Gimenez-Gomez, P.; Baldi, A.; Ayora, C.; Fernandez-Sanchez, C. Automated Determination of As(III) in Waters with an Electrochemical Sensor Integrated into a Modular Microfluidic System. *ACS Sens.* **2019**, *4*, 3156–3165. [CrossRef] [PubMed]
88. Trachioti, M.G.; Karantzalis, A.E.; Hrbac, J.; Prodromidis, M.I. Low-cost screen-printed sensors on-demand: Instantly prepared sparked gold nanoparticles from eutectic Au/Si alloy for the determination of arsenic at the sub-ppb level. *Sens. Actuators B Chem.* **2019**, *281*, 273–280. [CrossRef]
89. Wang, W.; Bao, N.; Yuan, W.; Si, N.; Bai, H.; Li, H.; Zhang, Q. Simultaneous determination of lead, arsenic, and mercury in cosmetics using a plastic based disposable electrochemical sensor. *Microchem. J.* **2019**, *148*, 240–247. [CrossRef]
90. Zhao, G.; Liu, G. Electrochemical Deposition of Gold Nanoparticles on Reduced Graphene Oxide by Fast Scan Cyclic Voltammetry for the Sensitive Determination of As(III). *Nanomaterials* **2019**, *9*, 41. [CrossRef]
91. Bu, L.; Xie, Q.; Ming, H. Gold nanoparticles decorated three-dimensional porous graphitic carbon nitrides for sensitive anodic stripping voltammetric analysis of trace arsenic(III). *J. Alloys Compd.* **2020**, *823*, 153723. [CrossRef]
92. Chen, C.; Yu, S.; Jiang, S.; Liu, J.; Wang, Z.; Ye, B.-C. A novel and sensitive electrochemical sensor based on nanoporous gold for determination of As(III). *Microchim. Acta* **2020**, *187*, 395. [CrossRef]
93. Ismail, S.; Yusof, N.A.; Abdullah, J.; Abd Rahman, S.F. Development of Electrochemical Sensor Based on Silica/Gold Nanoparticles Modified Electrode for Detection of Arsenite. *IEEE Sens. J.* **2020**, *20*, 3406–3414. [CrossRef]
94. Kongor, A.; Panchal, M.; Athar, M.; Vora, M.; Verma, N.; Pandya, A.; Jha, P.C.; Bhadresha, K.; Rawal, R.; Jain, V. Colorimetric and electrochemical sensing of As(III) using calix 4 pyrrole capped gold nanoparticles and evaluation of its cytotoxic activity. *J. Incl. Phenom. Macrocycl. Chem.* **2020**, *98*, 29–41. [CrossRef]
95. Sullivan, C.; Lu, D.; Brack, E.; Drew, C.; Kurup, P. Voltammetric codetection of arsenic(III) and copper(II) in alkaline buffering system with gold nanostar modified electrodes. *Anal. Chim. Acta* **2020**, *1107*, 63–73. [CrossRef] [PubMed]
96. Udayan, A.P.M.; Kachwala, B.; Karthikeyan, K.G.; Gunasekaran, S. Ultrathin quasi-hexagonal gold nanostructures for sensing arsenic in tap water. *RSC Adv.* **2020**, *10*, 20211–20221. [CrossRef]
97. Jijana, A.N.; Mphuthi, N.; Shumbula, P.; Vilakazi, S.; Sikhwivhilu, L. The Ultra-sensitive Electrochemical Detection of As(III) in Ground Water Using Disposable L-cysteine/Lipoic Acid Functionalised Gold Nanoparticle Modified Screen-Printed Electrodes. *Electrocatalysis* **2021**, *12*, 310–325. [CrossRef]

98. Li, C.-Y.; Wei, Y.-Y.; Shen, W.; Dong, X.; Yang, M.; Wei, J. Ultrahigh sensitivity electroanalysis of trace As(III) in water and human serum via gold nanoparticles uniformly anchored to Co_3O_4 porous microsheets. *Electrochim. Acta* **2021**, *368*, 137605. [CrossRef]
99. Nodehi, M.; Baghayeri, M.; Veisi, H. Preparation of GO/Fe_3O_4@PMDA/AuNPs nanocomposite for simultaneous determination of As^{3+} and Cu^{2+} by stripping voltammetry. *Talanta* **2021**, *230*, 122288. [CrossRef]
100. Zhu, X.; Zhu, G.; Ge, Y.; Zhang, B.; Yang, J.; Hu, B.; Liu, J. AU(nano)/Fe-MOF hybrid electrode for highly sensitive determination of trace As(III). *J. Electroanal. Chem.* **2021**, *899*, 115642. [CrossRef]
101. Tabibi, Z.; Massah, J.; Vakilian, K.A. A biosensor for the sensitive and specific measurement of arsenite using gold nanoparticles. *Measurement* **2022**, *187*, 110281. [CrossRef]
102. Nguyen, T.S.V.; Huynh, T.M.; Nguyen, L.D.; Doan, T.C.D.; Baptist, R.; Dang, C.M. Sensitive electrochemical measurement of As(III) using Nafion modified platinum electrode via anode stripping voltammetry. *Int. J. Nanotechnol.* **2020**, *17*, 659–672. [CrossRef]
103. Xu, H.; Zeng, L.; Xing, S.; Shi, G.; Chen, J.; Man, Y.; Jin, L. Highly ordered platinum-nanotube arrays for oxidative determination of trace arsenic(III). *Electrochem. Commun.* **2008**, *10*, 1893–1896. [CrossRef]
104. Xu, H.; Zeng, L.; Xing, S.; Xian, Y.; Jin, L. Microwave-irradiated synthesized platinum nanoparticles/carbon nanotubes for oxidative determination of trace arsenic(III). *Electrochem. Commun.* **2008**, *10*, 551–554. [CrossRef]
105. Sitorus, R.; Saefumillah, A.; Ivandini, T.A. Arsenic and arsenate detection with anodic stripping voltammetry technique using boron-doped diamond modified by platinum. *IOP Conf. Ser. Mater. Sci. Eng.* **2020**, *763*, 012013. [CrossRef]
106. Kempegowda, R.; Antony, D.; Malingappa, P. Graphene-platinum nanocomposite as a sensitive and selective voltammetric sensor for trace level arsenic quantification. *Int. J. Smart Nano Mater.* **2014**, *5*, 17–32. [CrossRef]
107. Hrapovic, S.; Liu, Y.; Luong, J.H.T. Reusable platinum nanoparticle modified boron doped diamond microelectrodes for oxidative determination of arsenite. *Anal. Chem.* **2007**, *79*, 500–507. [CrossRef]
108. Dai, X.; Compton, R.G. Detection of As(III) via oxidation to As(V) using platinum nanoparticle modified glassy carbon electrodes: Arsenic detection without interference from copper. *Analyst* **2006**, *131*, 516–521. [CrossRef]
109. Li, P.-H.; Yang, M.; Song, Z.-Y.; Chen, S.-H.; Xiao, X.-Y.; Lin, C.-H.; Huang, X.-J. Highly Sensitive and Stable Determination of As(III) under Near-Neutral Conditions: Benefit from the Synergetic Catalysis of Pt Single Atoms and Active S Atoms over Pt-1/MoS2. *Anal. Chem.* **2021**, *93*, 15115–15123. [CrossRef]
110. Han, D.-D.; Liu, Z.-G.; Liu, J.-H.; Huang, X.-J. The size effect of Pt nanoparticles: A new route to improve sensitivity in electrochemical detection of As(III). *RSC Adv.* **2015**, *5*, 38290–38297. [CrossRef]
111. Aguirre, M.d.C.; Rivas, B.L.; Basaez, L.; Pena-Farfal, C. Electrochemical Detection of Arsenite with Silver Electrodes in Inorganic Electrolyte and Natural System Mixtures. *J. Braz. Chem. Soc.* **2011**, *22*, 2362–2370. [CrossRef]
112. Sonkoue, B.M.; Tchekwagep, P.M.S.; Nanseu-Njiki, C.P.; Ngameni, E. Electrochemical Determination of Arsenic Using Silver Nanoparticles. *Electroanalysis* **2018**, *30*, 2738–2743. [CrossRef]
113. Prakash, S.; Chakrabarty, T.; Singh, A.K.; Shahi, V.K. Silver nanoparticles built-in chitosan modified glassy carbon electrode for anodic stripping analysis of As(III) and its removal from water. *Electrochim. Acta* **2012**, *72*, 157–164. [CrossRef]
114. Dar, R.A.; Khare, N.G.; Cole, D.P.; Karna, S.P.; Srivastava, A.K. Green synthesis of a silver nanoparticle-graphene oxide composite and its application for As(III) detection. *RSC Adv.* **2014**, *4*, 14432–14440. [CrossRef]
115. Wen, S.-H.; Liang, R.-P.; Zhang, L., Qiu, J. D. Multimodal Assay of Arsenite Contamination in Environmental Samples with Improved Sensitivity through Stimuli-Response of Multiligands Modified Silver Nanoparticles. *ACS Sustain. Chem. Eng.* **2018**, *6*, 6223–6232. [CrossRef]
116. Agustiany, T.; Khalil, M.; Einaga, Y.; Jiwanti, P.K.; Ivandini, T.A. Stable iridium-modified boron-doped diamond electrode for the application in electrochemical detection of Arsenic(III). *Mater. Chem. Phys.* **2020**, *244*, 122723. [CrossRef]
117. Mafakheri, E.; Salimi, A.; Hallaj, R.; Ramazani, A.; Kashi, M.A. Synthesis of Iridium Oxide Nanotubes by Electrodeposition into Polycarbonate Template: Fabrication of Chromium(III) and Arsenic(III) Electrochemical Sensor. *Electroanalysis* **2011**, *23*, 2429–2437. [CrossRef]
118. Gupta, R.; Gamare, J.S.; Pandey, A.K.; Tyagi, D.; Kamat, J.V. Highly Sensitive Detection of Arsenite Based on Its Affinity Toward Ruthenium Nanoparticles Decorated on Glassy Carbon Electrode. *Anal. Chem.* **2016**, *88*, 2459–2465. [CrossRef]
119. Dash, S.; Munichandraiah, N. Electroanalysis of As(III) at nanodendritic Pd on PEDOT. *Analyst* **2014**, *139*, 1789–1795. [CrossRef]
120. Alam, M.M.; Rashed, M.A.; Rahman, M.M.; Rahman, M.M.; Nagao, Y.; Hasnat, M.A. Electrochemical oxidation of As(III) on Pd immobilized Pt surface: Kinetics and sensing performance. *RSC Adv.* **2018**, *8*, 8071–8079. [CrossRef]
121. Murali Mohan, J.; Amreen, K.; Javed, A.; Dubey, S.K.; Goel, S. Electrochemical Mini-Platform with Thread-Based Electrodes for Interference Free Arsenic Detection. *IEEE Trans. Nanobiosci.* **1800**, *21*, 117–124. [CrossRef]
122. Sanllorente-Mendez, S.; Dominguez-Renedo, O.; Arcos-Martinez, M.J. Determination of Arsenic(III) Using Platinum Nanoparticle-Modified Screen-Printed Carbon-Based Electrodes. *Electroanalysis* **2009**, *21*, 635–639. [CrossRef]
123. Zhang, Y.; Li, D.; Compton, R.G. Arsenic(III) Detection with Underpotential Deposition and Anodic Stripping Voltammetry. *ChemElectroChem* **2021**, *8*, 3707–3715. [CrossRef]
124. Mendez Cortes, S.P.; Andres Galan-Vidal, C.; Rodriguez Avila, J.A.; Alvarez Romero, G.A.; Paez-Hernandez, M.E. Square Wave Anodic Stripping Voltammetry Determination of Arsenic(III) onto Carbon Electrodes by Means of Co-deposition with Silver. *J. Mex. Chem. Soc.* **2018**, *62*, 314–322. [CrossRef]

125. Salimi, A.; Hyde, M.E.; Banks, C.E.; Compton, R.G. Boron doped diamond electrode modified with iridium oxide for amperometic detection of ultra trace amounts of arsenic(III). *Analyst* **2004**, *129*, 9–14. [CrossRef]
126. Ivandini, T.A.; Sato, R.; Makide, Y.; Fujishima, A.; Einaga, Y. Electrochemical detection of arsenic(III) using iridium-implanted boron-doped diamond electrodes. *Anal. Chem.* **2006**, *78*, 6291–6298. [CrossRef]
127. Touilloux, R.; Tercier-Waeber, M.-L.; Bakker, E. Direct arsenic(III) sensing by a renewable gold plated Ir-based microelectrode. *Analyst* **2015**, *140*, 3526–3534. [CrossRef]
128. Gumpu, M.B.; Veerapandian, M.; Krishnan, U.M.; Rayappan, J.B.B. Simultaneous electrochemical detection of Cd(II), Pb(II), As(III) and Hg(II) ions using ruthenium(II)-textured graphene oxide nanocomposite. *Talanta* **2017**, *162*, 574–582. [CrossRef] [PubMed]
129. Gumpu, M.B.; Veerapandian, M.; Krishnan, U.M.; Rayappan, J.B.B. Amperometric determination of As(III) and Cd(II) using a platinum electrode modified with acetylcholinesterase, ruthenium(II)-tris(bipyridine) and graphene oxide. *Microchim. Acta* **2018**, *185*, 297. [CrossRef]
130. Moghimi, N.; Mohapatra, M.; Leung, K.T. Bimetallic Nanoparticles for Arsenic Detection. *Anal. Chem.* **2015**, *87*, 5546–5552. [CrossRef] [PubMed]
131. Lan, Y.; Luo, H.; Ren, X.; Wang, Y.; Liu, Y. Anodic stripping voltammetric determination of arsenic(III) using a glassy carbon electrode modified with gold-palladium bimetallic nanoparticles. *Microchim. Acta* **2012**, *178*, 153–161. [CrossRef]
132. Zhang, Q.-X.; Yin, L.-B. Electrochemical performance of heterostructured Au-Pd bimetallic nanoparticles toward As(III) aqueous media. *Electrochem. Commun.* **2012**, *22*, 57–60. [CrossRef]
133. Bu, L.; Liu, J.; Xie, Q.; Yao, S. Anodic stripping voltammetric analysis of trace arsenic(III) enhanced by mild hydrogen-evolution at a bimetallic Au-Pt nanoparticle modified glassy carbon electrode. *Electrochem. Commun.* **2015**, *59*, 28–31. [CrossRef]
134. Yang, M.; Guo, Z.; Li, L.-N.; Huang, Y.-Y.; Liu, J.-H.; Zhou, Q.; Chen, X.; Huang, X.-J. Electrochemical determination of arsenic(III) with ultra-high anti-interference performance using Au-Cu bimetallic nanoparticles. *Sens. Actuators B Chem.* **2016**, *231*, 70–78. [CrossRef]
135. Melinte, G.; Hosu, O.; Lettieri, M.; Cristea, C.; Marrazza, G. Electrochemical Fingerprint of Arsenic(III) by Using Hybrid Nanocomposite-Based Platforms. *Sensors* **2019**, *19*, 2279. [CrossRef] [PubMed]
136. Mushiana, T.; Mabuba, N.; Idris, A.O.; Peleyeju, G.M.; Orimolade, B.O.; Nkosi, D.; Ajayi, R.F.; Arotiba, O.A. An aptasensor for arsenic on a carbon-gold bi-nanoparticle platform. *Sens. Bio-Sens. Res.* **2019**, *24*, 8–12. [CrossRef]
137. Yadav, R.; Kushwah, V.; Gaur, M.S.; Bhadauria, S.; Berlina, A.N.; Zherdev, A.V.; Dzantiev, B.B. Electrochemical aptamer biosensor for As^{3+} based on apta deep trapped Ag-Au alloy nanoparticles-impregnated glassy carbon electrode. *Int. J. Environ. Anal. Chem.* **2020**, *100*, 623–634. [CrossRef]
138. Sedki, M.; Zhao, G.; Ma, S.; Jassby, D.; Mulchandani, A. Linker-Free Magnetite-Decorated Gold Nanoparticles (Fe_3O_4-Au): Synthesis, Characterization, and Application for Electrochemical Detection of Arsenic(III). *Sensors* **2021**, *21*, 883. [CrossRef]
139. Shin, S.-H.; Hong, H.-G. Anodic Stripping Voltammetric Detection of Arsenic(III) at Platinum-Iron(III) Nanoparticle Modified Carbon Nanotube on Glassy Carbon Electrode. *Bull. Korean Chem. Soc.* **2010**, *31*, 3077–3083. [CrossRef]
140. Xu, W.-H.; Wang, L.; Wang, J.; Sheng, G.-P.; Liu, J.-H.; Yu, H.-Q.; Huang, X.-J. Superparamagnetic mesoporous ferrite nanocrystal clusters for efficient removal of arsenite from water. *CrystEngComm* **2013**, *15*, 7895–7903. [CrossRef]
141. Devi, P.; Sharma, C.; Kumar, P.; Kumar, M.; Bansod, B.K.S.; Nayak, M.K.; Singla, M.L. Selective electrochemical sensing for arsenite using rGO/Fe_3O_4 nanocomposites. *J. Hazard. Mater.* **2017**, *322*, 85–94. [CrossRef]
142. Hu, H.; Lu, W.; Liu, X.; Meng, F.; Zhu, J. A High-Response Electrochemical As(III) Sensor Using Fe_3O_4-rGO Nanocomposite Materials. *Chemosensors* **2021**, *9*, 150. [CrossRef]
143. Chimezie, A.B.; Hajian, R.; Yusof, N.A.; Woi, P.M.; Shams, N. Fabrication of reduced graphene oxide-magnetic nanocomposite (rGO-Fe_3O_4) as an electrochemical sensor for trace determination of As(III) in water resources. *J. Electroanal. Chem.* **2017**, *796*, 33–42. [CrossRef]
144. Gao, C.; Yu, X.-Y.; Xiong, S.-Q.; Liu, J.-H.; Huang, X.-J. Electrochemical Detection of Arsenic(III) Completely Free from Noble Metal: Fe_3O_4 Microspheres-Room Temperature Ionic Liquid Composite Showing Better Performance than Gold. *Anal. Chem.* **2013**, *85*, 2673–2680. [CrossRef]
145. Huang, H.-Q.; Li, Y.-Y.; Chen, S.-H.; Liu, Z.-G.; Cui, Y.-M.; Li, H.-Q.; Guo, Z.; Huang, X.-J. Noble-metal-free Fe_3O_4/Co_3S_4 nanosheets with oxygen vacancies as an efficient electrocatalyst for highly sensitive electrochemical detection of As(III). *Anal. Chim. Acta* **2022**, *1189*, 339208. [CrossRef] [PubMed]
146. Ren, B.; Sudarsanam, P.; Kandjani, A.E.; Hillary, B.; Amin, M.H.; Bhargava, S.K.; Jones, L.A. Electrochemical Detection of As(III) on a Manganese Oxide-Ceria (Mn_2O_3/CeO_2) Nanocube Modified Au Electrode. *Electroanalysis* **2018**, *30*, 928–936. [CrossRef]
147. Ponnaiah, S.K.; Periakaruppan, P.; Selvam, M.; Muthupandian, S.; Jeyaprabha, B.; Selvanathan, R. Clinically Pertinent Manganese Oxide/Polyoxytyramine/Reduced Graphene Oxide Nanocomposite for Voltammetric Detection of Salivary and Urinary Arsenic. *J. Clust. Sci.* **2020**, *31*, 877–885. [CrossRef]
148. Salimi, A.; Manikhezri, H.; Hallaj, R.; Soltanian, S. Electrochemical detection of trace amount of arsenic(III) at glassy carbon electrode modified with cobalt oxide nanoparticles. *Sens. Actuators B Chem.* **2008**, *129*, 246–254. [CrossRef]
149. Santhosh, C.; Velmurugan, V.; Jacob, G.; Jeong, S.K.; Grace, A.N.; Bhatnagar, A. Role of nanomaterials in water treatment applications: A review. *Chem. Eng. J.* **2016**, *306*, 1116–1137. [CrossRef]
150. Jiang, T.-J.; Guo, Z.; Liu, J.-H.; Huang, X.-J. Gold electrode modified with ultrathin SnO_2 nanosheets with high reactive exposed surface for electrochemical sensing of As(III). *Electrochim. Acta* **2016**, *191*, 142–148. [CrossRef]

151. Bhanjana, G.; Mehta, N.; Chaudhary, G.R.; Dilbaghi, N.; Kim, K.-H.; Kumar, S. Novel electrochemical sensing of arsenic ions using a simple graphite pencil electrode modified with tin oxide nanoneedles. *J. Mol. Liq.* **2018**, *264*, 198–204. [CrossRef]
152. Karthika, A.; Selvarajan, S.; Karuppasamy, P.; Suganthi, A.; Rajarajan, M. A novel highly efficient and accurate electrochemical detection of poisonous inorganic Arsenic(III) ions in water and human blood serum samples based on $SrTiO_3$/beta-cyclodextrin composite. *J. Phys. Chem. Solids* **2019**, *127*, 11–18. [CrossRef]
153. Ndlovu, T.; Mamba, B.B.; Sampath, S.; Krause, R.W.; Arotiba, O.A. Voltammetric detection of arsenic on a bismuth modified exfoliated graphite electrode. *Electrochim. Acta* **2014**, *128*, 48–53. [CrossRef]
154. Mafa, P.J.; Idris, A.O.; Mabuba, N.; Arotiba, O.A. Electrochemical co-detection of As(III), Hg(II) and Pb(II) on a bismuth modified exfoliated graphite electrode. *Talanta* **2016**, *153*, 99–106. [CrossRef] [PubMed]
155. Durai, L.; Badhulika, S. Ultra-selective, trace level detection of As^{3+} ions in blood samples using PANI coated $BiVO_4$ modified SPCE via differential pulse anode stripping voltammetry. *Mater. Sci. Eng. C Mater. Biol. Appl.* **2020**, *111*, 110806. [CrossRef] [PubMed]
156. Bhanjana, G.; Dilbaghi, N.; Chaudhary, S.; Kim, K.-H.; Kumar, S. Robust and direct electrochemical sensing of arsenic using zirconia nanocubes. *Analyst* **2016**, *141*, 4211–4218. [CrossRef]
157. Noskova, G.N.; Zakharova, E.A.; Kolpakova, N.A.; Kabakaev, A.S. Electrodeposition and stripping voltammetry of arsenic(III) and arsenic(V) on a carbon black-polyethylene composite electrode in the presence of iron ions. *J. Solid State Electrochem.* **2012**, *16*, 2459–2472. [CrossRef]
158. Saha, S.; Sarkar, P. Differential pulse anodic stripping voltammetry for detection of As(III) by Chitosan-$Fe(OH)_3$ modified glassy carbon electrode: A new approach towards speciation of arsenic. *Talanta* **2016**, *158*, 235–245. [CrossRef]
159. Xie, Z.; Xu, J.; Xie, F.; Xiong, S. Electrochemical Detection of As(III) by a rGO/Fe_3O_4-modified Screen-Printed Carbon Electrode. *Anal. Sci.* **2016**, *32*, 1053–1058. [CrossRef] [PubMed]
160. Zhang, Z.; Ji, H.; Song, Y.; Zhang, S.; Wang, M.; Jia, C.; Tian, J.-Y.; He, L.; Zhang, X.; Liu, C.-S. Fe(III)-based metal-organic framework-derived core-shell nanostructure: Sensitive electrochemical platform for high trace determination of heavy metal ions. *Biosens. Bioelectron.* **2017**, *94*, 358–364. [CrossRef]
161. Gounden, D.; Khene, S.; Nombona, N. Electroanalytical detection of heavy metals using metallophthalocyanine and silica-coated iron oxide composites. *Chem. Pap.* **2018**, *72*, 3043–3056. [CrossRef]
162. Li, S.-S.; Zhou, W.-Y.; Jiang, M.; Guo, Z.; Liu, J.-H.; Zhang, L.; Huang, X.-J. Surface Fe(II)/Fe(III) Cycle Promoted Ultra-Highly Sensitive Electrochemical Sensing of Arsenic(III) with Dumbbell-Like Au/Fe_3O_4 Nanoparticles. *Anal. Chem.* **2018**, *90*, 4569–4577. [CrossRef]
163. Zhang, R.; Hong, S.; Zhang, D.; Li, S.; Zhao, Z.; Bai, H. Nano alpha-FeOOH Modified Carbon Paste Electrode for Arsenic Determination in Natural Waters. *Mater. Sci. Medzg.* **2018**, *24*, 355–361. [CrossRef]
164. Yogeeshwari, R.T.; Shreenivasa, L.; Krishna, R.H.; Prashanth, S.A.; Banuprakash, G.; Ashoka, S. Synthesis of acid resistant $Fe_2V_4O_{13}$-polypyrrole nanocomposite: Its application towards the fabrication of disposable electrochemical sensor for the detection of As(III). *Mater. Res. Express* **2019**, *6*, 126448. [CrossRef]
165. Zhao, Z.; Li, C.; Wu, H. Reduced graphene oxide nanosheets modified with plasmonic gold-based hybrid nanostructures and with magnetite (Fe_3O_4) nanoparticles for cyclic voltammetric determination of arsenic(III). *Microchim. Acta* **2019**, *186*, 226. [CrossRef] [PubMed]
166. Saleemi, A.S.; Hafeez, M.; Munawar, A.; Akhtar, N.; Abbas, W.; Mazhar, M.E.; Shafiq, Z.; Davis, A.P.; Lee, S.-L. Synthesis and sensing efficiency of CN-wrapped $ZnFe_2O_4$ microsphere-ionic liquid composites towards ultra-high sensitive arsenic(iii) monitoring of ground drinking water. *J. Mater. Chem. C* **2020**, *8*, 12984–12992. [CrossRef]
167. Zhou, S.-F.; Han, X.-J.; Fan, H.-L.; Zhang, Q.-X.; Liu, Y.-Q. Electrochemical detection of As(III) through mesoporous $MnFe_2O_4$ nanocrystal clusters by square wave stripping voltammetry. *Electrochim. Acta* **2015**, *174*, 1160–1166. [CrossRef]
168. Devi, P.; Bansod, B.; Kaur, M.; Bagchi, S.; Nayak, M.K. Co-electrodeposited rGO/MnO_2 nanohybrid for arsenite detection in water by stripping voltammetry. *Sens. Actuators B Chem.* **2016**, *237*, 652–659. [CrossRef]
169. Yang, M.; Chen, X.; Jiang, T.-J.; Guo, Z.; Liu, J.-H.; Huang, X.-J. Electrochemical Detection of Trace Arsenic(III) by Nanocomposite of Nanorod-Like alpha-MnO_2 Decorated with similar to 5 nm Au Nanoparticles: Considering the Change of Arsenic Speciation. *Anal. Chem.* **2016**, *88*, 9720–9728. [CrossRef] [PubMed]
170. Jiang, T.; Qi, L.; Lu, X.; Hou, C.; Qin, W. Selective determination of Arsenic(III) using a Nafion/alpha-MnO_2@polydopamine modified electrode. *J. Electroanal. Chem.* **2020**, *878*, 114562. [CrossRef]
171. Gumpu, M.B.; Mani, G.K.; Nesakumar, N.; Kulandaisamy, A.J.; Babu, K.J.; Rayappan, J.B.B. Electrocatalytic nanocauliflower structured fluorine doped CdO thin film as a potential arsenic sensor. *Sens. Actuators B Chem.* **2016**, *234*, 426–434. [CrossRef]
172. Jimana, A.; Peleyeju, M.G.; Tshwenya, L.; Pillay, K.; Arotiba, O.A. Voltammetric Analysis of As(III) at a Cobalt Nanoparticles/Reduced Graphene Oxide Modified Exfoliated Graphite Electrode. *Int. J. Electrochem. Sci.* **2018**, *13*, 10127–10140. [CrossRef]
173. Kumar, P.; Devi, P.; Jain, R.; Saini, A.; Noetzel, R. Electrochemical Detection of Trace Arsenic(III) by functionalized $In_{0.38}Ga_{0.62}N$/Si(111) electrode. *Mater. Lett.* **2019**, *236*, 587–590. [CrossRef]
174. Ramesha, G.K.; Sampath, S. In-situ formation of graphene-lead oxide composite and its use in trace arsenic detection. *Sens. Actuators B Chem.* **2011**, *160*, 306–311. [CrossRef]
175. Dai, X.A.; Compton, R.G. Direct electrodeposition of gold nanoparticles onto indium tin oxide film coated glass: Application to the detection of arsenic(III). *Anal. Sci.* **2006**, *22*, 567–570. [CrossRef] [PubMed]

176. El-Said, W.A.; Hajjar, D.; Makki, A.A.; Elshehy, E.A.; Choi, J.-W. A reusable Gemini surfactant-based electrochemical sensor for As(III) detection. *Int. J. Environ. Anal. Chem.* **2021**, *101*, 1–12. [CrossRef]
177. Pereira, F.J.; Vazquez, M.D.; Deban, L.; Aller, A.J. Inorganic arsenic speciation by differential pulse anodic stripping voltammetry using thoria nanoparticles-carbon paste electrodes. *Talanta* **2016**, *152*, 211–218. [CrossRef]
178. Zhang, X.; Zeng, T.; Hu, C.; Hu, S.; Tian, Q. Studies on fabrication and application of arsenic electrochemical sensors based on titanium dioxide nanoparticle modified gold strip electrodes. *Anal. Methods* **2016**, *8*, 1162–1169. [CrossRef]
179. Pourbeyram, S.; Asadi, S. Time Resolved Direct Determination of Arsenate in the Presence of Arsenite on Pencil Graphite Electrode Modified by Graphene Oxide and Zirconium. *Electroanalysis* **2018**, *30*, 154–161. [CrossRef]
180. Pandey, S.K.; Singh, P.; Singh, J.; Sachan, S.; Srivastava, S.; Singh, S.K. Nanocarbon-based Electrochemical Detection of Heavy Metals. *Electroanalysis* **2016**, *28*, 2472–2488. [CrossRef]
181. Kumar, S.; Bhanjanaa, G.; Dilbaghi, N.; Kumar, R.; Umar, A. Fabrication and characterization of highly sensitive and selective arsenic sensor based on ultra-thin graphene oxide nanosheets. *Sens. Actuators B Chem.* **2016**, *227*, 29–34. [CrossRef]
182. Yang, M.; Jiang, T.-J.; Wang, Y.; Liu, J.-H.; Li, L.-N.; Chen, X.; Huang, X.-J. Enhanced electrochemical sensing arsenic(III) with excellent anti-interference using amino-functionalized graphene oxide decorated gold microelectrode: XPS and XANES evidence. *Sens. Actuators B Chem.* **2017**, *245*, 230–237. [CrossRef]
183. Gumpu, M.B.; Veerapandian, M.; Krishnan, U.M.; Rayappan, J.B.B. Electrochemical sensing platform for the determination of arsenite and arsenate using electroactive nanocomposite electrode. *Chem. Eng. J.* **2018**, *351*, 319–327. [CrossRef]
184. Yuan, Y.-H.; Zhu, X.-H.; Wen, S.-H.; Liang, R.-P.; Zhang, L.; Qiu, J.-D. Electrochemical assay for As(III) by combination of highly thiol-rich trithiocyanuric acid and conductive reduced graphene oxide nanocomposites. *J. Electroanal. Chem.* **2018**, *814*, 97–103. [CrossRef]
185. Baghayeri, M.; Ghanei-Motlagh, M.; Tayebee, R.; Fayazi, M.; Narenji, F. Application of graphene/zinc-based metal-organic framework nanocomposite for electrochemical sensing of As(III) in water resources. *Anal. Chim. Acta* **2020**, *1099*, 60–67. [CrossRef] [PubMed]
186. Deshmukh, S.; Banerjee, D.; Bhattacharya, G.; Fishlock, S.J.; Barman, A.; McLaughlin, J.; Roy, S.S. Red Mud-Reduced Graphene Oxide Nanocomposites for the Electrochemical Sensing of Arsenic. *ACS Appl. Nano Mater.* **2020**, *3*, 4084–4090. [CrossRef]
187. Nunez, C.; Jose Trivino, J.; Segura, R.; Arancibia, V. Development of a fast and sensitive method for the determination of As(III) at trace levels in urine by differential pulse anodic voltammetry using a simple graphene screen-printed electrode. *Microchem. J.* **2020**, *159*, 105393. [CrossRef]
188. Liu, Y.; Wei, W. Jointly modified single-walled carbon nanotubes on low resistance monolayer modified electrode for arsenic(III) detection. *J. Electroanal. Chem.* **2008**, *624*, 299–304. [CrossRef]
189. Liu, Y.; Wei, W. Layer-by-layer assembled DNA functionalized single-walled carbon nanotube hybrids for arsenic(III) detection. *Electrochem. Commun.* **2008**, *10*, 872–875. [CrossRef]
190. Daud, N.; Yusof, N.A.; Tee, T.W.; Abdullah, A.H. Electrochemical Sensor for As(III) Utilizing CNTs/Leucine/Nafion Modified Electrode. *Int. J. Electrochem. Sci.* **2012**, *7*, 175–185.
191. Yusof, N.A.; Daud, N.; Tee, T.W.; Abdullah, A.H. Electrocatalytic Characteristic of Carbon Nanotubes/Glutamine/Nafion Modified Platinum Electrode in Development of Sensor for Determination of As(III). *Int. J. Electrochem. Sci.* **2011**, *6*, 2385–2397.
192. Gamboa, J.C.M.; Cornejo, L.; Squella, J.A. Vibrating screen printed electrode of gold nanoparticle-modified carbon nanotubes for the determination of arsenic(III). *J. Appl. Electrochem.* **2014**, *44*, 1255–1260. [CrossRef]
193. Wang, Y.; Wang, P.; Wang, Y.; He, X.; Wang, K. Single strand DNA functionalized single wall carbon nanotubes as sensitive electrochemical labels for arsenite detection. *Talanta* **2015**, *141*, 122–127. [CrossRef]
194. Roy, E.; Patra, S.; Madhuri, R.; Sharma, P.K. Europium doped magnetic graphene oxide-MWCNT nanohybrid for estimation and removal of arsenate and arsenite from real water samples. *Chem. Eng. J.* **2016**, *299*, 244–254. [CrossRef]
195. RasulKhan, B.; Ponnaiah, S.K.; Periakaruppan, P.; Venkatachalam, G.; Balasubramanian, J. A new CQDs/f-MWCNTs/GO nanocomposite electrode for arsenic (10^{-12} M) quantification in bore-well water and industrial effluents. *New J. Chem.* **2020**, *44*, 18149–18156. [CrossRef]
196. Baghayeri, M.; Amiri, A.; Karimabadi, F.; Di Masi, S.; Maleki, B.; Adibian, F.; Pourali, A.R.; Malitesta, C. Magnetic MWCNTs-dendrimer: A potential modifier for electrochemical evaluation of As(III) ions in real water samples. *J. Electroanal. Chem.* **2021**, *888*, 115059. [CrossRef]
197. Zhang, W.; Chen, Z.; Guan, Y.; Liu, C.; Zheng, K.; Zou, X. Aptamer-functionalized screen-printed electrode coupled with graphene oxide and methylene blue nanocomposite as enhanced signal label for total arsenic determination in shellfish. *Sens. Actuators B Chem.* **2021**, *335*, 129383. [CrossRef]
198. Arora, K.; Prabhakar, N.; Chand, S.; Malhotra, B.D. Immobilization of single stranded DNA probe onto polypyrrole-polyvinyl sulfonate for application to DNA hybridization biosensor. *Sens. Actuators B Chem.* **2007**, *126*, 655–663. [CrossRef]
199. Labuda, J.; Bubnicova, K.; Kovalova, L.; Vanickova, M.; Mattusch, J.; Wennrich, R. Voltammetric detection of damage to DNA by arsenic compounds at a DNA biosensor. *Sensors* **2005**, *5*, 411–423. [CrossRef]
200. Wen, S.; Zhang, C.; Liang, R.; Chi, B.; Yuan, Y.; Qiu, J. Highly sensitive voltammetric determination of arsenite by exploiting arsenite-induced conformational change of ssDNA and the electrochemical indicator Methylene Blue. *Microchim. Acta* **2017**, *184*, 4047–4054. [CrossRef]

201. Mao, K.; Zhang, H.; Wang, Z.; Cao, H.; Zhang, K.; Li, X.; Yang, Z. Nanomaterial-based aptamer sensors for arsenic detection. *Biosens. Bioelectron.* **2020**, *148*, 111785. [CrossRef]
202. Baghbaderani, S.S.; Noorbakhsh, A. Novel chitosan-Nafion composite for fabrication of highly sensitive impedimetric and colorimetric As(III) aptasensor. *Biosens. Bioelectron.* **2019**, *131*, 1–8. [CrossRef]
203. Cui, L.; Wu, J.; Ju, H. Label-free signal-on aptasensor for sensitive electrochemical detection of arsenite. *Biosens. Bioelectron.* **2016**, *79*, 861–865. [CrossRef]
204. Fuku, X.; Iftikar, F.; Hess, E.; Iwuoha, E.; Baker, P. Cytochrome c biosensor for determination of trace levels of cyanide and arsenic compounds. *Anal. Chim. Acta* **2012**, *730*, 49–59. [CrossRef]
205. Hwang, J.-H.; Pathak, P.; Wang, X.; Rodriguez, K.L.; Park, J.; Cho, H.J.; Lee, W.H. A novel Fe-Chitosan-coated carbon electrode sensor for in situ As(III) detection in mining wastewater and soil leachate. *Sens. Actuators B Chem.* **2019**, *294*, 89–97. [CrossRef]
206. Wen, S.-H.; Wang, Y.; Yuan, Y.-H.; Liang, R.-P.; Qiu, J.-D. Electrochemical sensor for arsenite detection using graphene oxide assisted generation of prussian blue nanoparticles as enhanced signal label. *Anal. Chim. Acta* **2018**, *1002*, 82–89. [CrossRef] [PubMed]
207. Sawan, S.; Hamze, K.; Youssef, A.; Bouhadir, K.; Errachid, A.; Maalouf, R.; Jaffrezic-Renault, N. The Use of Voltammetry for Sorption Studies of Arsenic(III) Ions by Magnetic Beads Functionalized with Nucleobase Hydrazide Derivatives. *Electroanalysis* **2021**, *33*, 1789–1799. [CrossRef]
208. Chen, L.; Zhou, N.; Li, J.; Chen, Z.; Liao, C.; Chen, J. Synergy of glutathione, dithiothreitol and N-acetyl-L-cysteine self-assembled monolayers for electrochemical assay: Sensitive determination of arsenic(III) in environmental and drinking water. *Analyst* **2011**, *136*, 4526–4532. [CrossRef] [PubMed]
209. Shah, A. L-tryptophan modified glassy carbon electrode for the picomolar detection of As(III). *J. Electrochem. Soc.* **2020**, *167*, 117509. [CrossRef]
210. Zaib, M.; Saeed, A.; Hussain, I.; Athar, M.M.; Iqbal, M. Voltammetric detection of As(III) with Porphyridium cruentum based modified carbon paste electrode biosensor. *Biosens. Bioelectron.* **2014**, *62*, 242–248. [CrossRef]
211. Irvine, G.W.; Tan, S.N.; Stillman, M.J. A Simple Metallothionein-Based Biosensor for Enhanced Detection of Arsenic and Mercury. *Biosensors* **2017**, *7*, 14. [CrossRef]
212. del Torno-de Roman, L.; Asuncion Alonso-Lomillo, M.; Dominguez-Renedo, O.; Julia Arcos-Martinez, M. Dual Biosensing Device for the Speciation of Arsenic. *Electroanalysis* **2015**, *27*, 302–308. [CrossRef]
213. Gu, H.; Yang, Y.; Chen, F.; Liu, T.; Jin, J.; Pan, Y.; Miao, P. Electrochemical detection of arsenic contamination based on hybridization chain reaction and RecJ(f) exonuclease-mediated amplification. *Chem. Eng. J.* **2018**, *353*, 305–310. [CrossRef]
214. Nunez, C.; Jose Trivino, J.; Arancibia, V. A electrochemical biosensor for As(III) detection based on the catalytic activity of Alcaligenes faecalis immobilized on a gold nanoparticle-modified screen-printed carbon electrode. *Talanta* **2021**, *223*, 121702. [CrossRef] [PubMed]
215. Cortes-Salazar, F.; Beggah, S.; van der Meer, J.R.; Girault, H.H. Electrochemical As(III) whole-cell based biochip sensor. *Biosens. Bioelectron.* **2013**, *47*, 237–242. [CrossRef] [PubMed]
216. Sanchez, S.; McDonald, M.; Silver, D.M.; de Bonnault, S.; Chen, C.; LeBlanc, K.; Hicks, E.C.; Mayall, R.M. The Integration of Whole-Cell Biosensors for the Field-Ready Electrochemical Detection of Arsenic. *J. Electrochem. Soc.* **2021**, *168*, 067508. [CrossRef]
217. Sciuto, E.L.; Petralia, S.; van der Meer, J.R.; Conoci, S. Miniaturized electrochemical biosensor based on whole-cell for heavy metal ions detection in water. *Biotechnol. Bioeng.* **2021**, *118*, 1456–1465. [CrossRef] [PubMed]
218. Vega-Figueroa, K.; Santillan, J.; Ortiz-Gomez, V.; Ortiz-Quiles, E.O.; Quinones-Colon, B.A.; Castilla-Casadiego, D.A.; Almodovar, J.; Bayro, M.J.; Rodriguez-Martinez, J.A.; Nicolau, E. Aptamer-Based Impedimetric Assay of Arsenite in Water: Interfacial Properties and Performance. *ACS Omega* **2018**, *3*, 1437–1444. [CrossRef]
219. Ismail, S.; Yusof, N.A.; Abdullah, J.; Abd Rahman, S.F. Electrochemical Detection of Arsenite Using a Silica Nanoparticles-Modified Screen-Printed Carbon Electrode. *Materials* **2020**, *13*, 3168. [CrossRef]
220. Rahman, M.M.; Hussein, M.A.; Aly, K.I.; Asiri, A.M. Thermally stable hybrid polyarylidene(azomethine-ether)s polymers (PAAP): An ultrasensitive arsenic(III) sensor approach. *Des. Monomers Polym.* **2018**, *21*, 82–98. [CrossRef]
221. Ma, W.; Chang, Q.; Zhao, J.; Ye, B.-C. Novel electrochemical sensing platform based on ion imprinted polymer with nanoporous gold for ultrasensitive and selective determination of As^{3+}. *Microchim. Acta* **2020**, *187*, 571. [CrossRef]

nanomaterials

Article

A Light-Driven Integrated Bio-Capacitor with Single Nano-Channel Modulation

Jie Lin [1,†], Yu-Jia Lv [2,3,†], Lei Han [1], Kuan Sun [1], Yan Xiang [2,*], Xiao-Xing Xing [1,*] and Yu-Tao Li [1,*]

1 College of Information Science and Technology, Beijing University of Chemical Technology, Beijing 100029, China; jaylin44@126.com (J.L.); 2020210529@mail.buct.edu.cn (L.H.); heaod1998@163.com (K.S.)
2 Beijing Key Laboratory of Bio-inspired Energy Materials and Devices, School of Space and Environment, Beihang University, Beijing 100191, China; lvyj16@buaa.edu.cn
3 Beijing Institute of Nanoenergy and Nanosystems Chinese Academy of Sciences, Beijing 100083, China
* Correspondence: yt-li@mail.buct.edu.cn (Y.-T.L.); xxing@mail.buct.edu.cn (X.-X.X.); xiangy@buaa.edu.cn (Y.X.)
† These authors contributed equally to this work.

Abstract: Bioelectronics, an emerging discipline formed by the biology and electronic information disciplines, has maintained a state of rapid development since its birth. Amongst the various functional bioelectronics materials, bacteriorhodopsin (bR), with its directional proton pump function and favorable structural stability properties, has drawn wide attention. The main contents of the paper are as follows: Inspired by the capacitive properties of natural protoplast cell membranes, a new bio-capacitor based on bR and artificial nanochannels was constructed. As a point of innovation, microfluidic chips were integrated into our device as an ion transport channel, which made the bio-capacitor more stable. Meanwhile, a single nanopore structure was integrated to improve the accuracy of the device structure. Experiments observed that the size of the nanopore affected the ion transmission rate. Consequently, by making the single nanopore's size change, the photocurrent duration time (PDT) of bR was effectively regulated. By using this specific phenomenon, the original transient photocurrent was successfully transformed into a square-like wave.

Keywords: bacteriorhodopsin; photoelectric conversion; bioelectronics; nanopore; microfluidic

Citation: Lin, J.; Lv, Y.-J.; Han, L.; Sun, K.; Xiang, Y.; Xing, X.-X.; Li, Y.-T. A Light-Driven Integrated Bio-Capacitor with Single Nano-Channel Modulation. *Nanomaterials* **2022**, *12*, 592. https://doi.org/10.3390/nano12040592

Received: 21 December 2021
Accepted: 3 February 2022
Published: 9 February 2022

1. Introduction

With the rapid development of bioelectronics [1–7], increasing numbers of bioelectronic devices have been developed and widely used. Bacteriorhodopsin (bR) extracted from *Halobacterium halobium* is a new type of biological material, which is widely used in bioelectronic devices. Moreover, bR is a novel photosynthetic system with a more controlled proton transport than chlorophyll in plants [8]. This consequently makes it easier to utilize in bioelectronic devices. The proton pump function of bR is the biophysical basis for designing novel bioelectronic devices. The pump function of bR would enable protons to be actively transported from the intracellular space to the extracellular under induced light [9,10]. Thus, an electrochemical potential difference inside and outside the cell membrane is formed, realizing photoelectric conversion [11–18]. Moreover, bR has many excellent material properties such as a higher light cycle stability [19], pH tolerance [20], thermal stability [19], and chemical stability [21], which improves its functional activity in the fabrication of bioelectronic devices outside the cell. Based on these material properties and the advantages of bR described above, it has been applied in a variety of bioelectronic devices, such as artificial retinal prosthesis [22], optical storage [23–25], photovoltaic cells [26], and pH sensors [27–29]. These applications are mainly based on the sensing, collection, and conversion of bR intrinsic photocurrent, while there is limited research on the shape regulation of photocurrent. The regulation of photocurrent shape can be achieved by some basic electronic components (e.g., capacitors), which greatly expands the application field of bR-based bioelectronic devices and lays a foundation for the construction of the next generation of intelligent bioelectronic systems.

As a typical electronic component, capacitors can easily output standard square waves with a controllable duty cycle through electronic circuits. A square wave can be used as the basic digital logic in a digital circuit, so it plays an important role in digital circuits [30]. However, in real life, most bio-photoelectric signals exist as a transient wave, resulting in limited applications. Therefore, transforming the transient wave to generate square waves by bio-capacitors will lead to the further development of bioelectronic digital logic and many related applications [31]. Some studies have used rhodopsin to fabricate biological capacitors to obtain waveform conversion functions. For example, Rao et al. used proteus rhodopsin and artificial anodized aluminum nanochannels to construct a light-powered bio-capacitor [32]; subsequently, they determined the influence of nanopore aperture on the photocurrent duration time (PDT) of the photocurrent of the bio-capacitor. However, the porous structure of the nanopore integrated into the device has a deviation from the size of nanopores. The maximum deviation could reach about 30% (171.6 $\pm$ 25.1 nm), which possibly makes the result inaccurate. An improved proposal is to replace the porous structure with a single-pore structure by novel nanopore fabrication technology. Meanwhile, in the bio-capacitor designed by Rao et al., the Ag/AgCl electrode used in the test system was unstable under induced light. Integrating the microfluidic chip may be a potential solution to enhance the device stability, and an electrode without light instability can make a device more stable. The design of these integrated devices is expected to improve the stability and accuracy of device performance and provide support for the performance improvement of bio-capacitors.

In this work, microfluidic chip, bR, and Si/SiO_2-based size-controllable Si_3N_4 single nanopore are combined to construct a light-driven bio-capacitor. After that, a test system is constructed to verify the capacitance properties of the entire system. It is found that the PDT is affected by the change in the nanopore size. Quantitative theoretical analysis based on the capacitor discharge process is proposed, as well as its capacitive property. Finally, this bio-capacitor is used to successfully convert the transient spike signal waveform of the bR into a square wave-like photocurrent waveform. This work is of great significance for the development of novel bR-based electronic components and the development of intelligent bioelectronics.

2. Materials and Methods

2.1. Modeling

Figure 1a shows a bionic schematic diagram of biological capacitance and it is the basis of the biological model of our device with capacitor property. This biological model consists of three parts: the membrane unit, ion pump protein (corresponding to the bR in our bio-capacitor), and the ion channel protein (corresponding to the nanopore in our bio-capacitor) [33–35]. Amongst them, the membrane unit (consisting of blue balls and yellow curves) is used to construct the purple membrane skeleton and support ion pump protein and ion channel protein. Ion pump proteins (yellow channel) can enable ions to reach the outside of the cell through active transport, thus forming a transmembrane electrochemical potential difference that is consistent with the function of bR. In addition, this process can simulate the charging process of the capacitor (represented by the yellow line in the voltage–time curve). The ion channel proteins (purple channel) and the nanopore structure in the device are functionally the same. They both can maintain the balance of ion concentration inside and outside the cell by transferring ions, exhibiting a discharging feature (represented by the purple line in the voltage–time curve). Therefore, nanopores can be analogous to ion channel proteins in the device. The cooperation of the ion pump protein (yellow channel) and the ion channel protein (purple channel) leads to the capacitance characteristic of the model. The purpose of the as-designed bio-capacitor is to simulate and regulate this biophysical process.

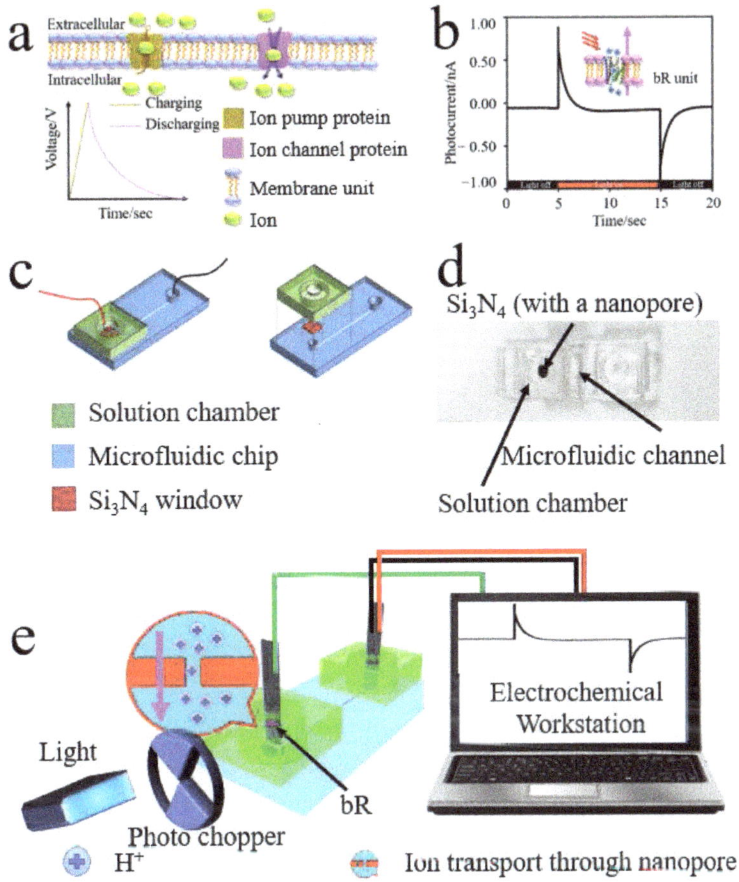

Figure 1. (**a**) The biological model of our device. The cooperation of ion-pump and ion-channel proteins makes the membrane act as a capacitor. (**b**) Photocurrent–time image of bR. (**c**) As-designed microfluidic chip. (**d**) Optical image of the device. (**e**) Schematic diagram of the complete test system.

Figure 1b depicts a basic photocurrent curve of bR and it can be seen that the shape of the curve is in accordance with the charging process of the plasma membrane. Therefore, it can be inferred that bR can be used to simulate ion pump protein by designing a reasonable device structure. When the induced light is shining on bR, the retinal of bR is isomerized (from the all-trans form to the 13-cis form). Through this process, H^+ ions are transferred from the intracellular to the extracellular, thereby generating potential differences between the inside and outside of the membrane. Based on this model and bR's property, we use bR as the part of the device structure that acts as an ion pump protein, while the fabricated nanopore is used to act as an ion channel protein.

By designing the proper microfluidic chip, the nanopore can be integrated with the bR purple membrane into one device. Figure 1c displays the structure and fabrication process of the microfluidic chip, while Figure 1d demonstrates the optical image of the chip. The entire structure is on the left, and the individual structure is on the right. The entire chip is made up of three parts: the blue polydimethylsiloxane (PDMS) substrate, the red Si_3N_4 wafer with a nanopore, and the green solution chamber. Amongst them, the PDMS substrate contains a microfluidic channel (a semi-cylinder with a height of 1.8 cm and a radius of 162 μm), which links both sides of the solution to connect the outside electrode. During testing, 0.05 mol/L of Na_2SO_4 solution fills the microfluidic channel to make sure the ion can smoothly and stably transfer. The Si_3N_4 wafer with a nanopore (3 nm or 8 nm in diameter) can enable H^+ to transfer from one side to the other to simulate the discharge process of ion channel proteins. The solution chamber has a hollow cylindrical structure (0.5 cm in diameter × 0.5 cm in height), which is used to store the Na_2SO_4 solution so that ITO glass with bR purple membrane can be in full contact with the solution. The entire microfluidic device is used to provide channels for H^+ ion movement and support the Si_3N_4 window with good sealing. After the production of bio-capacitors, subsequent testing and analysis can be carried out.

During the test, we first connected various test equipment to the system (shown in Figure 1e). The inducing light provided by the light source (tungsten lamp PLX-SXE250) in the test system was 86,600 $\mu W/cm^2$ and this was cut off by a photo chopper with a specific cutting frequency before shining on the bio-capacitor. A three-wire connection method was adopted to link the bio-capacitor with an electrochemical workstation. The green, yellow, and red lines in Figure 1e represent the working electrode, counter electrode, and reference electrode, respectively. Under the excitation by a flicker of light at a specific flickering frequency, bR transferred H^+ ions from the intracellular to extracellular space and generated an electrochemical potential difference. Then, by controlling the size of the nanopore, the ion diffusion rate was controlled to reduce the potential difference. The photocurrent was collected by an electrochemical workstation (CHI660E, biasing condition is 0 V) and the photocurrent–time (I–t) curve was obtained. The connection mode of the electrochemical workstation was as follows: the working electrode was connected with the ITO electrode deposited with bR and the counter/reference electrode was connected with the pure ITO electrode.

2.2. Device Construction

It should be noted that the nanopores used in the experiment have two different sizes: 3 nm or 8 nm diameter. This is of great significance for understanding the effect of nanopores size on device performance, in addition to improving the device accuracy. Figure 2a demonstrates the fabrication process used for nanopores: (1) deposition of Si_3N_4 film on both sides of the silicon wafer by low-pressure chemical vapor deposition; (2) the square area is lithographed and etched from the Si_3N_4 film on the back where the nanopore needs to be drilled; (3) the Si in the specific area is removed with wet anisotropic etching method; (4) a nano-scale hole in the Si_3N_4 film on the front is drilled with an electron beam.

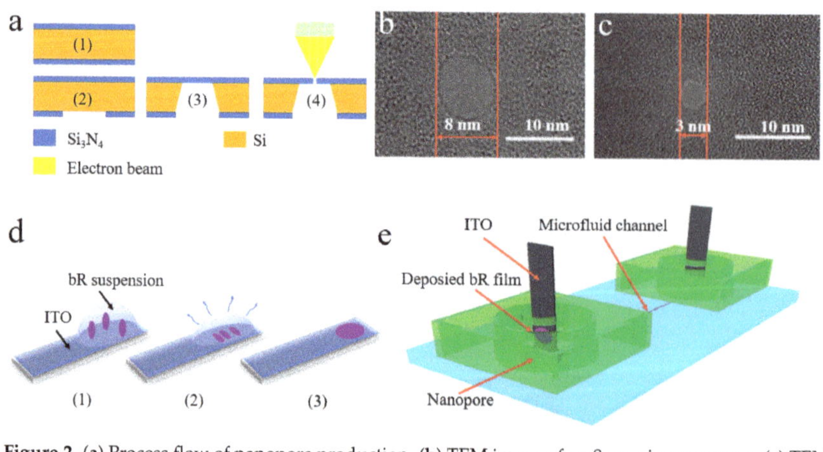

Figure 2. (**a**) Process flow of nanopore production. (**b**) TEM image of an 8 nm-size nanopore. (**c**) TEM image of 3 nm-size nanopore. (**d**) Process of evaporative deposition. (**e**) Complete device schematic. The PDMS substrate contains a microfluidic channel (a semi-cylinder with a height of 1.8 cm and a radius of 162 μm).

The TEM characterization [36] of two devices with different sizes of nanopores is shown in Figure 2b,c (the scale bar is 10 nm). Both holes appear to be quasi-circular. Two parallel red lines are used to highlight the size of the nanopore (8 nm on the left and 3 nm on the right). These two devices with different size nanopores are made to study the effect of the size in a subsequent test.

While there is a three-wire connection method in our test system, an ITO glass deposited with bR is used as a working electrode and the pure one is used as a counter/reference electrode. Figure 2d shows the following process: In step (1), 12 μL of bR suspension with a concentration of 1 mg/mL is dripped on the conductive side of ITO glass, which is then heated at 40 °C in step (2) to cause the excess solution to evaporate. Furthermore, step (3) demonstrates the modified electrode.

The manufacturing process of the microfluidic chip with a nanopore can be divided into three parts: The first involves making the SU-8 mold, the second involves making the microfluidic chip, and the third involves integrating the PDMS and Si_3N_4 wafer. The specific steps are as follows:

1. SU-8 mold making:

(1) Prepare a clean silicon wafer with SU-8 photoresist spin-coating. Place the suspended silicon wafer on the homogenizer. (2) Transfer the static silicon wafer to a heating plate and then use a patterned glass chrome plate for UV exposure. (3) Transfer the exposed silicon wafer to a heating plate to stabilize the model. (4) Soak the silicon wafer in the developer for a while and then clean the remaining developer. (5) Heat the developed silicon wafer to harden the film. At this time, the SU-8 mold can be obtained.

2. Microfluidic chip production:

(1) Mix the PDMS and the supporting coagulant aid in a specific ratio then stir. (2) Place the processed container in the vacuum machine. (3) Use an aluminum paper box to keep the vacuumed PDMS in, flatten the mold to make sure that it is in full contact with the aluminum paper, then heat them. (4) Cut out the required cleaned model and perform ultrasonic cleaning. (5) Use the PLASMA CLEANER to treat the ultrasonically cleaned PDMS mold for viscous treatment. (6) Glue the mold components together and let them stand overnight.

3. Integration of PDMS and Si_3N_4 wafer:

Si_3N_4 wafer is bonded onto PDMS by the O_2 plasma method. Turn the interface to be bonded upward and put it into the plasma cleaner machine (Harric PDC-002-HP), then work at 15 V power for one minute. After that, join the interface together, gently press, and leave it overnight.

Construction of the test platform: (1) Cut out 1.5 cm × 5 cm ITO glass with a glasscutter and have it cleaned. (2) Extract an appropriate amount of the bR solution, shake the solution evenly on the vortex oscillator. (3) Transfer the 12 μm bR solution to the ITO glass conductive surface. (4) Install in on the prepared microfluidic device. Subsequently, install the entire device into the shield box. (5) Place an opaque partition between the shielding box and the xenon light source. Thus, the whole test platform is constructed.

A complete schematic of the manufactured device is shown in Figure 2e, which shows the bR, the microfluidic channel, and the nanopores as well as their functions. The bR deposited in ITO is used as the working electrode in the test system and to simulate the ion pump protein. A microfluidic channel is used to transport H^+ ions, while fabricated nanopores are integrated into the device upon the microfluidic channel to control the rate of diffusion of H^+ ions.

3. Results

Figure 3a demonstrates the comparison of two different fabricated devices' photocurrent curves. It is clear that the device integrated with 3 nm nanopores has a longer sustainable photocurrent duration than another one (integrated with 8 nm nanopores). To make the quantitative difference more intuitive, photocurrent duration time (PDT) is defined and shown in Figure 3b. Figure 4 demonstrates that the photocurrent has been successfully converted.

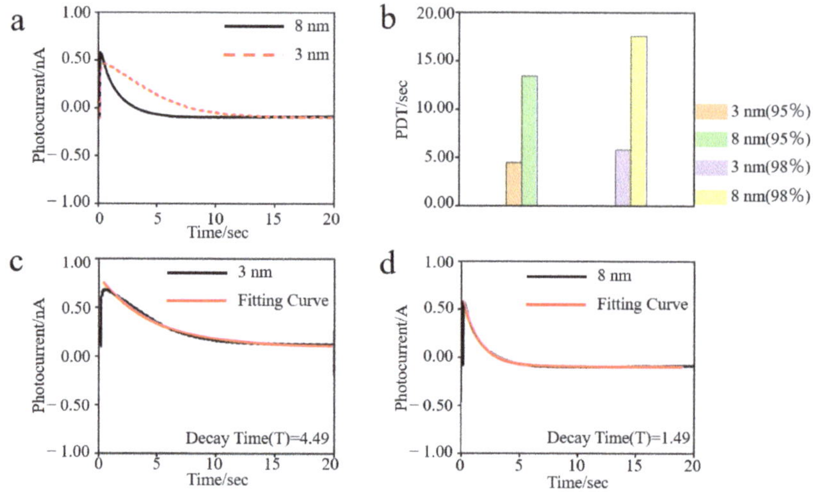

Figure 3. (**a**) The PDT is greatly affected by the changed nanopores. (**b**) The PDT in different indexes. (**c**) The fitting curve of 3 nm with a decay time of 4.49 s. (**d**) Fitting curve of 8 nm with a decay time of 1.49 s.

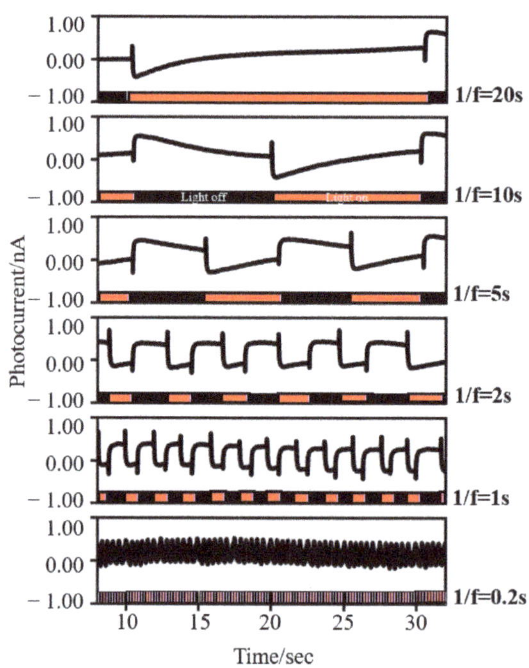

Figure 4. The photocurrent waveform changes with the light flickering frequency. f represents the flickering frequency of the induced light.

PDT [32] contains the light cycle time of bR and the decay time of the photocurrent. The decay time of the photocurrent reaches the seconds level, which is much longer than the light cycle time of bR (microsecond level). Thus, PDT could be simply defined as the decay time of the photocurrent. The decay time of the photocurrent shows a descending tendency in the i–t image; hence, an exponential decay function is used to fit the line. In addition, the time parameter (T) of the function is used to evaluate the PDT, which is defined as the time when the fitting line descends 95 percent. This is demonstrated by the following function:

$$Y = Y_0 + Ae^{-X/T}$$

where Y represents the instantaneous photocurrent, Y_0 represents the initial photocurrent, A represents the amplitude of photocurrent, X represents the X-axis (time), and T represents the decay time. By this equation, the data from the peak photocurrent amplitude down to 5% photocurrent amplitude (the photocurrent amplitude descends 95 percent) are selected for the fitting calculation as T. Then, the decay time (T) is obtained and PDT is calculated. Figure 3c,d demonstrate that the fitting result is in accordance with the experimental results.

4. Discussion

Figure 5 illuminates the mechanism underlying the photocurrent. First, since the ion pump function of bR protein can transport H^+ ions to the extracellular (EC) side, there will be an ion concentration gradient on both sides of the nanopore. Then, the concentration gradient descent rate is regulated by the size of the nanopore. H^+ ions passively diffuse to the other side through the nanopore, and this charge movement process forms a current. The size of the nanopore will influence the amount of H^+ ions passing through the pore per unit time, affecting the equilibrium time of the potential difference at both sides. Then, the PDT becomes larger under a smaller nanopore size, resulting in a slowly descending photocurrent. Controlling the light flickering frequency causes the chopping to occur in a period where the photocurrent changes are relatively flat, meaning that the photocurrent waveform can be adjusted and converted from peak signal to square wave signal.

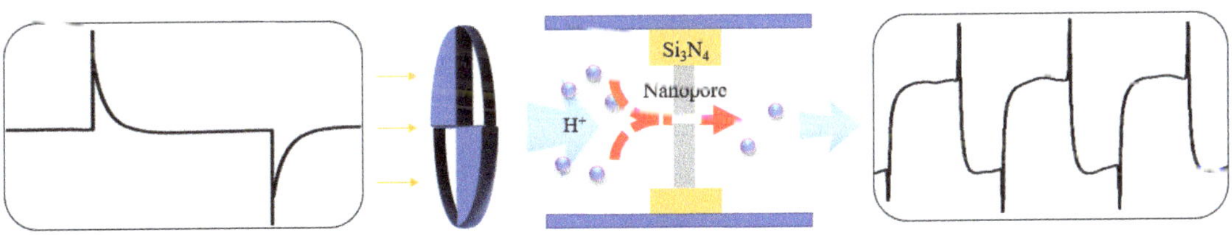

Figure 5. Schematic diagram of the mechanism underlying the photocurrent.

It can be seen from Figure 3 that the smaller the nanopore is, the longer the PDT is, which represents a longer charging and discharging time in the bio-capacitors under a different process parameter. The decay times (that is, PDT) of different nanopore sizes are 4.49 s and 1.49 s, corresponding to 3 nm and 8 nm, respectively. As shown in Figure 3, the PDT of a 3 nm device is about 3 times that of the 8 nm device. Compared with the previous data [32], they both show the same linear relationship—that is, the PDT becomes longer as the size of the nanopore decreases.

Since this study was conducted under a low frequency condition, bR has sufficient time to complete its whole photocycle process [37]. Therefore, in our experiment, the amplitude of the photocurrent varies little under different light flickering frequency, which is also confirmed by previous studies [32].

Based on bR's specific phenomenon, we believe that the PDT can also be regulated by the light flickering frequency. Through this, the bio-capacitor's charging and discharging time can be regulated. As shown in Figure 4, the transient spike current waveform can be successfully converted into a square wave-like waveform with the change in the light flickering frequency. In the bio-capacitor system, the photocurrent response changes with the increase in the light flickering frequency; the main characteristic of this is the descending tendency of the photocurrent to be flat. When the light flickering frequency reaches 0.5 Hz, the transient spike waveform is transformed into a square wave-like waveform. Theoretically, this phenomenon will lead to a longer PDT, verifying the successful regulation of the charging and discharging time of a build-up bio-capacitor.

The other potential application for regulating the photocurrent is that the square waveform can be seen everywhere in digital electronic circuits and expressed as binary "0" and "1" through analog-to-digital conversion. Simultaneously, because the bio-capacitor can form a square-like waveform while the light is on, a selective filter can be made by controlling the frequency of flickering illumination.

5. Conclusions

In this paper, the design and construction of a bR bio-photoelectric system is carried out and a bR-based bio-capacitor system is built. The photocurrent–time curve of this bio-capacitor system is measured in the electrochemical workstation and analyzed. The system with 3 nm-diameter nanopores has a PDT of about 13.45 s, while another device with 8 nm-diameter nanopores has one of about 4.47 s. Obviously, the PDT of the bR is regulated by the changing of the size of the nanopore channel. The i–t curve proves that bR has bio-capacitor properties. Finally, through optimizing the control parameters and the light flickering frequency, the original transient spike photocurrent waveform of the bR can be successfully converted into a square wave signal that is more suitable for the combination of micro-device circuits. As a newly developed bio-device, the nano-channel modulated bio-capacitor has shown great potential in the application of biological devices in the future.

Author Contributions: Conceptualization, Y.-T.L. and J.L.; methodology, Y.-T.L. and X.-X.X.; validation, Y.-T.L., J.L. and Y.-J.L.; resources, X.-X.X., Y.-J.L. and Y.X.; data curation, Y.-T.L. and J.L.; investigation, L.H., K.S. and J.L.; writing—original draft preparation, J.L.; writing—review and editing, Y.-T.L.; funding acquisition, Y.-T.L. All authors have read and agreed to the published version of the manuscript.

Funding: This research was supported by the Central University Fund of China (buctrc202108) and Open Research Fund Program of Beijing National Research Center for Information Science and Technology (BNR2021KF0100).

Institutional Review Board Statement: Not applicable.

Informed Consent Statement: Not applicable.

Data Availability Statement: The data presented in this study are available on request from the corresponding author. The data are not publicly available due to privacy concerns.

Acknowledgments: Equipment supported by Beijing Key Laboratory of Bio-inspired Energy Materials and Devices, School of Space and Environment, Beihang University.

Conflicts of Interest: The authors declare no conflict of interest.

References

1. Berggren, M.; Richter-Dahlfors, A. Organic bioelectronics. *Adv. Mater.* **2007**, *19*, 3201–3213. [CrossRef]
2. Someya, T.; Bao, Z.; Malliaras, G.G. The rise of plastic bioelectronics. *Nature* **2016**, *540*, 379–385. [CrossRef] [PubMed]
3. Szent-Györgyi, A. Bioelectronics: Intermolecular electron transfer may play a major role in biological regulation, defense, and cancer. *Science* **1968**, *161*, 988–990. [CrossRef] [PubMed]
4. Willner, I.; Katz, E. *Bioelectronics: From Theory to Applications*; John Wiley & Sons: Hoboken, NJ, USA, 2006.
5. Willner, I.; Willner, B. Biomaterials integrated with electronic elements: En route to bioelectronics. *Trends Biotechnol.* **2001**, *19*, 222–230. [CrossRef]
6. Willner, I.; Baron, R.; Willner, B. Integrated nanoparticle–biomolecule systems for biosensing and bioelectronics. *Biosens. Bioelectron.* **2007**, *22*, 1841–1852. [CrossRef]
7. Zhang, A.; Lieber, C.M. Nano-bioelectronics. *Chem. Rev.* **2016**, *116*, 215–257. [CrossRef]

8. Stoeckenius, W.; Kunau, W.H. Further characterization of particulate fractions from lysed cell envelopes of Halobacterium halobium and isolation of gas vacuole membranes. *J. Cell Biol.* **1968**, *38*, 337–357. [CrossRef]
9. Birge, R.R. Photophysics and molecular electronic applications of the rhodopsins. *Annu. Rev. Phys. Chem.* **1990**, *41*, 683–733. [CrossRef]
10. Trissl, H.-W.; Montal, M. Electrical demonstration of rapid light-induced conformational changes in bacteriorhodopsin. *Nature* **1977**, *266*, 655–657. [CrossRef]
11. Birge, R.R. A nonlinear proton pump. *Nature* **1994**, *371*, 659–660. [CrossRef]
12. Drachev, L.; Kaulen, A.; Skulachev, V. Correlation of photochemical cycle, H^+ release and uptake, and electric events in bacteriorhodopsin. *FEBS Lett.* **1984**, *178*, 331–335. [CrossRef]
13. Edman, K.; Nollert, P.; Royant, A.; Belrhali, H.; Pebay-Peyroula, E.; Hajdu, J.; Neutze, R.; Landau, E.M. High-resolution X-ray structure of an early intermediate in the bacteriorhodopsin photocycle. *Nature* **1999**, *401*, 822–826. [CrossRef] [PubMed]
14. Han, B.-G.; Vonck, J.; Glaeser, R.M. The bacteriorhodopsin photocycle: Direct structural study of two substrates of the M-intermediate. *Biophys. J.* **1994**, *67*, 1179–1186. [CrossRef]
15. Luecke, H.; Richter, H.-T.; Lanyi, J.K. Proton transfer pathways in bacteriorhodopsin at 2.3 angstrom resolution. *Science* **1998**, *280*, 1934–1937. [CrossRef]
16. Oesterhelt, D.; Stoeckenius, W. Rhodopsin-like protein from the purple membrane of Halobacterium halobium. *Nat. New Biol.* **1971**, *233*, 149–152. [CrossRef]
17. Polland, H.-J.; Franz, M.; Zinth, W.; Kaiser, W.; Kölling, E.; Oesterhelt, D. Early picosecond events in the photocycle of bacteriorhodopsin. *Biophys. J.* **1986**, *49*, 651–662. [CrossRef]
18. Shichida, Y.; Matuoka, S.; Hidaka, Y.; Yoshizawa, T. Absorption spectra of intermediates of bacteriorhodopsin measured by laser photolysis at room temperatures. *Biochim. Biophys. Acta (BBA)-Bioenerg.* **1983**, *723*, 240–246. [CrossRef]
19. Ranaghan, M.J.; Shima, S.; Ramos, L.; Poulin, D.S.; Whited, G.; Rajasekaran, S.; Stuart, J.A.; Albert, A.D.; Birge, R.R. Photochemical and thermal stability of green and blue proteorhodopsins: Implications for protein-based bioelectronic devices. *J. Phys. Chem. B* **2010**, *114*, 14064–14070. [CrossRef]
20. Brouillette, C.G.; McMichens, R.B.; Stern, L.J.; Khorana, H.G. Structure and thermal stability of monomeric bacteriorhodopsin in mixed pospholipid/detergent micelles. *Proteins Struct. Funct. Bioinform.* **1989**, *5*, 38–46. [CrossRef]
21. Rao, S.; Guo, Z.; Liang, D.; Chen, D.; Wei, Y.; Xiang, Y. A proteorhodopsin-based biohybrid light-powering pH sensor. *Phys. Chem. Chem. Phys.* **2013**, *15*, 15821–15824. [CrossRef]
22. Frydrych, M.; Silfsten, P.; Parkkinen, S.; Parkkinen, J.; Jaaskelainen, T. Color sensitive retina based on bacteriorhodopsin. *Biosystems* **2000**, *54*, 131–140. [CrossRef]
23. Fischer, T.; Neebe, M.; Juchem, T.; Hampp, N.A. Biomolecular optical data storage and data encryption. *IEEE Trans. Nanobiosci.* **2003**, *2*, 1–5. [CrossRef] [PubMed]
24. Imhof, M.; Rhinow, D.; Hampp, N. Two-photon polarization data storage in bacteriorhodopsin films and its potential use in security applications. *Appl. Phys. Lett.* **2014**, *104*, 081921. [CrossRef]
25. Yu, X.; Yao, B.; Lei, M.; Gao, P.; Ma, B. Femtosecond laser-induced permanent anisotropy in bacteriorhodopsin films and applications in optical data storage. *J. Mod. Opt.* **2013**, *60*, 309–314. [CrossRef]
26. Allam, N.K.; Yen, C.-W.; Near, R.D.; El-Sayed, M.A. Bacteriorhodopsin/TiO 2 nanotube arrays hybrid system for enhanced photoelectrochemical water splitting. *Energy Environ. Sci.* **2011**, *4*, 2909–2914. [CrossRef]
27. Lv, Y.; Yang, N.; Li, S.; Lu, S.; Xiang, Y. A novel light-driven pH-biosensor based on bacteriorhodopsin. *Nano Energy* **2019**, *66*, 104129. [CrossRef]
28. Lin, Y.-C.; Lin, C.-Y.; Chen, H.-M.; Kuo, L.-P.; Hsieh, C.-E.; Wang, X.-H.; Cheng, C.-W.; Wu, C.-Y.; Chen, Y.-S. Direct and label-free determination of human glycated hemoglobin levels using bacteriorhodopsin as the biosensor transducer. *Sensors* **2020**, *20*, 7274. [CrossRef]
29. Partridge, M.; Korposh, S.; James, S.; Tatam, R. Long period grating sensors response to photosensitive bacteriorhodopsin coating. In Proceedings of the Fifth Asia-Pacific Optical Sensors Conference, Jeju, Korea, 20–22 May 2015; p. 965535.
30. Srinivasan, K.; Thamilmaran, K.; Venkatesan, A. Effect of nonsinusoidal periodic forces in Duffing oscillator: Numerical and analog simulation studies. *Chaos Solitons Fractals* **2009**, *40*, 319–330. [CrossRef]
31. Yuan, L.; Xiao, X.; Ding, T.; Zhong, J.; Zhang, X.; Shen, Y.; Hu, B.; Huang, Y.; Zhou, J.; Wang, Z.L. Paper-based supercapacitors for self-powered nanosystems. *Angew. Chem. Int. Ed.* **2012**, *51*, 4934–4938. [CrossRef]
32. Rao, S.; Lu, S.; Guo, Z.; Li, Y.; Chen, D.; Xiang, Y. A Light-Powered Bio-Capacitor with Nanochannel Modulation. *Adv. Mater.* **2014**, *26*, 5846–5850. [CrossRef]
33. Ashrafuzzaman, M.; Tuszynski, J.A. *Membrane Biophysics*; Springer Science & Business Media: Berlin/Heidelberg, Germany, 2012.
34. Gadsby, D.C. Ion channels versus ion pumps: The principal difference, in principle. *Nat. Rev. Mol. Cell Biol.* **2009**, *10*, 344–352. [CrossRef] [PubMed]
35. Gouaux, E.; MacKinnon, R. Principles of selective ion transport in channels and pumps. *Science* **2005**, *310*, 1461–1465. [CrossRef] [PubMed]

36. Wu, S.; Cao, F.; Zheng, H.; Sheng, H.; Liu, C.; Liu, Y.; Zhao, D.; Wang, J. Fabrication of faceted nanopores in magnesium. *Appl. Phys. Lett.* **2013**, *103*, 243101. [CrossRef]
37. Lanyi, J.K. Proton transfers in the bacteriorhodopsin photocycle. *Biochim. Biophys. Acta (BBA)-Bioenerg.* **2006**, *1757*, 1012–1018. [CrossRef] [PubMed]

MDPI
St. Alban-Anlage 66
4052 Basel
Switzerland
Tel. +41 61 683 77 34
Fax +41 61 302 89 18
www.mdpi.com

Nanomaterials Editorial Office
E-mail: nanomaterials@mdpi.com
www.mdpi.com/journal/nanomaterials

www.ingramcontent.com/pod-product-compliance
Lightning Source LLC
Chambersburg PA
CBHW061435210926
51850CB00008B/1520
9783036558844